THE ECONOMICS OF BARGAINING

THE ECONOMICS OF BARGAINING

Edited by

Ken Binmore
and Partha Dasgupta

Basil Blackwell

First published 1987

Basil Blackwell Ltd
108 Cowley Road, Oxford OX4 1JF, UK

Basil Blackwell Inc.
432 Park Avenue South, Suite 1503
New York, NY 10016, USA

British Library Cataloguing in Publication Data

The Economics of bargaining.
1. Collective bargaining – Mathematical models
I. Binmore, K. G. II. Dasgupta, Partha
331.89'0724 HD6483
ISBN 0-631-14254-1

Library of Congress Cataloging in Publication Data

The Economics of bargaining.
Bibliography: p.
Includes index.
1. Game theory. 2. Negotiation – Mathematical models.
I. Binmore, Ken. II. Dasgupta, Partha.
HB144.E29 1986 381 86-17169
ISBN 0-631-14254-1

Typeset in 10 on 12pt Press Roman by Unicus Graphics Ltd,
Horsham, West Sussex

Printed in Great Britain by TJ Press Ltd, Padstow

Contents

Preface

The Suntory-Toyota International Centre for Economics and Related Disciplines (ICERD) was established at the London School of Economics in 1978. One of its earliest initiatives was the Economic Theory Workshop which was set up as a forum for the interchange of ideas in economic theory and provided with funds from the Suntory-Toyota Foundation to finance international visits and a discussion paper series. Since that time, the Economic Theory Workshop has entertained many visitors, witnessed much heated debate and produced large numbers of discussion papers. A central theme in much of this work has been the theory of games and its applications in economics. In this and the companion volume *Economic Organizations as Games* we offer a selection of papers written by members of the workshops and our visitors on this theme. For each volume, we have written a lengthy, but we hope useful, introduction which is intended to explain the significance of the papers chosen and their place in the general development of the subject.

We should like to express our gratitude to Suntory and Toyota for their vision in supporting this truly international project and to the staff at ICERD for their assistance and guidance in running the workshop. All royalties for this and the companion volume will be donated to ICERD.

Ken Binmore
London School of Economics

Partha Dasgupta
St John's College, Cambridge

Acknowledgements

The publishers and editors acknowledge with thanks permission to reproduce in this volume material previously published in *Econometrica*. Chapter 3 appeared in vol. 50 (1982), 207–11; chapter 6 in vol. 52 (1984), 1351–64; chapter 7 in vol. 50 (1982), 607–38; chapter 9 in vol. 53 (1985), 1151–72; and chapter 10 in vol. 49 (1981), 597–617.

1

Nash Bargaining Theory: An Introduction

K. Binmore and P. Dasgupta

1 BACKGROUND

In spite of early hopes, it is only in recent years that the von Neumann and Morgenstern theory of games has begun to be genuinely fruitful in economic analysis. In retrospect, it seems clear that the delay was due only in part to the incompleteness of the theory, since much of what is now being used has been available in essence from the early 1950s. A larger stumbling block has been the problem of determining under what circumstances the available theory is (or is not) applicable.

When seeking to employ a game-theoretic analysis, one is faced with two fundamental questions:

a What is the appropriate game?
b How should it be analysed?

It will typically be impractical to incorporate into the formal model all aspects of the potential interaction between the agents. It is therefore necessary to decide which of these aspects should be formally modelled and which should be taken account of informally. Where and how this line is drawn will be highly significant to the question of what type of analysis is appropriate. If all, or almost all, of the essential strategic structure is adequately formalized, then the conceptual problems involved in determining what constitutes a solution to the game may well be easy. But this will seldom also be true of the technical problems which arise in calculating the solution. It seems doubtful, for example, that it will ever be possible actually to compute the solutions of such games as Chess or 'Go' (although Poker and its relatives are reasonably amenable to analysis). On the other hand, if little of the strategic structure is formalized, then the technical problems are likely to be slight once a solution concept has been chosen. Under the heading of a 'solution concept' it is common to include a very wide variety of ideas. We shall use the term to signify a function which maps from the given formal structure of the game to subsets (or, sometimes, sets of subsets) of possible outcomes. The less structure incorporated in the formal description of a game, the less complex such functions can be; and therefore the less likely it is that any mathematical difficulties encountered will turn out to be intractable. Equilibrium ideas, such as those discussed in the companion volume *Economic Organizations as Games*, make no sense unless the formal description of the game includes a detailed specification of the strategic opportunities open to the players. Such a specification is often complicated and the corresponding equilibrium analyses are therefore accordingly seldom straightforward. On the other hand, solution concepts like the 'core' or the 'Shapley value' are relatively easy to deal with because their definition requires comparatively little formal structure on the part of a game.

The problem at this end of the spectrum is that it is not clear which of the many available solution concepts, if any, is the 'right' concept for the problem in hand. This question is a difficult one and it is not surprising that authors should be reluctant to provide tightly argued justifications for their answers. But to fudge the question is to risk using game theory as a Procrustean bed on which the economics can be chopped or stretched at will to suit one's computational convenience.[1] This danger is particularly acute in those cases, which frequently occur in a game-theoretic context, where the results of an analysis are surprising or downright counterintuitive. For example, in two-person, pot-limit, straight Poker it is optimal for the opening player *always* to bet on the first round with a ten-high 'bust' (or worse), *never* to bet with a pair of twos and to bet incredibly conservatively with a low straight flush (Cutler 1975). Untutored common sense is therefore not necessarily a reliable guide.

These remarks are not intended to suggest that there is no room in game-theoretic modelling for good judgement or intuitive insight. Quite the contrary is the case. But intuition needs to be trained. In particular, it is important to adopt a critical attitude towards any general assumptions about optimal behaviour which may be advanced and to seek for ways of testing their validity.

The introduction to the companion volume *Economic Organizations as Games*, was insistent on the importance of the idea of a Nash equilibrium. But this idea is only one of several important contributions made by Nash during the short time he devoted to game theory. Perhaps an idea of equal importance is embodied in what has now become known as the 'Nash program'. This provides a conceptual framework within which it is possible to evaluate general assumptions about optimal behaviour in a unified and disciplined manner. The fundamental point is that a choice of solution concept for a game-theoretically based economic model needs to be considered

1 In a one-seller, two-buyer market, outcomes in which the buyers form an unbreakable coalition lie outside the core. But this is hardly a reason for excluding them as possible 'economic outcomes' (see Weintraub 1975, p. 70).

very carefully. In so far as bargaining theory is concerned, we shall observe that the appropriate solution concept depends very strongly on the informational and institutional properties of the negotiation arrangements. These, of course, will seldom be deducible from what would usually be regarded as an adequate 'economic' description of a problem. A choice of solution concept therefore implicitly generates assumptions about the 'non-economic' characteristics of the situation the model is intended to describe. It is necessary, at the very least, to be conscious of the fact that such implicit assumptions are being made and to have some feeling for their nature. What is certainly not the case is that the choice of solution concept is simply, or even largely, a matter of taste.[2] If an inappropriate solution concept is grafted onto a model, one has to expect that the results will be grotesquely distorted. The purpose of the Nash program is to minimize the risk of generating such distortions.

2 THE NASH PROGRAM

In the introduction to the companion volume, we introduced the idea of a formal game and commented upon the analysis of such games *in the absence of* a cooperative infrastructure. In this volume, however, our interest centres precisely on those situations in which a cooperative infrastructure *is present.* The necessary considerations are therefore very different.

Firstly, we should clarify what we mean by a 'cooperative infrastructure'. Suppose that it is recognized that a certain situation has game-like characteristics and an attempt is made to construct a formal model G which captures these characteristics. For the purposes of the current discussion it does not matter greatly whether G exhibits all of the properties required of a formal game, because, even if it did, a knowledge of the formal structure of G would still not usually be adequate to analyse the game. In general, it will also be necessary to know something about the *unformalized*[3] activity which may precede the play of the game. It is such unformalized pre-play activity that we have in mind when using the term 'cooperative infrastructure'. (Note the contrast with *contests* which were the focus of attention in the introduction to the companion volume.)

Where possible, for example, players of a game will usually wish to *coordinate* their strategy choices.[4] If some players are reluctant to cooperate perhaps they should be bribed or threatened. If so, how big should the

2 As, for example, is suggested explicitly by Friedman in his otherwise instructive book (1977).

3 Where the opportunities for pre-play activity open to the players are sufficiently rich, it can happen that the precise nature of the *formal* strategies for G becomes almost irrelevant. Descriptions of the formal strategies are then typically omitted.

4 As in the theory of *correlated equilibria* (see Myerson 1984 for a topical account).

bribes be? What threats will be effective? Such pre-play activity is not always permitted. For example, bribery in the species of game played by two contractors submitting tenders to a public official is illegal. Similarly, anti-trust laws are designed to prevent the formation of certain coalitions for mutual advantage. Perhaps of most vital importance is the question of whether any pre-play agreements are *binding* on the players. The interest of the parlour game 'Diplomacy', for example, lies in the judicious breaking of agreements, while, in the commercial world, contracts are supposedly legally enforceable. On a more fundamental level, it may be that the players cannot communicate with each other at all (e.g. if one does not know the other's whereabouts) or that meetings of three or more players are impossible (e.g. if they communicate by telephone).[5] Where self-binding threats are possible, the order in which the players can make their statements becomes highly relevant. This brings us to the general question of commitment. Commitments made physically (for example, by investing in a plant) are easily modelled; but what of verbal commitments? To what extent do these bind those who make them? Can we suppose that commitments the players make are credible because of the impact that a failure to keep the commitment would have on their reputations? If so, what is the mechanism? (For a more detailed discussion of these and related questions on trust, see Dasgupta 1986.)

These considerations are listed in an attempt to make it clear that what happens in a game will in general depend on what may or may not have happened *before* the formal game is played (or on what is anticipated might happen *after* the formal game is played). But solution concepts simply consist of mappings from a formal game G (often suppressing much of the structure of G) to outcomes of G. It follows that the choice of a solution concept must necessarily entail a whole mix of implicit assumptions about the nature of the cooperative infrastructure within which the game is thought to operate. And, to carry the same point further, the existence of a battery of such solution concepts necessarily implies the existence of a *classification* of cooperative infrastructures – each different concept being appropriate for a different infrastructure.

The standard classification, illustrated below, is almost shamefully primitive.

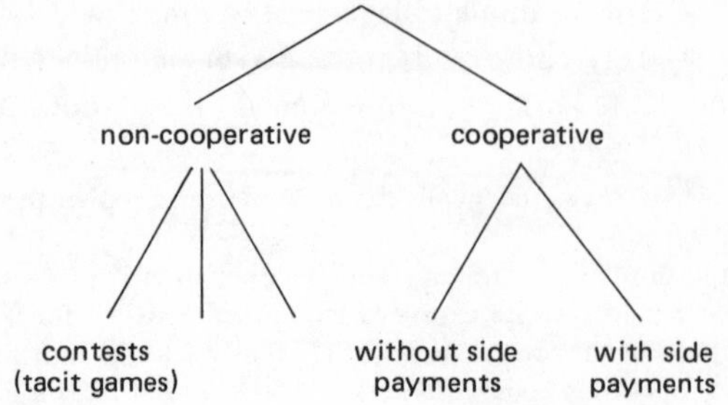

Before strategies are chosen in a formal game played cooperatively, it is supposed that players may communicate costlessly and without restriction and that they may enter into any agreements whatsoever that they choose. Most important of all, it is assumed that mechanisms exist for enforcing such agreements. A side payment is to be thought of as a sum of money which may be paid by one player to another in order to facilitate (or hinder) the signing of a pre-play contract. In principle, the notion of a non-cooperative game covers a wide spectrum of possibilities but attention is usually addressed to what Harsanyi calls 'tacit games' and we prefer to call contests. It is to be understood that, before the play of such a game, the players have no opportunities, explicit or implicit, for any type of communication at all.

This classification is too crude to provide more than preliminary assistance in evaluating solution concepts. For this reason certain concepts for cooperative games come equipped with characterizing 'axiom systems'. Chapter 2 of this volume, for example, provides an alternative axiom system to that usually offered in support of the Nash bargaining solution. But such axiom systems are typically couched in rather abstract terms and there seem to be no cases at all where the appropriate interpretation is not controversial.

The Nash program (outlined in a few sentences by Nash 1951) provides a very much more satisfactory framework within which to consider these matters. His basic view was that the most fundamental type of game is what we have called a *contest.* We have discussed the game-theoretic analysis of such games in the companion volume. Briefly, the notion of a *Nash equilibrium* provides a firm foundation upon which the theory can rest. This is not to say that problems do not exist nor that refinements of the Nash equilibrium idea are not essential.[6] On the contrary, there are numerous difficulties to which we shall find it necessary to return. But these difficulties are of a secondary nature compared with the conceptual problems resolved for contests by the idea of a Nash equilibrium but which remain largely unresolved for other types of game.

We now turn to games other than contests - i.e. games which are to be analysed on the assumption that some measure of pre-play interaction between the players is possible.[7] Nash proposed to deal with the problem of pre-play activity in the following way. If G is a formal game, imagine the various possible steps in the negotiations which precede its play as moves in a larger 'negotiation game' N - i.e. formalize the negotiation procedure. A strategy for the formal game N is a statement of how to conduct the negotiations

5 This consideration can matter enormously to market outcomes. (See Binmore 1983, Rubinstein and Wolinsky 1984 and Binmore and Herrero 1985.)

6 We have in mind problems of equilibrium selection and the difficulties which arise in extensive-form games or in games of incomplete information.

7 For simplicity, we leave aside the question of post-play interaction although there will be many situations where this is very significant.

under all possible eventualities and how finally to choose a strategy for G contingent on the course the negotiations took. The negotiation game N is then analysed as a *contest.* The solution of N as a contest then provides a 'solution concept' for the original game G. Nash maintains that *all* 'solution concepts' for G should be regarded this way – i.e. as the solutions of associated negotiation contests.

An obvious objection to this line of attack lies, of course, in the difficulty of finding a model for the negotiation procedure which is both realistic and sufficiently simple to be amenable to analysis. One response is that difficulties cannot successfully be evaded by pretending that they do not exist. However, matters are not so hopeless as they may seem at first sight. In considering appropriate negotiation models, it is necessary to focus on those factors which have, or appear to have, a genuine strategic relevance to the situation. This certainly does not apply to the bulk of manoeuvres common in real-life negotiations. Under this latter heading, for example, come flattery, abuse, the inducement of boredom and other more subtle attempts to put the opponent at a psychological disadvantage. These factors would certainly be of the greatest importance in a behavioural analysis but they have no place in a game-theoretic analysis. A rational player will simply ignore such irrelevancies.[8]

It is as well to enlarge upon this point since it is an issue on which the literature has a tendency to fudge. (See also Section 2 of the introduction to the companion volume.) Consider, for example, a two-person cooperative game. Von Neumann and Morgenstern (1944) originally suggested that the 'solution' of such a game be identified with a whole continuum of possible outcomes called the 'negotiation set' or 'core' of the game. To identify a particular outcome within the core, they suggested, is an enterprise outside the scope of the game theorist on the grounds that the outcome on which the players finally agree will depend on their relative 'bargaining powers' and such *psychological* factors elude adequate formalization. Nash himself makes reference to a similar behavioural notion of 'bargaining skill' in 1950 but explicitly corrects himself in the later paper (1953) where he asserts that, '*with people who are sufficiently intelligent and rational there should not be any question of "bargaining ability", a term which suggests something like skill in duping the other fellow*'. An analogy may be of some help here. Imagine two chimpanzees playing chess. Perhaps in the circumstances, the players might find it difficult to formalize the rules of the game and it will certainly be true that the outcome of the game will depend principally on psychological factors. But these facts are irrelevant to a game theorist who is concerned with what *would* be best for the players if only they knew it.

8 Assuming his concern was only with the outcome of G obtained and not with the manner in which it was achieved.

This information will remain the same even if one of our chess-playing chimpanzees is replaced by Capablanca provided only that both have the same preferences over the set of outcomes. A behavioural analysis would be quite a different matter and would doubtless take account of such factors as 'chess playing power' etc. An ethical analysis would similarly require taking into account numerous issues which are irrelevant to a game-theoretic analysis. This is not to say that a behavioural or an ethical analysis might not be more useful than a game-theoretic analysis: only that it is as well not to be confused about what sort of analysis is actually being attempted.[9]

These ideas carry over to bargaining games. If the game-theoretic optimal strategy involves playing like Attila the Hun, then this will remain true no matter who the player might be. It may be true that St Francis of Assisi might find this strategy hard to implement but this is equally true of a chimpanzee attempting the Sicilian Defence. Alternatively, one may object that St Francis would not *want* to behave in this way on the grounds that the 'end does not justify the means'. But this is an ethical question to be dealt with, if at all, by altering the payoffs.[10]

We have been arguing that the Nash program for formalizing negotiation procedures is less formidable than it at first appears provided that one adheres firmly to a game-theoretic viewpoint. Nevertheless, it remains the case that, when the negotiation procedure has been formalized, one must expect to be faced with a structure whose analysis may be technically difficult. For this reason and others, Nash was very far from suggesting that cooperative game theory should be abandoned. To do so would be like attempting to study the movement of billiard balls using only quantum theory. On the contrary, Nash thought of cooperative and non-cooperative theory as complementary and mutually supporting prongs of a pincer attack upon the problem of negotiated games. This view is sometimes classified along with Hegelian dialectic as being impenetrably paradoxical, but we see little justification for such an attitude since the idea, once stated, seems quite simple.

As stressed in Section 1, a mathematical modeller in economic theory (or anything else) has to exercise his judgement in deciding how much of a situation to represent *formally* within his model. His decision will depend on what aspects of the situation are of particular interest to him and on his knowledge of the situation under study. In practice, the extent to which he is successful will depend on how cleverly he draws the line between the formal and the informal so as to maximize the effectiveness of the analytical tools at his

9 The reason for attempting a game-theoretic analysis rather than a behavioural analysis is that it seems sensible to solve the simple problems before seeking to attack the more difficult ones.

10 In a formal game as defined by von Neumann and Morgenstern, the ends are inseparable from the means in that each terminal mode can be reached by only one path through the game tree.

disposal and thus to minimize his dependence on speculation and guesswork. In so far as bargaining is concerned, two immediate issues arise:

(1) The modeller is unlikely to be immediately interested in the details of the negotiations per se. What he will care about is how the *result* of the negotiations is likely to change as relevant economic parameters alter. In many cases he will be seeking only *qualitative* rather than quantitative conclusions on this issue.

(2) The modeller is unlikely to have hard information about a whole spectrum of issues relevant to the detailed conduct of the negotiations. It is remarkable how little of the empirical work on market institutions is directed at issues which a game theorist would regard as significant.

Both these considerations militate in favour of consigning the details of the negotiation process to the area to be dealt with informally by intelligent speculation and informed guesswork. In practical terms, this means choosing an appropriate 'cooperative solution concept' under which heading can be included such notions as Walrasian equilibrium, the Shapley value, the core or a von Neumann and Morgenstern solution set. Within the Nash program, each of these and other cooperative solution concepts are seen as an attempt to describe the outcome of a non-cooperative negotiation game *without formalizing the negotiation procedure.* To be useful, of course, a cooperative solution procedure must be applicable (or thought to be applicable) to a fairly wide class of negotiation procedures. In some cases, such as the Shapley value, the concept is best seen as an 'average' of what is to be expected over such a class of negotiation procedures. In other cases, such as the core or a von Neumann and Morgenstern solution set, the concept does not supply a unique outcome and one is free to attribute this indeterminacy to the breadth of the class of procedures to which the concept is intended to be applicable.

These last remarks make it clear that the Nash viewpoint is valuable as a philosophical prop in thinking about cooperative solution concepts. But the real value of the Nash program lies in the point to be made next. In the physical sciences, experiment provides a final court to which critics of exercises in speculation and guesswork can appeal. In the social sciences, conclusive appeals to experimental evidence are seldom possible - partly because of Murphy's Law[11] and partly because it is seldom feasible to impose the necessary controls. This makes it all the more important in the social sciences to exercise restraint in making assumptions - particularly those which are not stated explicitly in the description of the formal model but appear implicitly as a result of the choice of a structure for the formal model and whose justification, if any, is based on informal considerations. One form of self-discipline available to a mathematical modeller in both the physical and the social

11 Under carefully controlled experimental conditions, laboratory animals do what they damn well please.

sciences is the mind experiment[12] and it is the idea of a mind experiment which is at the heart of the non-cooperative side of the Nash program.

The purpose of constructing negotiation contests in which the details of the negotiation procedure are spelled out explicitly and without omission is *not* because it is thought that such models will displace the use of cooperative solution concepts. The purpose is to *test* cooperative solution concepts.[13] If a certain concept is claimed to apply in the presence of a particular type of negotiation set-up, one may ask the claimant for his reasons. Typically, the reasons offered will constitute a defence not only for the negotiation set-up originally contemplated but also for other related procedures. The argument can then be examined by formalizing the simplest of these procedures and analysing the resulting contest non-cooperatively. Such a method allows the exposure of silly theories (of which there seems no lack) and the opportunity to refine incomplete theories.[14] *The hallmark and the essence of the Nash program is therefore the imperative to test abstract or informal reasoning with simple but specific negotiation models.* And an important duty for games theorists is to provide a battery of suitable models to facilitate such testing.

As an example, consider the use that De Menil (1971) and others have made of the Nash bargaining solution in discussing union power and related issues. From the point of view of the Nash program, such work is suspect because it does not seriously address the question of *why* the Nash bargaining solution is to be used rather than some other cooperative notion. Like ready-made suits, solution concepts should not be bought without first trying them on for size. As it happens, recent work, based on an important paper by Rubinstein (chapter 3 in the current volume), indicates that such suspicion is justified. The difficulty concerns 'outside options' – by which we mean the utility levels which bargainers can achieve by leaving the bargaining table and abandoning the negotiations. Traditional wisdom takes for granted that these should be dealt with by identifying the 'status quo' point with the 'outside option' point. However, an examination of the appropriate extension of Rubinstein's very natural, non-cooperative bargaining model shows that, while this identification may have merit in some special circumstances, it is *not* the right way to proceed in general.[15] The point is that

12 This point is made at greater length in the Introduction to the companion volume, *Economic Organizations as Games.*

13 Or, on occasion, to assist in the formulation of new cooperative solution concepts.

14 By the type of process, long familiar to mathematicians, described by Lakatos in his *Proofs and Refutations* (1976).

15 See Binmore (1983) or Shaked and Sutton (1984). A more widely applicable procedure is to locate the 'status quo' at the point which would result if nobody left the table but there was perpetual disagreement and to observe that 'outside options' will affect the bargaining outcome if and only if it assigns one or more of the bargainers less than his outside option utility level.

traditional wisdom, which is intuitively quite compelling on this question, would not have been challenged, let alone altered, without the intervention of a testing mind experiment as advocated by the Nash program.

3 THE FORMALIZATION OF NEGOTIATION PROCESSES

At first sight, it is surprising how difficult it can be to identify the crucial features of a bargaining process. One would think that the very considerable experience we all have of bargaining at one level or another would be more useful in this respect than it seems to be.[16] This apparent failure of our intuitive insight makes the availability of the von Neumann and Morgenstern formulation of an extensive game particularly valuable. The attempt to fit a negotiation process into their framework forces attention to be paid to features of the process which otherwise might be dismissed as irrelevant. One of the more significant discoveries of modern bargaining theory has been the identification of *discontinuities* in the bargaining outcome as certain 'frictions' or 'imperfections' in the bargaining process are allowed to tend to zero. In the limit, a unique outcome is obtained in many cases. On the other hand, if the frictions are set equal to zero a priori, then a whole range of outcomes appears possible. What seemed at the time to be an entirely natural and straightforward idealization therefore now appears as a possible root cause for the indeterminacy problem which has plagued bargaining theory ever since Edgeworth wrote on the subject. (See chapters 4, 5 and 8.)

There is still much to be learned about bargaining processes but it is now fairly well-established that one cannot usually offer a sensible estimate of what is likely to happen without having a view on the roles to be ascribed to

a commitment;
b time;
c information

within the bargaining process. We shall take these one at a time.

16 It is easy, if not particularly profitable, to speculate on the reasons. The simplest explanation is that psychological issues dominate at the expense of rational issues (as in poker games among novices). Possibly, social evolution has not equipped us well for dealing with bargaining situations in isolation since we shall usually have to deal with our bargaining partners in the future on a variety of different issues. Possibly the problem is simply that we are reluctant to think too closely about our bargaining behaviour in case we find a second-hand car-dealer lurking inside our skulls. Where personal issues are concerned, we certainly have a strong tendency to pretend that we have not been bargaining at all. A more encouraging hypothesis for a game theorist is that our bargaining behaviour is shaped by what *would happen* if we did not bargain as we do. Such an explanation requires taking account of events which are not normally observed and are unlikely to be properly understood.

3.1 Commitment

In much economic theory it is taken for granted that agents can write whatever contracts they choose and that these contracts will be totally binding on the players - i.e. that the basic hypothesis for a cooperative game is satisfied. Two defences of the hypothesis are usually advanced. The first cites the existence of a legal system and the second refers to the importance of maintaining a reputation for good faith in so far as future transactions are concerned. In many cases both considerations will be relevant. Whatever the reasons, all the essays in this volume take for granted that the aim of bargaining is to provide a contract describing the agreed transaction and that all agents regard such a signed contract as equivalent to the transaction itself. Of course, there will be situations for which this is a bad assumption[17] as the current disarray in the world banking system amply demonstrates.[18] However, the hypothesis that the final contract reached will be honoured does not seem an unreasonable assumption from which to begin.

But it does not follow that, because it is reasonable to assume that one type of contract will be honoured, then *any* type of contract will be honoured. In particular, it is easy to overestimate what can be done via the legal system in this context. We shall take up the issue of contracts which relate to the actual transaction again under the heading of 'information'. For the moment, however, we wish to concentrate on commitments (unilateral or multilateral) that the players may wish to make about how they propose to conduct the negotiations in the future. Schelling (1960) has emphasized both how useful it is to players if they can leave their opponent with a genuine 'take it or leave it' problem and, simultaneously, how hard it is in practical terms to make the necessary commitments stick. Perhaps the best known example, although in a somewhat different context, is that of Ellsberg (1975). He cites the kidnapping victim who would dearly wish to make a commitment not to reveal the identity of the kidnapper in order to escape murder but has no way to do so.

Of course, it does not follow that, because the legal system is unlikely to be effective in enforcing in-bargaining commitments, that such commitments may not be enforced because of the damage that a failure to carry through on a commitment may incur in respect of a player's reputation. Obviously, a reputation for 'toughness' can be very valuable if bargaining with an individual with a reputation for 'softness'. This is another matter we shall take up again under the heading of 'information'. For the moment,

17 One could, of course, regard the players as bargaining over their valuations of the possible contracts or restrict attention to 'self-enforcing' contracts in which goods are exchanged a little at a time (as in illicit drug dealing).

18 Or, less obviously, the Westinghouse contracts on uranium supply.

however, we wish only to register that assumptions about in-bargaining commitment should not be made without careful thought. In so far as they are sustainable by appeals to reputation ideas, (Rosenthal and Landau 1979, Kreps and Wilson 1982, Wilson 1983), it is necessary that the bargaining take place in an environment that makes the necessary long-term considerations credible. But then one has the difficulty that these long-term considerations are likely to intrude to such an extent that it becomes unreasonable to suppose that the players are motivated only by the utility they will obtain as a result of the deal currently in hand.

Obviously the best way to deal with these conceptual difficulties is to abandon the attempt to model the impact of reputation considerations through hypothesis about commitment in single bargaining games in favour of a model in which commitment is *not* assumed but in which the players plan their negotiation strategies not only for the current bargaining game in which they are involved but also for the bargaining games in which they may potentially be involved in the future. As always, what is most satisfactory from a conceptual point of view is the least satisfactory from the practical standpoint of obtaining results. Here, as on the broader issues, the spirit of the Nash program applies. Clearly, we ought to make assumptions about in-bargaining commitment where our judgement suggests that this is appropriate. But, at the same time, we ought to *test* such assumptions by constructing larger and more detailed models in which these assumptions are *not* made.

Having made all these reservations, we now turn to bargaining models in which in-bargaining commitment *is* possible. In trying to formalize the negotiation procedure, one may begin by assuming that the players are equipped with a common formal language which they use to encode a sequence of messages of finite length (signals). These messages are then transmitted to the opponent according to a pre-determined timetable. Binding commitments may be modelled by postulating the existence of a stake-holding third party (a lawyer perhaps) with whom the players may elect to lodge bonds on the understanding that these will be forfeit should the commitments not be honoured. No real existence need be attributed to such a lawyer. He serves simply as a substitute for a more complex mechanism which is not to be modelled, just as the Walrasian auctioneer substitutes for an unmodelled tâtonnement process. Commitments need not be absolute. They may be made conditional on appropriate commitments by other players. Thus threats, promises and multilateral agreements may be incorporated. Nor need the commitments be absolutely binding since the element of compulsion is included in the penalty clauses for non-completion.

No mathematical analysis is necessary to see that, within such a framework, bargaining activity is likely to be reduced to a race to be the first to make a commitment[19] and that this commitment would be as all-embracing as

19 See Schelling (1960, 1966) for such commitment battles in a less abstract context.

possible so as to leave the opponent with no room for manoeuvre. If one of the players is able to win this race, then we will be left with a Stackelberg 'leader–follower' game. A more interesting situation will arise when the commitment race is not winnable in that each player's opening statement will be published simultaneously. In an extreme case, these opening statements might consist of a complex, interlocking system of conditional clauses for the unravelment of which an army of lawyers would be required. The role of the lawyers would be to search for a joint course of action for the players which would avoid triggering any penalty clauses. In essence, each player would be announcing a response function[20] to the choices of the other players and we would end up with a 'follower–follower' game to analyse.

Clearly it is not going to be easy to say anything very definitive about the latter situation beyond the observation that all activity is likely to be compressed into the opening instant of the bargaining period. However, a not unreasonable conjecture is that, for a fairly wide class of processes of the type we are considering, the bargaining game obtained will not differ in its intrinsic character from Nash's simple demand game. This begins and ends with each player simultaneously making a commitment to a utility level beneath which he will not consider an agreement.

Nash's simple demand game presents a difficulty in so far as a non-cooperative analysis is concerned. The reason is that the game has many equilibria. In particular, any individually rational, Pareto-efficient outcome is an equilibrium of the game and hence a candidate for being regarded as the solution of the game. Nash dealt with this problem by asking which of the many equilibria are stable in the presence of small 'trembles' in the information that the agents share. In this, he anticipated the essence of Selten's (1975) notion of a 'trembling-hand' equilibrium which we mention again in Section 4. As the 'trembles' are made arbitrarily small, the stable equilibria converge on a single outcome; namely, the *symmetric* Nash bargaining solution as described in Section 5.

Nash is laconic on this subject (as he is on other subjects). The first part of chapter 4 is therefore devoted to a close analysis of a very specific model of the demand game with informational trembling. Within this model, it is possible to justify Nash's conclusions in a precise manner. A more abstract justification appears in the early part of chapter 8.

A very different defence of Nash's choice of an equilibrium as the solution of his demand game appears in the later part of chapter 2. When many equilibria exist with no obvious way of distinguishing between them, players of a

20 A viable abstraction would be to identify the set of commitments open to a player with a suitably restricted set of formulae with as many free variables as there are other players. Each free variable would then be interpreted as the Gödel number of the formula chosen by another player.

game are faced with a *coordination* problem. In real-life, this is often solved by using an essentially arbitrary *convention* (e.g. 'ladies-first', 'drive on the right'). Schelling (1960) gives many instructive examples. Obviously, a convention is of no great use unless it is employed consistently. Chapter 2 explores the consequences of requiring consistency of the convention used to resolve the coordination problems which arise in Nash-type demand games. It is argued that a consistent convention must coordinate play on the *symmetric* Nash bargaining solution as before.

Given that irrevocable commitment is possible, the Nash demand game provides a satisfying means of resolving the indeterminacy problem in bargaining situations. But what if there is uncertainty about the extent to which commitments are genuinely irrevocable? This point is taken up in chapter 7 by Vincent Crawford. He modifies Nash's game so that a demand no longer represents an unconditional take-it-or-leave-it offer. Instead, a demand sets in motion a process which makes it costly, *to an uncertain extent*, for the demanding agent to later accept less than his demand. Crawford studies a simple two-stage game of incomplete information which embodies this structure. Among other things, he demonstrates that rational agents with a mutually profitable deal available may fail to agree. (See also Chatterjee and Samuelson 1983.)

3.2 Time

The relevance of Nash's demand game depends heavily on the possibility of making irrevocable commitments (as did earlier work such as that of Hicks 1934). This is also true of Nash's (1953) later 'variable threats' game. As soon as the possibility that demands may not be unconditionally binding is entertained, it becomes necessary to take account of time as a significant variable. For this reason, although seeking the simplest model with enough structure to make his point, Crawford (chapter 7) finds it necessary to use a *two-stage* model. Without commitment assumptions at all (or, rather, with minimal commitment assumptions), time assumes a dominant role. As Cross (1969) observes[21]

. . . the passage of time has a cost in terms of both dollars and the sacrifice of utility which stems from the postponement of consumption, and it will be precisely this cost which motivates the whole bargaining process. If it did not matter when the parties agreed, it would not matter whether they agreed at all.

21 For an appreciation of Cross's work and the related work of Foldes (1964) and Hicks (1934), see the appendix to Ståhl (1972). The quotation which follows is reproduced from Cramton's instructive thesis 'Time and Information in Bargaining' (1984).

In assigning the entire motivation for bargaining behaviour to the effects of the passage of time, perhaps Cross overstates the case. In the absence of informational problems, however, it does seem natural to focus on models in which commitments can only be made to stick for short time periods. In such models, a player may threaten to delay agreement (perhaps for ever). But such a threat must be credible if it is not to be dismissed as idle bombast. It is therefore natural to follow Ståhl (1972) in framing a non-cooperative analysis of such a model in terms of the notion of a subgame-perfect equilibrium. (See Section 4 and Selten 1975.)

Ståhl (1972) looked at finite-horizon models in which two players alternate in making proposals until agreement is reached (if ever). In chapter 8, a finite-horizon model is studied in which the players make simultaneous proposals at a succession of times until a pair of compatible proposals is made (if ever). However, the most significant (and certainly the most elegant) model of this type is that of Rubinstein (chapter 3). In this model, players alternate in making proposals, over a shrinking 'cake' with no exogenous bound on the length of time that they may bargain. Under mild conditions, Rubinstein shows that his model has a *unique* subgame-perfect equilibrium. Chapter 5 introduces a simple, graphical technique for dealing with such models. With this technique, it is shown that Rubinstein's conclusion is stable under a variety of modifications of his basic assumptions. Chapter 6 provides a particularly simple proof of Rubinstein's result when the underlying structure is stationary with respect to time as well as offering an interesting application to the economics of involuntary unemployment.

Rubinstein's result has been criticized as being very special in that the bargaining procedure is too specific to find ready application. But this criticism misses an important point: namely, that the result should not be viewed in isolation but as a contribution to the Nash program. In particular, it provides evidence as to the range of application of the (asymmetric) Nash bargaining solution described in Section 5. In the latter part of chapter 4 (see also chapter 5), it is shown that, if the players have preferences in respect of time which can be described by discount factors δ_1 and δ_2, then the bargaining outcome converges on an asymmetric Nash bargaining solution (in which the 'bargaining powers' β_1 and β_2 are given functions of δ_1 and δ_2) as the time interval between successive proposals is allowed to diminish to zero. Considering this limiting case corresponds to studying a bargaining situation in which bargaining frictions (i.e. the cost of delaying agreement for a fixed number of periods) are negligible compared with what is being traded. It is important that this result is *not* special to the specific model studied by Rubinstein. A very wide class of variants of his model yield the *same* conclusion (although β_1 and β_2 may be different). Chapter 5 examines some of these (see also MacLennan 1980). What we therefore have is evidence for the value of an asymmetric Nash bargaining solution in certain situations when transaction costs are negligible. It is true that data about the mini-micro structure of bargaining institutions in use are unlikely to be sufficiently specific to allow a theoretical calculation of β_1 and β_2. But this does not mean that β_1 and β_2 cannot be estimated empirically from observations of other deals using the same institutions.

This has a bearing on what has come to be called the 'Coase theorem' (Coase 1960). In chapter 3 (see also chapter 5), Rubinstein shows that *Pareto-inferior*, subgame-perfect equilibria exist for a 'fixed-cost' model in which it costs each player ct $(c > 0)$ to delay agreement by time t. This result survives when the time interval between successive bids becomes negligible and so transaction costs cease to be significant.

This result and, more significantly, that described in the previous paragraph, show that the impact of time-dependent costs on the strategic behaviour of players does *not* necessarily become negligible as the costs themselves become negligible. Some care is therefore necessary in interpreting the results in models in which such costs are set equal to zero a priori.

Actually the point is more fundamental. Suppose that the elements of the description of a formal game are listed under the headings

a structure of the game tree;
b preferences;
c information.

Time is a significant factor in each of these categories. In the quotation offered above, Cross (1969) assigns a dominating role to the manner in which time affects the preferences of the players via their attitudes to time-dependent costs. Even if informational questions are left aside, there remains the role that time plays in the description of the game tree representing the institutional framework within which the players bargain - i.e. who can do what when. Rubinstein's work (and that of others)[22] makes it clear that a classification of cooperative bargaining concepts must necessarily be founded on an appreciation of the different varieties of institutional framework within which bargaining takes place. One might say that, with perfect information, the institutional framework determines the *qualitative* nature of the result while the costs determine the result *quantitatively* (even though they may be small in an absolute sense).

22 Consider, for example, the result of removing the apparently harmless assumption of Shaked and Sutton in chapter 6 of this volume that a firm must always listen to an 'inside' worker's counteroffer before switching to an 'outsider'. Without this very reasonable assumption, a Walrasian outcome will always be obtained. See also Binmore (1983), Shaked and Sutton (1984), Rubinstein and Wolinsky (1984), Binmore and Herraro (1985), Gale (1986).

The lesson is that, without commitment, one cannot afford to be casual about the institutional assumptions made. Items which seem of little importance at first sight can sometimes be of considerable significance and we repeat our observations of Section 2 that little empirical work has been done which is of immediate use to a game theorist on this issue. Having chosen an institutional framework, an analysis is then necessary which is based on the time-dependent costs faced by the bargainers. This raises a number of spectres. The first is the general problem of modelling information which we treat briefly below. The other problems are theoretical problems in game theory. What is the correct equilibrium idea to use in an extensive form game without commitment? What is to be done when there are multiple equilibria? These and other questions are too large to be treated in more than a cursory fashion in this introduction (even if we thought we could provide adequate answers). We shall, however, make some comments in Section 4.

3.3 Information

Until fairly recently, pieces on bargaining theory had little or nothing to say about informational questions. Presumably this was because Harsanyi's theory of 'games of incomplete information' needed time to percolate. However, it is obvious that what the players know or do not know matters a great deal in practice. At the same time there is the prospect that the bargaining equivalent of the 'fog of war' may serve to simplify the class of cooperative solution concepts that we need to take seriously. Clearly these concepts cannot depend sharply on parameters which are only known very approximately. Ignorance may not be as successful a leveller as death, but it has its uses in this direction. On the other hand, the introduction of informational questions can only complicate the formal, non-cooperative analysis of bargaining contests. As far as the Nash program is concerned, this means that we have to work much harder but with the hope of a less diverse class of results.

Information may be lacking about a variety of factors in bargaining problems. This volume contains chapters which treat uncertainty about the extent to which commitment is possible (chapter 7), uncertainty about the preferences the other players may have over the possible deals that can be reached (chapter 8) and uncertainty over time-dependent costs (chapters 8 and 9). This work depends on Harsanyi's theory of 'games of incomplete information' as discussed in the companion volume. The theory deals with differences in the information the players have by introducing ideal 'chance moves' whose outcome is observed by some players but not by others. This allows players to have information which is secret from the other players. However, it is important for the analysis of a 'game of incomplete information' that this structure of the chance moves themselves (i.e. what can happen and with what probabilities) *not* be secret. The theory therefore requires the existence of an informational pool in the background which is common knowledge (Aumann 1976, Mertens and Zamir 1983, Myerson 1984). The basic idea is reasonably simple but theoretical difficulties intrude concerning how beliefs are to be updated once time enters the picture.

Chapter 8 contains an analysis of Nash's demand game in an incomplete information context (Sections 5, 6 and 7). Since commitment is assumed in this analysis, time is not a significant factor and so technical problems are minimized. It is shown that a version of the asymmetric Nash bargaining solution introduced by Harsanyi and Selten (1972) is relevant in this context. Chapter 7 introduces a two-stage model, but difficulties with time-dependent beliefs are not at issue. The later part of chapter 8 (Section 8 et seq) represents an early attempt to grapple with these difficulties without the later theoretical framework which became available - notably the Kreps-Wilson (1982) notion of a sequential equilibrium. This is employed by Rubinstein in chapter 9 where he extends his discounting model of chapter 3 to the case in which there is incomplete information over the discount factors.

These three chapters cover a fairly wide spectrum, but it will be appreciated that the basic problem is too large to be more than just touched upon in this volume. Numerous articles[23] have been written which treat different aspects of the problem, but we are still just feeling our way in so far as the formal analysis of bargaining contests of incomplete information is concerned. Certainly it would seem premature to pontificate on the shape that the general theory will take other than to observe that it will be a theory of great significance.

Instead, we turn from the situation in which players have information which is secret from each other. This returns us to the domain of games of complete information in which the rules of the game and the tastes and beliefs of the players are common knowledge. But this return to safer ground does not eliminate informational questions altogether. Because an item of information is common knowledge, it does not necessarily follow that it is *public knowledge.* Under the latter heading, we propose to include all information which can be taken for granted in the drafting of contracts. This includes such matters as the language in which the players communicate and everyday facts about the world but we are concerned with information about the *players* which can be said to be public. In the last resort, this means information which can be established in a court of law during an action for breach of contract.

In a formal treatment of this idea, the lawyers would be modelled not as creatures with motives of their own, but as mindless automata executing a

23 To mention just a few, Chatterjee and Samuelson (1983), Cramton (1984), Fudenberg et al. (1983), Matthews (1983), Mookherjee (1983), Myerson (1979, 1982), Samuelson (1984), Sobel and Takahashi (1983).

fixed procedure by rote.[24] Here, as previously, the legal process serves as a surrogate for a wider class of procedures including that via which reputations may be established. Public knowledge is anything which may serve as an input to this procedure. In general, public knowledge will be a subset of common knowledge in that public knowledge will be common not only to the players but to outsiders too.

Any contract at which the players arrive after bargaining must rely *only* on data which are public knowledge for how else could it be enforced? It may be, for example, that the bargaining contest faced by the players is one of complete information in that they face no doubts about each other's von Neumann and Morgenstern utility functions. But it does not follow from the fact that the utility functions are common knowledge among the players that they are public knowledge. To establish such information in a court of law (or the world at large) would require an appeal to the theory of revealed preference.[25] But then a player would have an incentive to reveal false preferences[26] and, although the other players might be aware of the deception, they would be helpless to prevent its implementation. Similar considerations, of course, apply to other pieces of information which may not be public knowledge where the players have the opportunity to distort the available evidence.

Suspicion should therefore be directed at cooperative solution concepts which take for granted that common knowledge is public knowledge. As chapter 11 makes clear in the context of simple bartering, the fact that contracts have to be written in terms of what can actually be weighed or measured can make a very great deal of difference[27] - in fact, all the difference between a Walrasian outcome and the Nash bargaining solution.

Finally we wish to enlarge on the point made above about the misrepresentation of utility functions. How does information about preferences become incorporated in a body of common or public knowledge? Presumably, the story is that players acquire this information by observing the actions of others and observing that others are carrying out similar observations. Possibly, social statistics may be available for examination. However, in most of the activities being observed, either directly or indirectly, the individuals under observation will not regard themselves as faceless and will be well aware that

24 And, if any pretence at realism is to be maintained, at very great cost.

25 Is this theory itself public knowledge?

26 The large literature on incentive compatibility and social choice is relevant here. See Dasgupta, Hammond and Maskin (1979) and Myerson (1984).

27 This is only apparently at variance with what Harsanyi (1977) and others have to say about restricting attention to solution concepts which depend only on genuinely 'game-theoretic variables'. Such assertions are made on the implicit assumption that the game is played *in isolation* - i.e. independently of external institutions. Thus Harsanyi quotes telephone numbers as an extraneous variable although it is obvious that situations exist for which knowledge of a certain telephone number may be the crucial factor.

their current behaviour may influence the games they have to play in the future with those around them. The fact that they are observed will therefore change their behaviour since they will need to strike a balance between current gains and potential future losses. Where the future outweighs the present, a 'true' picture of the preferences of a population may never emerge. Indeed, if an agent systematically reveals 'false' preferences, then, from an operational point of view, these *are* his preferences. Even the agent himself (especially if a politician) may not be able to distinguish at all clearly between his 'true' preferences and those he has chosen to reveal over a long period.

Chapters 10 and 11 examine such models in the context of simple exchange economies. The models are very simple, problems of commitment, time and incomplete information being ruthlessly truncated. But their structure is adequate to demonstrate that, when public knowledge is restricted to the commodity bundles actually bartered (rather than the von Neumann and Morgenstern utilities attached by the agents to these bundles), then the bargaining process is likely to zero in on a Walrasian equilibrium - even in the case when there are only two agents and so neither is negligible compared with the total size of the 'market'. Chapter 11 glosses these results by providing an axiomatic justification of a Walrasian equilibrium which mimics Nash's axiomatic defence of the Nash bargaining solution *except* in respect of the nature of the information needed to compute the bargaining outcome.

4 THE ANALYSIS OF NEGOTIATION CONTESTS

Recall that we defined a contest as a game which is to be analysed in the absence of any pre-play communication at all. A minimal necessary condition that any candidate for the solution of such a game must satisfy is that it be a *Nash equilibrium.* This remains true for 'games of incomplete information' provided these are analysed using Selten's version of the Harsanyi theory. This requires postulating an initial chance move which selects the players who actually participate in the game from known populations of potential players, each of whom is characterized by a formal statement of his preferences and beliefs. The purpose of this device is to convert a 'game of incomplete information' H with a few 'players' into a game[28] of imperfect information G with many players. The players for the game G are the potential players in the original game H. A solution for H is then identified with a solution for G and hence is to be sought for amongst the Nash equilibria of G. Such a Nash equilibrium is mathematically the same object as what is usually called a Bayesian equilibrium[29] for H.

28 Strictly a 'game of incomplete information' is not a game at all in the sense of von Neumann and Morgenstern.

29 Although it seems harsh that the Reverend Bayes should be blamed for so much of which he was quite innocent.

A Nash equilibrium is essentially a static notion - i.e. it is satisfactory without refinement[30] only for games in which time is irrelevant because each player may be regarded as making a single choice in ignorance of the choices of the other players. As we have observed earlier, such static games arise naturally in a bargaining context when commitment is possible during the negotiations. However, in static bargaining contests there is an unfortunate tendency for the set of Nash equilibria to be embarrassingly large. A *convention* is then necessary to determine which of these equilibria is to be selected as the solution (see chapter 2).

For dynamic games[31] (or for games which are formally static but which are to be analysed with some implicit dynamic story in mind), Selten (1975) has shown that refinements of the Nash equilibrium idea are necessary. Consideration of these refinements has proved very profitable in bargaining theory. The crucial paper on this topic is that of Rubinstein (chapter 3 of this volume). In short, the use of refinements of the Nash equilibrium idea which explicitly take into account the dynamics of the negotiation game can resolve altogether the problem of multiple equilibria for a fairly wide class of two-person bargaining problems. Three dynamic equilibrium ideas which have been widely used are listed below.

1 *Subgame-perfect equilibrium* (Selten 1975). A Nash equilibrium which induces a Nash equilibrium in every subgame.

2 *Trembling-hand-perfect equilibrium* (Selten 1975). A limit of Nash equilibria of perturbed games in which each agent is supposed to tremble in making decisions so that each available action is chosen with a minimum probability $\epsilon > 0$ which is then allowed to recede to zero.

3 *Sequential equilibrium* (Kreps and Wilson 1982). To specify a sequential equilibrium it is necessary to state not only the strategies which the players propose to use but also their beliefs, i.e. the probabilities to be assigned to nodes at each information set. Natural consistency requirements are imposed on these beliefs and strategy choices are then required to be optimal at each information set given the beliefs held at that information set and the strategies employed elsewhere.

A lengthy disquisition on these ideas would be out of place. We shall therefore only briefly mention some well-known difficulties. Subgame-perfect equilibria (often simply referred to as perfect equilibria in this volume) work well for some games of perfect information but do not get

30 Stability considerations require postulating some dynamics even if these lurk only implicitly in the wings. This is not to say that it may not be a good idea to postulate some background dynamics: only that it is necessary to be aware that such a postulate has been made and to be ready to defend it.

31 By which we mean a game which is not static.

their teeth sufficiently into the dynamics of games of imperfect information to provide much assistance with the problem of multiple equilibria. Trembling-hand equilibria are technically difficult to deal with and there are difficulties of interpretation when the results depend on the relative frequencies with which agents make trembles. Sequential equilibria present the problem that one has to explain where the beliefs came from since the results can be very sensitive to beliefs held at information sets which are never reached in equilibrium. Rubinstein (1983) is instructive on this point in so far as this relates to his chapter 9.

It may not be out of order to speculate that some at least of these difficulties are due to a reluctance to face the issue of how players find their way to an equilibrium. At least two routes can be distinguished which are described as *eductive* and *evolutive* in Binmore (1985). The former refers to the traditional route described by von Neumann and Morgenstern in which the equilibrating process takes place inside the players' heads. The latter refers to situations in which equilibrium is reached via an adjustment process of trial and error. As we emphasized in the introduction to the companion volume *Economic Organizations as Games*, game theorists normally defend their analysis eductively and much confusion can (and has) resulted from attempting an evolutive analysis without treating the background dynamics adequately. But, in the context of the Nash program, an eductive approach does not always carry conviction because it often implicitly assumes some measure of pre-play communication.[32] But the spirit of the Nash program is to model such pre-play communication formally. Once this has been done, one is left with nothing else to work with *except* an evolutive approach. However, recent progress in mathematical biology suggests that this may not be such a bad thing provided one is not tempted to cheat on the dynamics (see Maynard-Smith 1982, Selten 1983).

5 THE NASH BARGAINING SOLUTION

We have concentrated so far on the non-cooperative side of the Nash program because that is where recent progress has been made and where the prospects for immediate advances seem to be. However, something needs to be said to explain the apparent pervasiveness of the Nash bargaining solution. For a reasonably comprehensive account of cooperative solution concepts tailored to the bargaining problem, see Roth (1979).

32 The fact that two players know that both of them are using the same game theory book may be regarded as a form of implicit communication. But often players are explicitly assumed to meet 'in the bar the previous night'.

Suppose that n players bargain over the division of a compact, convex cake $\mathscr{X}$ in utility space. Given a reference point,[33] $\xi \in \mathscr{X}$ and positive 'bargaining powers' $\beta_1, \beta_2, \ldots, \beta_n$, we can define an (asymmetric) Nash bargaining solution as the point $\sigma \in \mathscr{X}$ at which

$$(x_1 - \xi_1)^{\beta_1}(x_2 - \xi_2)^{\beta_2} \ldots (x_n - \xi_n)^{\beta_n}$$

is maximized subject to the constraints $x \geqslant \xi$ and $x \in \mathscr{X}$. This notion has been characterized by Roth (1979) with a set of three axioms:

Axiom 1: strong individual rationality;
Axiom 2: independence of utility calibration;
Axiom 3: independence of irrelevant alternatives.

A function which maps from pairs $(\mathscr{X}, \xi)$ to points of $\mathscr{X}$ and satisfies these three axioms is necessarily an asymmetric Nash bargaining solution for some set of bargaining powers $\beta_1, \beta_2, \ldots, \beta_n$. With the addition of a fourth axiom postulating symmetry, the classical Nash bargaining solution is obtained (with $\beta_1 = \beta_2 = \ldots = \beta_n$).

The Nash bargaining solution is pervasive in this volume but excessive significance should not be attached to this[34] as its constant reappearance can be attributed to the highly simplified nature of the models studied. Note in particular the informational assumption that the outcome of negotiations depends *only* on $\mathscr{X}$ and ξ. This heavy assumption has received little attention presumably because it is not listed as a formal axiom. But, as we argued in Section 3, public knowledge is unlikely to include precise information on utilities.

Something also needs to be said about axiom 3 which has been the subject of much debate although seldom in the context of the Nash program. Consider a static bargaining contest (such as would be obtained with no constraints on commitment). Suppose that the players choose strategies from sets $S_1, S_2, \ldots, S_n$ which are independent of $\mathscr{X}$ and ξ. A function $\phi : S_1 \times \ldots \times S_n \to \mathbb{R}^n$ determines 'agreement' payoffs. The payoff vector is $\phi(s)$ provided that $\phi(s) \in \mathscr{X}$. Otherwise 'disagreement' results producing a payoff vector which we may take to be ξ for simplicity (although this is not strictly necessary for the argument). Assume that this bargaining contest always has an unequivocal solution in the sense that the solution is a Nash equilibrium

33 In this chapter, we follow Roth (1979) in avoiding the misleading description of ξ as the 'status quo' point. See the concluding remarks to Section 2. For the same reason, we avoid speaking of a model with 'fixed threats'. Both terms invite misleading implicit assumptions about commitment possibilities.

34 No significance at all should be assigned to the assertion that an ethical notion from social choice theory has emerged from a strategic analysis. Nash (1950) did not intend his notion as an ethical idea and there seems no good reason for seeking to re-interpret axiom 3 in this direction.

which Pareto-dominates any other Nash equilibria. Suppose that the solution s^* of the bargaining contest based on $\mathscr{X}$ and ξ yields the agreement outcome $\phi(s^*) > \xi$. What happens if $\mathscr{X}$ is replaced by a subset $\mathscr{Y}$ containing ξ and $\phi(s^*)$ as contemplated in axiom 3? Obviously, s^* remains a Nash equilibrium which Pareto-dominates any other Nash equilibria in the new game and therefore is the solution of this game and hence we have a justification of axiom 3. In dynamic bargaining contests (i.e. where time plays a role), the argument above needs to be adapted but still survives in a modified form.[35]

Finally, it is necessary to comment on the fact that none of the non-cooperative bargaining models which have been studied implement the Nash bargaining solution *exactly*. In each case, the implementation is *approximate* (or exact only in the limit).

35 One needs that unequivocal solutions to subgames involving the shrunken cake $\mathscr{X}_t$ available at time t exist and are simply related to the solution at time 0. This is *not* true of the Rubinstein model (chapter 3) because the solution depends on which player goes first except in the limit as the length of the time intervals is allowed to recede to zero.

REFERENCES

Aumann, R. 1976: Agreeing to disagree. *Am. Stat.*, **4**, 1236–9.

Binmore, K. 1977: Bargaining, barter and competitive equilibrium, I and II. Mimeo, London School of Economics.

Binmore, K. 1983: Bargaining and coalitions. ICERD discussion paper 83/71, London School of Economics.

Binmore, K. 1985: Modelling rational players. Forthcoming as ICERD paper, London School of Economics.

Binmore, K. and Dasgupta P. (ed.), 1986: *Economic Organizations as Games.* Basil Blackwell, Oxford.

Binmore, K. and Herrero, M. 1985: Matching and bargaining in dynamic markets. ICERD discussion paper 85/117, London School of Economics.

Chatterjee, K. and Samuelson, W. 1983: Bargaining under incomplete information. *Op. Research*, **31**, 835–51.

Coase, R. 1960: The problem of social cost. *J. Law Econ.*, 1.

Cramton, P. 1984: Time and information in bargaining. PhD thesis, Stanford University, California.

Cross, J. 1969: *The Economics of Bargaining.* Basic Books, New York.

Cutler, W. 1975: An optimal strategy for pot-limit poker. *Am. Math. Monthly*, **82**, 368–76.

Dasgupta, P. 1986: Trust as a commodity, in D. Gambetta (ed.) *Trust and Agency: Making and Breaking Cooperative Relations.* Basil Blackwell, Oxford, forthcoming (1987).

Dasgupta, P., Hammond, P. and Maskin, E. 1979: The implementation of social choice rules: some general results in incentive compatibility. *Review of Economic Studies*, **46**, 185–216.

De Menil, G. 1971: *Bargaining: Monopoly Power Versus Union Power.* MIT Press, Cambridge, Massachusetts.

Ellsberg, D. 1975: The theory and practice of blackmail. *Bargaining: Formal Theories of Negotiation.* University of Illinois Press, Urbana, Illinois.

Foldes, L. 1964: A determinate model of bilateral monopoly. *Economica*, **31**, 117–31.

Friedman, J. 1977: *Oligopoly and the Theory of Games.* North Holland, Amsterdam.

Fudenberg, D., Levine, D. and Tirole, J. 1983: Infinite-horizon models of bargaining with one-sided incomplete information. Forthcoming in Roth, A. (ed.) *Game-Theoretic Models of Bargaining.* CUP, Cambridge.

Gale, D. 1986: Bargaining and competition, I and II. *Econometrica*, **54**, 785–806, 807–18.

Harsanyi, J. 1977: *Rational Behaviour and Bargaining Equilibrium in Games and Social Situations.* CUP, Cambridge.

Harsanyi, J. and Selten, R. 1972: A generalised Nash solution for two person bargaining games with incomplete information. *Man. Sci.*, **18**, 80–106.

Hicks, J. 1934: *The Theory of Wages.* Macmillan, London.

Kreps, D. and Wilson, R. 1982a: Sequential equilibrium. *Econometrica*, **50**, 863–94.

Kreps, D. and Wilson, R. 1982b: Reputation and imperfect information. *J. Econ. Th.*, **27**, 253–79.

Lakatos, L. 1976: *Proofs and Refutations: The Logic of Mathematical Discovery.* CUP, Cambridge.

MacLennan, A. 1980: A general non-cooperative theory of bargaining. Working paper, University of Toronto, Ontario.

Mathews, S. 1983: Selling to risk averse buyers with unobservable tastes. *J. Econ. Th.*, 29.

Maynard Smith, J. 1982: *Evolution and the Theory of Games.* CUP, Cambridge.

Mertens, J. and Zamir, S. 1983: Formalisation of Harsanyi's notion of 'type' and 'consistency' in games with incomplete information. CORE discussion paper, Université Catholique de Louvain.

Mookherjee, D. 1983: One stage simultaneous bargaining with incomplete information. Research paper 392, Stanford Graduate School of Business, California.

Myerson, R. 1982: Cooperative games with incomplete information. Research paper 528, Northwestern University, Evanston, Illinois.

Myerson, R. 1984: An introduction to game theory. Northwestern University discussion paper 623, Evanston, Illinois.

Nash, J. 1950: The bargaining problem. *Econometrica*, **18**, 286–95.

Nash, J. 1951: Non-cooperative games. *Ann. Math.*, **54**, 289–95.

Nash, J. 1953: Two-person cooperative games. *Econometrica*, **21**, 128–40.

Rosenthal, R. and Landan, H. 1979: A game-theoretic analysis of bargaining with reputations. *J. Math. Psych.*, **20**, 233–55.

Roth, A. 1979: Axiomatic models of bargaining. *Lecture Notes in Economics and Mathematical Systems*, 170, Springer-Verlag, Berlin.

Rubinstein, A. 1982: Perfect equilibrium in a bargaining model. *Econometrica*, **50**, 97–109.

Rubinstein, A. 1983: Conjectures in bargaining models with incomplete information. Forthcoming in Roth, A. (ed.) *Game Theoretic Models of Bargaining.* CUP, Cambridge.

Rubinstein, A. and Wolinsky, A. 1984: Equilibrium in a market with sequential bargaining. ICERD discussion paper 84/91, London School of Economics.

Samuelson, W. 1984: Bargaining under asymmetric information. Forthcoming in *Econometrica.*

Schelling, T. 1960: *The Strategy of Conflict.* OUP, Oxford.

Schelling, T. 1966: *Arms and Influence*, Yale University Press, New Haven, Connecticut.

Selten, R. 1975: Reexamination of the perfectness concept for equilibrium points in extensive games. *Int. J. Game. Th.*, **4**, 25–55.

Selten, R. 1983: Evolutionary stability in extensive 2-person games. Bielefeld working papers 121, 122, Bielefeld.

Shaked, A. and Sutton, J. 1984: The semi-Walrasian economy. ICERD discussion paper 84/98, London School of Economics.

Sobel, J. and Takahashi, I. 1983: A multi-stage model of bargaining. *Rev. Econ. Studies*, **50**, 411–26.

Ståhl, I. 1972: *Bargaining Theory.* Economics Research Institute, Stockholm.

von Neumann, J. and Morgenstern O. 1944: *Theory of Games and Economic Behaviour.* Princeton University Press, Princeton, New Jersey.

Weintraub, R. 1975: *Conflict and Cooperation in Economics.* Macmillan, London.

Wilson, R. 1983: Reputations in games and markets. ISMSSS report 434, Stanford University, California.

2

Nash Bargaining Theory I

K. Binmore

1 INTRODUCTION

In three remarkable papers written in the early 1950s, Nash outlined an approach to the theory of games in general and to the theory of bargaining in particular which remains of the greatest significance. In this and subsequent chapters, it is intended to discuss and to extend the general approach to bargaining outlined in those papers.

In the current chapter, we propose to re-examine Nash's axiomatic derivation of the 'Nash bargaining solution'. Of the various axioms which Nash formulated, it is the so-called 'independence of irrelevant alternatives' on which criticism has been concentrated and with good reason. The main part of this chapter is therefore concerned with the formulation and justification of an alternative version of this axiom.

Although this chapter is devoted to the classical 'Nash bargaining solution' it should be emphasized that we do *not* regard this as being universally applicable to all two-person bargaining problems. We shall in fact argue that the 'Nash bargaining solution' is appropriate only for a rather special class of bargaining problems with a highly simplified informational structure. In a later chapter we shall discuss some bargaining problems with a somewhat less simplified information structure. In the case of bartering over commodities, for example, we shall argue that, with the natural informational structure in this context, the application of Nash's principles leads *not* to the 'Nash bargaining solution' but to a competitive equilibrium.[1]

1 This is a somewhat unconventional viewpoint. Harsanyi (1977) for example, has a special 'rationality postulate' B3 to exclude possibilities of this type. Note, incidentally, that although we are very close to Harsanyi on many issues, we do not find his rationality postulates, taken as a whole, particularly convincing. In particular, 'Zeuthen's principle', by means of which he justifies the 'Nash bargaining solution' seems less than adequate as a basis for such a justification.

We prefix the technical discussion with a brief account of Nash's general approach to bargaining theory as we understand it. This is a strictly *game-theoretic* approach – i.e. Nash is concerned with optimal play between rational players with individual goals. Like Harsanyi (1977), we feel that it is important to distinguish carefully between such an approach and *behavioural* theories (i.e. theories which seek to describe how individuals behave in practice) on the one hand and *ethical* theories (i.e. theories which prescribe outcomes which are optimal for society 'as a whole') on the other. Attempts have been made by various authors to interpret Nash's work from both of the latter viewpoints. In particular, it is common for the 'Nash bargaining solution' to be referred to as an 'arbitration scheme'. Such an interpretation seems to us of limited interest and was certainly not intended by Nash. In any case, this chapter is not concerned with this interpretation except for some brief comments in Section 4 (note 1).

2 NASH THEORY: GENERAL APPROACH

Nash felt that the most fundamental type of game is what we shall call a contest. By this is to be understood a formal two-person game in extensive form which is to be analysed on the assumption that no pre-play communication is permitted. The question as to what constitutes a solution to such a game is by no means clear-cut. It is, however, far less cloudy than the question of what constitutes a solution to a game in general. In particular, for a contest, one has available the concept of a Nash equilibrium (or, more generally, that of a perfect equilibrium).

In games other than contests, the nature of the possible pre-play interaction between players is often all-important. In a bargaining situation, for example, if one player is able to get in first with a pre-emptive, self-binding commitment, it will usually be to his advantage to do so. His opponent is then left with a 'take it or leave it' problem. Of course, if the roles of the players are reversed or if simultaneous opening commitments are possible, the situation becomes quite different.

Nash (1951) proposed to deal with the problem of pre-play interaction in the following way. Let G denote a formal game. One can imagine the various possible steps in the pre-play negotiation procedure as moves in a larger 'negotiation game' N – i.e. one envisages a formalization of the negotiation procedure. A strategy for the formal game N tells one how to conduct the negotiations under all possible eventualities and how to choose a (possibly mixed) strategy for G contingent upon the course which the negotiations took. The negotiation game N is then analysed as a contest and its solution leads to an outcome of the game G which we shall call the 'negotiated outcome' of G.

A *bargaining contest B* associated with a game *G* will be understood to be a negotiation contest in which any pair of negotiation strategies (one for each player) can be classified as compatible or incompatible. If compatible, the choice of these negotiation strategies by the players results in a binding agreement (contract) being signed which determines the strategies to be used by the players in the play of *G*. If incompatible, the choice of these negotiation strategies results in a pre-determined outcome of *G* which is called the status quo.

As an example we consider the standard two-person Edgeworth market game in which the first player has quantity *W* of wheat and the second has *F* of fish. We can make this into a formal game *G* by stipulating that the strategies for the two players consist of the unilateral dispatch of some proportion of their initial endowment to the other. It is assumed that each player has a utility function defined over the set of relevant commodity bundles and that this is known to the other. An opportunity for barter now exists and the question is: what will the result of this barter be? Following Nash, we take the view that this will depend on the nature of the negotiation strategies open to the players. If no binding agreements are possible, for example, then the 'negotiated outcome' of *G* will be the initial endowment point (for 'Prisoner's Dilemma' reasons). Binding agreements are therefore necessary for an interesting negotiated outcome. But what precisely this negotiated outcome will be depends on the nature of the negotiation procedure and what the players regard as constituting the solution of the resulting negotiation contest.

A simple bargaining procedure considered by Nash consists of each player simultaneously announcing a real number. The announcement of the real number d_1 by the first player represents a demand for an outcome of *G* which yields the first player a utility of d_1. Similarly for the announcement of d_2 by the second player. A pair of announcements is compatible if a pair of (possibly mixed) strategies for *G* exists whose implementation leads to each player receiving the utility demanded. In this case, a binding agreement to implement such a pair of strategies is signed. If the demands are incompatible, the status quo results. In this case the status quo is identified with the initial endowment (or 'no-trade') point.

The original game *G* preceded by this simple bargaining procedure constitutes a bargaining contest *B*. What can be said about the 'solution' of *B*?

The bargaining contest *B* has many Pareto-efficient Nash equilibria (and this is typical of bargaining contests). In fact, *every* Pareto-efficient outcome of *G* can be realized as the result of a Nash equilibrium of *B* under suitable concavity assumptions on the utility functions. Under such assumptions (and with free disposal) the set $\mathscr{X}$ of achievable utility pairs will be as illustrated in figure 2.1. It is clear that under these circumstances, the demands d_1 and d_2 together constitute a Nash equilibrium of *B* if and only if (d_1, d_2) is a

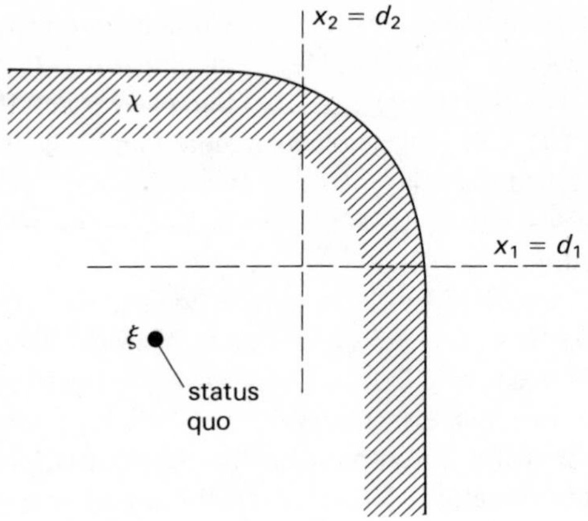

Figure 2.1

Pareto-efficient point of $\mathcal{X}$ (provided that trivial cases are excluded by constraining d_1 and d_2 to values which do not force disagreement independently of the opponent's action).

For the theory to make sense, the players must recognize one of these Nash equilibria as the solution of the game. For this purpose the players require a *convention*. Such a convention may appear entirely arbitrary if the play of a single game is considered by itself.[2] But a pattern will emerge when the play of several games is observed provided that

1 The players all subscribe to a common game theory
2 The games are played separately (i.e. the players do not treat the games as subgames of a larger super-game).

One would not, for example, have to observe many cars driving on the left to deduce that one is in a country where the conventional solution of the game of 'Chicken' is to drive on the left. The point is that if one knows the solution of one or more distinct games and also knows something about the possible conventions which may be in use, then one should be able to say something about the solution of a further distinct game whose play has not been observed.

We shall return to this point later. Assuming however that the players share a common convention, we can then solve the bargaining contest B. This solution then determines the negotiated outcome of the original game G. In

2 Although it should be noted that Nash (1950) manages an ingenious defence of the convention built into the 'Nash bargaining solution' based on its stability in the face of infinitesimal informational perturbations.

the particular case we have been considering, the negotiated outcome consists of a contract to exchange certain specified quantities of wheat and fish.

The above example was intended to illustrate the general approach advocated by Nash and to focus upon the important fact that the determination of the negotiated outcome will depend, in general, not only on a knowledge of the bargaining procedure to be employed but also on a knowledge of the conventions built into the game theory which the players will use to resolve the bargaining game *B*.

There are, of course, an unlimited number of bargaining procedures with which one could precede the formal play of *G*. Equally, there are numerous conventions which might be used in the solution of the resulting bargaining contests. Nash, however, felt that it was reasonable to expect that many of these seemingly disparate bargaining situations would have essentially the same strategic structure and hence lead to the same negotiated outcome. In support of this view, Nash (1950) formulated a system of axioms for the negotiated outcome of a game *G* with given status quo. These axioms implicitly incorporate assumptions about the bargaining procedure and the conventions to be used in solving the resulting bargaining contest. However, the axioms are sufficiently general that it is reasonable to suppose that they would be satisfied for a fairly wide class of bargaining problems. On the basis of his axioms, Nash was able to show that only *one* negotiated outcome (in terms of utilities) is possible. This is the 'Nash bargaining solution'.

An account of Nash's axiomatic argument is given in Section 4. In this account we point out that his axioms concerning Pareto efficiency and symmetry are unnecessary to obtain a result.[3] We then turn to Nash's much criticized 'independence of irrelevant alternatives'. This has been interpreted in various ways but we follow Nash (and Harsanyi) in regarding this as an 'institutional assumption' – i.e. it is an assumption about the convention the players are to use in resolving the bargaining game *G*. It is, however, an assumption which is not particularly easy to justify as it stands. Harsanyi (1977), for example, makes no attempt to do so and offers instead an alternative derivation of the 'Nash bargaining solution' using 'Zeuthen's principle'.

Our response is similar to Harsanyi's in that we do not feel that the 'independence of irrelevant alternatives' can be adequately justified on a priori grounds. We discuss this point in Section 5 and then go on to propose a modified condition which we call 'convention consistency'. This is a weaker assumption than the 'independence of irrelevant alternatives' and has the additional advantage that its status as a condition which requires that the same convention is to be used in resolving distinct games is quite explicit.

3 This work is not original. See, for example, Roth (1979) which contains further references.

We feel however quite strongly that a condition which asserts the consistent use of a convention cannot be expected to have a wide domain of applicability. Conventions are matters of convenience rather than necessity and a condition which leads to the 'Nash bargaining solution' can only be expected to be applicable when it is convenient for the players to ignore certain aspects of the bargaining situation. In particular, it is essential that they make no use of the structure of G except in so far as this is incorporated in the pair $(\mathcal{X}, \xi)$. In a later chapter, we shall argue that this and other information may very well be relevant to the bargaining process and we shall describe how the principles implicit in Nash's axiom system can be adapted in certain special cases to take account of such extra informational structure. This will lead to 'bargaining solutions' different to that given by Nash. At this point we wish only to stress that we do not accept the claims for universality which are sometimes advanced in respect of the 'Nash bargaining solution'.

3 TECHNICAL PRELIMINARIES

In the background we have a set of formal two-person games G given in extensive form with each terminal node (outcome) labelled with an element ω from an outcome space $\Omega = \Omega(G)$. We shall use the notation

$$\lambda = \begin{Bmatrix} \omega_1 & \omega_2 & \cdots & \omega_k \\ p_1 & p_2 & \cdots & p_k \end{Bmatrix}$$

to denote the lottery λ in which the prize ω_j is available with probability p_j. The set of all such lotteries with prizes in Ω will be denoted by $\mathcal{L} = \mathcal{L}(\Omega)$. Each player is assumed to be equipped with a preference relation defined on $\mathcal{L}$ which can be described by a von Neumann and Morgenstern utility function defined on Ω. The linear extension of these utility functions to $\mathcal{L}$ will be denoted by $\phi_1 : \mathcal{L} \to \mathbb{R}$ and $\phi_2 : \mathcal{L} \to \mathbb{R}$. We define $\phi : \mathcal{L} \to \mathbb{R}^2$ by

$$\phi(\lambda) = (\phi_1(\lambda), \phi_2(\lambda)).$$

The payoff region $\mathcal{X} = \mathcal{X}(G)$ for the game G is

$$\mathcal{X} = \phi(\mathcal{L})$$

The set $\mathcal{X}$ is a convex subset of $\mathbb{R}^2$. We shall also assume that $\mathcal{X}$ is compact.

The play of the game G is to be preceded by a period of bargaining. One of a set of formalized bargaining procedures is used, thus embedding G in a larger bargaining contest B. Two bargaining strategies (one for each player) are either compatible or incompatible. If incompatible, their play results in a pre-specified outcome q of $\mathcal{L}$ called the status quo. We use the notation

$$\boldsymbol{\xi} = \phi(q).$$

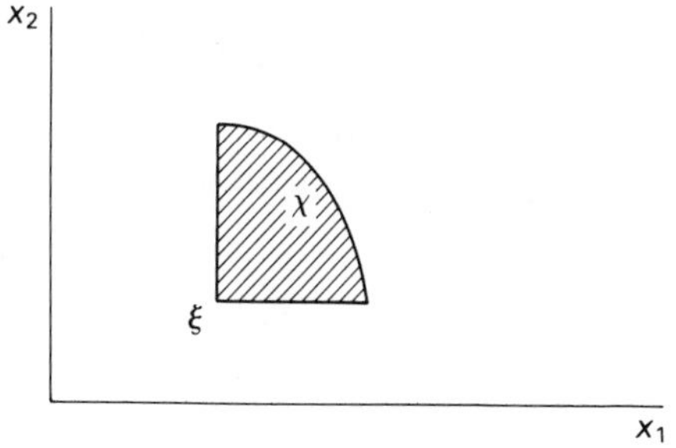

Figure 2.2

The following technical assumptions will be made. These assumptions are not essential and are made only to simplify the analysis somewhat (see figure 2.2).

a $\mathcal{X} \subseteq \{\mathbf{x} : \mathbf{x} \geqslant \boldsymbol{\xi}\}$
b $\mathcal{X} \cap \{\mathbf{x} : \mathbf{x} > \boldsymbol{\xi}\} \neq \phi$
c $(\boldsymbol{\xi} < \mathbf{x} \leqslant \mathbf{y}$ and $\mathbf{y} \in \mathcal{X}) \Rightarrow \mathbf{x} \in \mathcal{X}$.

The set of all pairs $(\mathcal{X}, \boldsymbol{\xi})$ with these properties will be denoted by D.

The set of axioms which Nash gave for his 'bargaining solution' is a list of restrictions on a function $f : D \to \mathbb{R}^2$ (see Section 4). Nash showed that his axioms specify a unique function $f : D \to \mathbb{R}^2$. But, in order for it to be sensible to interpret $f(\mathcal{X}, \boldsymbol{\xi})$ as a 'bargaining solution', further assumptions are necessary. These are implicit in Nash's paper (1950) but we shall state them explicitly.

Axiom A. *The bargaining contest B has a solution*

We denote the point of $\mathcal{L}$ which results from the play of the solution strategies by $\nu = \nu(B)$. Thus ν is the negotiated outcome of G.

Axiom B. *There exists a function $f : D \to \mathbb{R}^2$ such that*

$$f(\mathcal{X}, \boldsymbol{\xi}) = \phi(\nu).$$

Axiom B is an important assumption but it is one which is often taken for granted. As a universal principle, its validity seems doubtful. For example, it does not seem unreasonable that the negotiated outcome of the example of Section 2 should depend on the configuration of the Edgeworth box as well as on $\mathcal{X}$ and $\boldsymbol{\xi}$. We shall, however, postpone considerations of this nature to a later chapter and restrict our attention in this chapter to situations in which axiom B does apply. That is to say, we shall regard axiom B as a restriction on the class of bargaining problems to be considered rather than as a universal principle.

4 NASH'S AXIOM SYSTEM

The following is a list of somewhat freely adapted versions of Nash's axioms. These axioms are concerned with the mathematical properties of a function $f: D \to \mathbb{R}^2$.

Axiom 1 (*feasibility*)

$$\boldsymbol{\xi} < f(\mathcal{X}, \boldsymbol{\xi}) \in \mathcal{X}.$$

Axiom 2 (*invariance*)

For any strictly increasing, affine transformation $\alpha : \mathbb{R}^2 \to \mathbb{R}^2$,

$$f(\alpha\mathcal{X}, \alpha\boldsymbol{\xi}) = \alpha f(\mathcal{X}, \boldsymbol{\xi}).$$

Axiom 3 (*efficiency*)

$$\mathbf{y} > f(\mathcal{X}, \boldsymbol{\xi}) \Rightarrow \mathbf{y} \notin \mathcal{X}.$$

Axiom 4 (*independence of irrelevant alternatives*)

$$f(\mathcal{X}, \boldsymbol{\xi}) \in \mathcal{Y} \subseteq \mathcal{X} \Rightarrow f(\mathcal{Y}, \boldsymbol{\xi}) = f(\mathcal{X}, \boldsymbol{\xi}).$$

Axiom 5 (*symmetry*)

$$\textit{If } \tau : (x_1, x_2) \mapsto (x_2, x_1), \textit{ then}$$
$$f(\tau\mathcal{X}, \tau\boldsymbol{\xi}) = \tau f(\mathcal{X}, \boldsymbol{\xi}).$$

Nash shows that, with these assumptions, $f(\mathcal{X}, \xi)$ is the point $x \in \mathbb{R}^2$ at which

$$\max_{\substack{\mathbf{x} \in \mathcal{X} \\ \mathbf{x} > \xi}} (x_1 - \xi_1)(x_2 - \xi_2)$$

is achieved. The axioms therefore determine f uniquely. This uniquely determined function f is interpreted as in Section 3 and $f(\mathcal{X}, \boldsymbol{\xi})$ is called the 'Nash bargaining solution' of the game G.

We discuss the role of the axioms briefly in Section 5. The rest of this section is devoted to the proof of a version of Nash's theorem.

Given $\tau\,(0 < \tau < 1)$, define $g_\tau : D \to \mathbb{R}^2$ by

$$g_\tau(\mathcal{X}, \boldsymbol{\xi}) = T$$

where T is the point indicated in figure 2.3. In this diagram, RS is a supporting line to the convex set $\mathcal{X}$ and T is chosen so that

$$\frac{ST}{SR} = \tau.$$

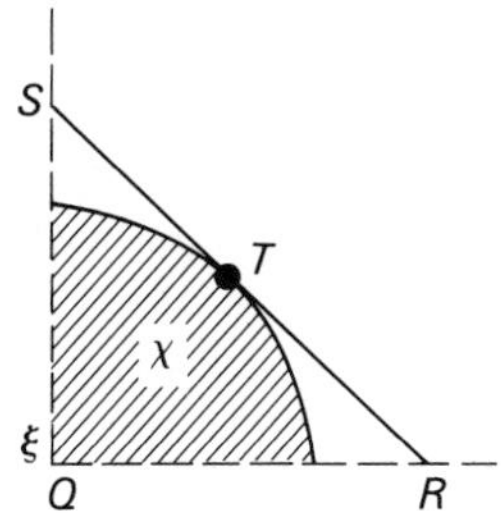

Figure 2.3

Alternatively, one may define $T = (x_1, x_2)$ to be the point at which

$$\max_{\substack{\mathbf{x} \in \mathscr{X} \\ \mathbf{x} > \boldsymbol{\xi}}} (x_1 - \xi_1)^{\tau}(x_2 - \xi_2)^{1-\tau}$$

is achieved. We refer to $g_\tau(\mathscr{X}, \boldsymbol{\xi})$ as an 'asymmetric Nash bargaining solution' of G and to τ and $1-\tau$ as the associated 'bargaining powers'.

It is trivial to show that g_τ satisfies axioms 1–4 inclusive and that g_τ satisfies axiom 5 if and only if $\tau = \frac{1}{2}$. The following theorem of Roth (1979) deserves a little more attention.

Theorem 1. Let $f: D \to \mathbb{R}^2$ satisfy axioms 1, 2 and 4. Then there exists a $\tau (0 < \tau < 1)$ such that

$$f = g_\tau$$

Proof. Let

$$\Delta = \{(x_1, x_2) : x_1 + x_2 \leqslant 1, x_1 \geqslant 0, x_2 \geqslant 0\}$$

and let $f(\Delta, \mathbf{0}) = \boldsymbol{\delta}$. By axiom 1, $\boldsymbol{\delta} > 0$. Suppose that $\boldsymbol{\delta}$ does not lie on the line segment AB (see figure 2.4).

By axiom 4, $f(\Delta_1, \mathbf{0}) = \boldsymbol{\delta}$. Let α be the affine transformation which maps $0 \to 0$, $B \to B$ and $D \to A$. Then, by similarity, $\alpha\boldsymbol{\delta}$ lies on AB and so $\alpha\boldsymbol{\delta} \neq \boldsymbol{\delta}$. By axiom 2, $f(\alpha\Delta_1, \alpha\mathbf{0}) = \alpha\boldsymbol{\delta}$. But $f(\alpha\Delta_1, \alpha\mathbf{0}) = f(\Delta, \mathbf{0}) = \boldsymbol{\delta}$. This is a contradiction and so $\boldsymbol{\delta}$ lies on AB as in figure 2.5.

Choose τ so that

$$\tau = \frac{BC}{BA}$$

in figure 2.5. Given any $(\mathscr{X}, \xi)$ in D construct the diagram of figure 2.3 with $T = g_\tau(\mathscr{X}, \boldsymbol{\xi})$. Apply the unique strictly increasing, affine transformation

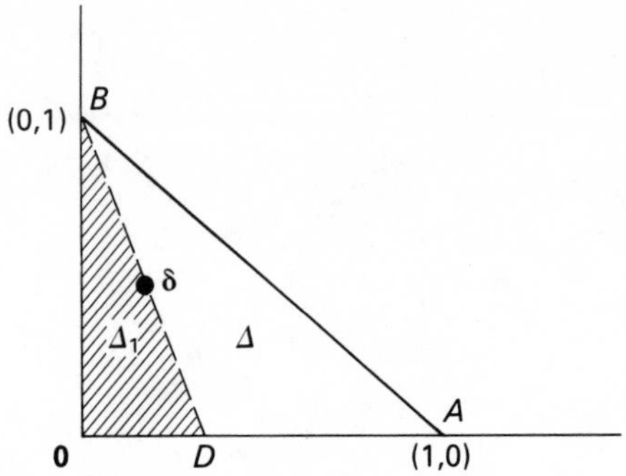

Figure 2.4

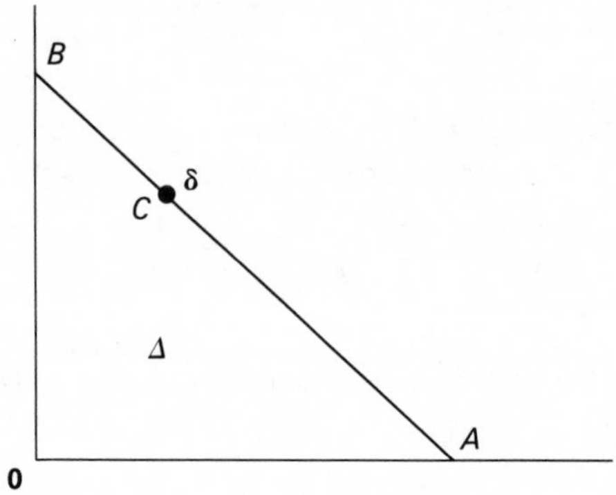

Figure 2.5

β: $\mathbb{R}^2 \to \mathbb{R}^2$ which maps $Q \to 0$, $S \to B$ and $R \to A$. By similarity, $T \to G$. Hence, by axiom 4,

$$f(\beta \mathscr{X}, \beta \boldsymbol{\xi}) = f(\Delta, \mathbf{0}) = \beta g_T(\mathscr{X}, \boldsymbol{\xi}).$$

Thus, by axiom 2,

$$f(\mathscr{X}, \boldsymbol{\xi}) = \beta^{-1} f(\beta \mathscr{X}, \beta \boldsymbol{\xi}) = g_T(\mathscr{X}, \boldsymbol{\xi}).$$

Note 1. The 'feasibility axiom' requires that $\boldsymbol{\xi} < f(\mathscr{X}, \boldsymbol{\xi})$. If we require only that $\boldsymbol{\xi} \leqslant f(\mathscr{X}, \boldsymbol{\xi})$, three further possibilities for f emerge. Those are the 'dictatorships' g_0 and g_1 and the 'apathy' solution $f(\mathscr{X}, \boldsymbol{\xi}) \equiv \boldsymbol{\xi}$.

Sometimes axiom 4 is treated as a condition of 'collective rationality' but this interpretation has a number of difficulties. One of these is that a result similar to Arrow's theorem holds. Recall that we are interpreting the points of $\mathbb{R}^2$ as pairs of von Neumann and Morgenstern utilities. The 'sure-thing principle' therefore requires that a collective preference relation should have straight line indifference curves if it is to be regarded as 'rational'. But the only such collective preference relations consistent with f as a choice function are the three degenerate cases mentioned in the first paragraph.

Note 2. The function $f: D \to \mathbb{R}^2$ is point-valued. If instead $f(\mathscr{X}, \boldsymbol{\xi})$ is a non-empty, compact, convex set in $\mathscr{X}$, the theorem remains true provided axiom 1 is altered to read $\boldsymbol{\xi} < f(\mathscr{X}, \boldsymbol{\xi}) \subseteq \mathscr{X}$ and axiom 4 is altered to read

$$\mathscr{Y} \subseteq \mathscr{X} \text{ and } \mathscr{Y} \cap f(\mathscr{X}, \boldsymbol{\xi}) \neq 0 \Rightarrow f(\mathscr{Y}, \boldsymbol{\xi}) \subseteq f(\mathscr{X}, \boldsymbol{\xi}).$$

5 INDEPENDENCE OF IRRELEVANT ALTERNATIVES

Theorem 1 invites us to focus our attention on axioms 1, 2 and 4. Axiom 1 has very little content. Axiom 2 asserts that the solution is to be independent of the calibration of the utility functions. In the context of axiom B, this seems an entirely innocent assumption. Indeed, in a treatment which worked directly in terms of preference relations on $\mathscr{L}$ (rather than utility functions on $\mathscr{L}$), axiom 2 would become redundant, its content having been absorbed into the appropriate version of axiom B.

It follows that, axiom B having been accepted, the full weight of theorem 1 falls on axiom 4 – i.e. the 'independence of irrelevant alternatives'. We see this as a *consistency* condition – i.e. it is a condition which asserts that the 'same' convention is to be used in solving two games. (Presumably this is what Nash meant by describing it as an 'institutional assumption'.) Given theorem 1, it is only necessary to observe the outcome of one bargaining contest to be able to predict the outcome of any other distinct bargaining contest. In particular, if a symmetric bargaining contest turns out to yield a symmetric solution, then the 'Nash bargaining solution' applies (within the range of validity of the analysis).

Some consistency condition is clearly necessary in order to obtain a result. What is not so clear is why the 'independence of irrelevant alternatives' is an appropriate consistency condition.

The usual defence offered of the 'irrelevance of independent alternatives' is a defence of the condition in the arbitration scheme context. We therefore offer a strictly game-theoretic alternative. This is a sort of 'perfect equilibrium' argument but applied to conventions rather than equilibria.

Suppose that a bargaining game is played in two stages. First a partial agreement is signed followed by a full agreement. We shall suppose that the partial agreement is conditional in the sense that the players are freed from any obligations under the partial agreement should it happen that the ensuing negotiations break down and no full agreement is reached. The status quo then results. On the other hand, if full agreement is reached, the provisions of the full agreement must be consistent with those of the partial agreement.

Suppose that this bargaining game specifies the pair $(\mathscr{X}, \boldsymbol{\xi})$. If the bargaining procedure given above falls within the scope of these procedures covered by axioms A and B, then the solution of the game will result in a payoff pair $\boldsymbol{\sigma}=f(\mathscr{X}, \boldsymbol{\xi})$. Now suppose that a partial agreement has been signed as a first step to implementing the solution strategies. Such a partial agreement will have the effect of restricting the possible payoff pairs available at the full agreement stage to a subset $\mathscr{Y}$ of $\mathscr{X}$. Clearly $\boldsymbol{\sigma} \in \mathscr{Y}$ since otherwise the solution would not be implementable by this route. At the second stage of the bargaining procedure, the players therefore have to consider the pair $(\mathscr{Y}, \boldsymbol{\xi})$ where $\boldsymbol{\sigma} \in \mathscr{Y}$. If the bargaining procedure of this subgame falls within the scope of those procedures covered by axioms A and B, then the solution of this subgame will result in a payoff pair $\boldsymbol{\tau}=f(\mathscr{Y}, \boldsymbol{\xi})$. But an inconsistency now results unless $\boldsymbol{\sigma}=\boldsymbol{\tau}$ and hence the 'independence of irrelevant alternatives'.

The weakness in this argument is the assumption that the subgame which comes into existence after partial agreement has been reached can be analysed without reference to the bargaining which leads up to the partial agreement. Even if commitments during the initial round of negotiations binding players to certain courses of action in the final round of negotiations are forbidden, it may well still be the case that the convention the players use in resolving the final subgame will depend on the signals they exchanged while negotiating the preliminary partial agreement or on the nature of the partial agreement itself.

It is difficult to see how this criticism can be answered in the general case. However, in certain special cases, a defence is readily available. Suppose that, in the concluding subgame, the players are able to evaluate the bargaining strategies available to them without reference to the partial agreement they have reached. For example, a brother and sister might be left a house, a car and a sum of money without instructions on the division of the inheritance. Assume their utilities linear in money and that they begin by agreeing (conditional on a subsequent final agreement being reached) on who gets the house and who gets the car. In general, this agreement will require the use of some random device. They now turn to the question of how to divide the money. Observe that this matter can be considered quite separately from the problem of the house and the car in the sense that both parties can put the negotiations over the money in the hands of their lawyers *without* informing the lawyers of the result of their preliminary partial agreement but the lawyers will still have all the information they need to negotiate properly on behalf of their principals. This would certainly not be the case, for example, if no money were involved and the preliminary partial agreement were concerned with the disposal of half the house and half the car and the final agreement were concerned with the disposal of the other halves of the house and car.

In situations where the negotiation of the final agreement can be left in the hands of lawyers who are ignorant of the preliminary partial agreement, it clearly makes a good deal more sense to apply axiom B to the concluding subgame than it does in the general case. The remainder of this chapter is devoted to expressing this notion in formal terms and then to showing that the appropriately weakened form of the 'independence of irrelevant alternatives' so obtained remains adequate for the proof of a version of theorem 1.

6 SEPARABLE GAMES

In section 7 we introduce a substitute for the independence of irrelevant alternatives which we call 'convention consistency'. This condition asserts that, if a bargaining game can be decomposed into two distinct subgames which it makes sense to analyse separately, then the convention built into the game theory in use will yield the same result when applied to the original bargaining game as when applied to each subgame separately. In this section, we describe formally what we mean by saying that a game is separable.

Two preference relations, $\lesssim_1$ and $\lesssim_2$, will be said to be *compatible* if, for any x and y,

$$x <_1 y \Rightarrow x \lesssim_2 y.$$

(It then follows that $x <_2 y \Rightarrow x \lesssim_1 y$.) If $\lesssim_1$ and $\lesssim_2$ are defined on the set $\mathscr{L} = \mathscr{L}(\Omega)$ of lotteries with prizes in Ω and satisfy the von Neumann and Morgenstern rationality postulates, then they may be described by linear utility functions $\psi_1 : \mathscr{L} \to \mathrm{I\!R}$ and $\psi_2 : \mathscr{L} \to \mathrm{I\!R}$. If the preference relations $\lesssim_1$ and $\lesssim_2$ are also compatible on $\mathscr{L}$, then

$$\psi_1 = A\psi_2 + B \text{ or } \psi_2 = A\psi_1 + B$$

where $A \geqslant 0$ and B are constants. (If $A > 0$, $\lesssim_1$ and $\lesssim_2$ are the same preference relation. If $A = 0$, at least one of $\lesssim_1$ or $\lesssim_2$ is complete indifference.)

Suppose now that Ω_1 and Ω_2 are two outcome spaces and that $\mathscr{L}_1 = \mathscr{L}(\Omega_1)$ and $\mathscr{L}_2 = \mathscr{L}(\Omega_2)$ are the associated spaces of lotteries. Put $\Omega = \Omega_1 \times \Omega_2$ and $\mathscr{L} = \mathscr{L}(\Omega)$. If λ and μ are lotteries in $\mathscr{L}_1$ and $\mathscr{L}_2$ respectively given by

$$\lambda = \begin{Bmatrix} x_1 & x_2 & \ldots \\ p_1 & p_2 & \ldots \end{Bmatrix}; \quad \mu = \begin{Bmatrix} y_1 & y_2 & \ldots \\ q_1 & q_2 & \ldots \end{Bmatrix}$$

we shall use the notation (λ, μ) to indicate the lottery

$$(\lambda, \mu) = \begin{Bmatrix} (x_1, y_1) & (x_1, y_2) & (x_2, y_1) & \ldots \\ p_1 q_1 & p_1 q_2 & p_2 q_1 & \ldots \end{Bmatrix}$$

in $\mathscr{L}$.

If $\lesssim$ is a preference relation defined on $\mathscr{L}$, we shall say that $\lesssim$ evaluates $\mathscr{L}_1$ and $\mathscr{L}_2$ separately if and only if,

$$(\lambda, \mu) < (\lambda, \mu') \Rightarrow (\lambda', \mu) \lesssim (\lambda', \mu')$$

and

$$(\lambda, \mu) < (\lambda', \mu) \Rightarrow (\lambda, \mu') \lesssim (\lambda', \mu')$$

for all $(\lambda, \lambda') \in \mathscr{L}_1^2$ and $(\mu, \mu') \in \mathscr{L}_2^2$.

Theorem 2. Let $\Phi : \mathscr{L} \to \mathbb{R}$ be a linear utility function which describes the preference relation $\lesssim$ on $\mathscr{L} = \mathscr{L}(\Omega)$. If $\lesssim$ evaluates $\mathscr{L}_1$ and $\mathscr{L}_2$ separately, then there exist linear functions $\alpha : \mathscr{L}_1 \to \mathbb{R}$ and $\beta : \mathscr{L}_2 \to \mathbb{R}$ such that

$$\Phi(\lambda, \mu) = a\, \alpha(\lambda)\beta(\mu) + b\, \alpha(\lambda) + c\, \beta(\mu) + d$$

for appropriate constants a, b, c and d.

Proof. Define the preference relation $\lesssim_\lambda$ on $\mathscr{L}_2$ by

$$\mu \lesssim_\lambda \mu' \Leftrightarrow (\lambda, \mu) \lesssim (\lambda, \mu')$$

and define $\lesssim_\mu$ on $\mathscr{L}_1$ similarly. If trivial cases are excluded, we may suppose that $\lesssim_{\lambda_0}$ and $\lesssim_{\mu_0}$ are not total indifference relations. Put

$$\alpha(\lambda) = \Phi(\lambda, \mu_0)$$

$$\beta(\mu) = \Phi(\lambda_0, \mu).$$

Since $\lesssim_\lambda$ is compatible with $\lesssim_{\lambda_0}$ and $\lesssim_\mu$ is compatible with $\lesssim_{\mu_0}$, it follows that

$$\Phi(\lambda, \mu) = A_\mu \alpha(\lambda) + B_\mu$$

$$\Phi(\lambda, \mu) = C_\lambda \beta(\mu) + D_\lambda$$

where $A_\mu \geqslant 0$ and $C_\lambda \geqslant 0$. If we normalize $\Phi(\lambda, \mu)$ by taking $\Phi(\lambda_0, \mu_0) = 0$, we can obtain that $B_\mu = \beta(\mu)$ and $D_\lambda = \alpha(\lambda)$. Thus

$$A_\mu \alpha(\lambda) + \beta(\mu) = C_\lambda \beta(\mu) + \alpha(\lambda)$$

$$\alpha(\lambda)(A_\mu - 1) = \beta(\mu)(C_\lambda - 1)$$

from which it follows that $A_\mu = a\, \beta(\mu) + 1$ and $C_\lambda = a\, \alpha(\lambda) + 1$ for some constant a. Thus

$$\Phi(\lambda, \mu) = a\alpha(\lambda)\beta(\mu) + \alpha(\lambda) + \beta(\mu)$$

and hence the result.

It is convenient to normalize so that, if (m_1, m_2) is the worst outcome of $\Omega_1 \times \Omega_2$ and (M_1, M_2) is the best outcome, then

$$\Phi(m_1, m_2) = 0; \quad \alpha(m_1) = 0; \quad \beta(m_2) = 0;$$

$$\Phi(M_1, M_2) = 1; \quad \alpha(M_1) = 1; \quad \beta(M_2) = 1.$$

We then obtain that

$$\Phi(\lambda, \mu) = (1-\alpha(\lambda), \alpha(\lambda)) \begin{pmatrix} 0 & c \\ b & 1 \end{pmatrix} \begin{pmatrix} 1-\beta(\lambda) \\ \beta(\lambda) \end{pmatrix}$$

where $b = \Phi(M_1, m_2)$ and $c = \Phi(m_1, M_2)$.

It is useful to think of the linear functions $\alpha : \mathscr{L}_1 \to \mathbb{R}$ and $\beta : \mathscr{L}_2 \to \mathbb{R}$ as von Neumann and Morgenstern utility functions of preference relations $\lesssim_1$ and $\lesssim_2$ given on $\mathscr{L}_1$ and $\mathscr{L}_2$ separately.

Now suppose that Ω is the outcome space of a bargaining contest B with status quo q (see section 3). Suppose that $\Omega = \Omega_1 \times \Omega_2$ and that $q = (q_1, q_2)$. We shall say that B is *separable* provided the following conditions are satisfied.

1 The players' preference relations on $\mathscr{L}$ both evaluate $\mathscr{L}_1$ and $\mathscr{L}_2$ separately.
2 $\phi(\mathscr{L}) = \phi(\mathscr{L}_1 \times \mathscr{L}_2)$.
3 For each $\lambda \in \mathscr{L}_1$ and each $\mu \in \mathscr{L}_2$, $\phi(q_1, q_2) = \phi(\lambda, q_2) = \phi(q_1, \mu)$.

The first condition means that if B is split into two bargaining contests B_1 and B_2 with respective outcome spaces Ω_1 and Ω_2, then it is feasible for the players to negotiate these games separately of each other. The second condition asserts that any outcome which can be achieved in the bargaining contest B can also be achieved by separate negotiations in B_1 and B_2. The third condition says that failure to agree in one of the three games is regarded by both players as being equivalent to failure to agree in all three games. The point of this latter condition is that it means that neither player gains any 'bargaining power' as a result of the completion of negotiations in B_1. Failure to agree in B_2 renders any agreement reached in B_1 worthless.

In this chapter, we are restricting our attention to circumstances under which axiom B holds. It is therefore necessary to translate the three conditions given above into conditions relating to pairs of the form $(\mathscr{X}, \boldsymbol{\xi})$. For this purpose, we require some further notation.

If $\mathbf{x} \in \mathbb{R}^2$ and $\mathbf{y} \in \mathbb{R}^2$, we define

$$\mathbf{x} * \mathbf{y} = (x_1 y_1, x_2 y_2).$$

Similarly, if $\mathscr{X}$ and $\mathscr{Y}$ are sets in $\mathbb{R}^2$, we define

$$\mathscr{X} * \mathscr{Y} = \{\mathbf{x} * \mathbf{y} : (\mathbf{x}, \mathbf{y}) \in \mathscr{X} \times \mathscr{Y}\}$$

Recall that D denotes the set of all admissible pairs $(\mathscr{X}, \boldsymbol{\xi})$. We shall use E to denote the set of all $(\mathscr{X}, \boldsymbol{\xi})$ in D such that $\boldsymbol{\xi} = 0$ and $x_1 = 1$ and $x_2 = 1$ are supporting lines for $\mathscr{X}$. We shall say that a pair $(\mathscr{X}, \boldsymbol{\xi})$ in E is *normalized* (see figure 2.6).

Theorem 3. Suppose that a bargaining contest B corresponding to the pair $(\mathscr{Z}, \mathbf{0})$ in E is separable and that the bargaining contests into which it separates correspond to the pairs $(\mathscr{X}, \mathbf{0})$ and $(\mathscr{Y}, \mathbf{0})$ in E. Then $\mathscr{Z} = \mathscr{X} * \mathscr{Y}$.

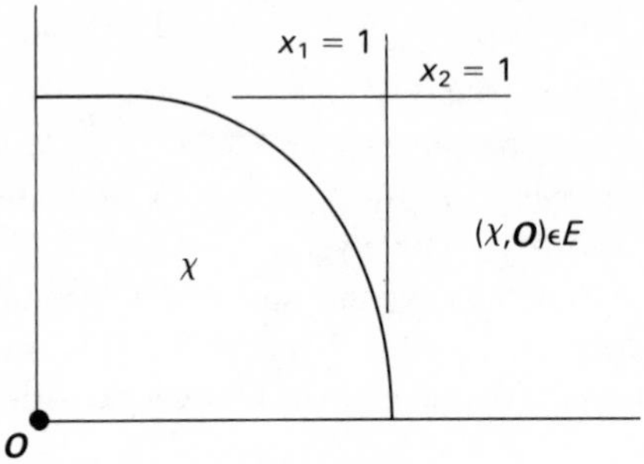

Figure 2.6

Proof. In the formula for normalized utility functions which follows theorem 2, take $(m_1, m_2) = (q_1, q_2)$. Condition 3 for separability assures that $b = c = 0$ for each player's utility function on $\mathscr{L}$. Thus

$$\left.\begin{aligned} \phi_1(\lambda, \mu) &= \alpha_1(\lambda)\beta_1(\mu) \\ \phi_2(\lambda, \mu) &= \alpha_2(\lambda)\beta_2(\mu) \end{aligned}\right\}$$

provided the utility functions are suitably normalized. But $\mathscr{Z} = \phi(\mathscr{L})$, $\mathscr{X} = \alpha(\mathscr{L}_1)$ and $\mathscr{Y} = \beta(\mathscr{L}_2)$. Using condition 2 for separability, we obtain that

$$\mathscr{Z} = \phi(\mathscr{L}) = \phi(\mathscr{L}_1 \times \mathscr{L}_2) = \alpha(\mathscr{L}_1) * \beta(\mathscr{L}_2) = \mathscr{X} * \mathscr{Y}.$$

In the context of axiom B it follows that the assertion that a bargaining contest B is separable into bargaining contests B_1 and B_2 translates into the assertion that a normalized pair $(\mathscr{Z}, \mathbf{0})$ may be written in the form

$$(\mathscr{Z}, \mathbf{0}) = (\mathscr{X} * \mathscr{Y}, \mathbf{0})$$

for appropriate normalized pairs $(\mathscr{X}, \mathbf{0})$ and $(\mathscr{Y}, \mathbf{0})$.

Note 3. Suppose in theorem 2 that (m_1, m_2) is the worst outcome in $\Omega = \Omega_1 \times \Omega_2$ and (M_1, M_2) is the best outcome. Then the sign of the number

$$\Delta = \Phi(m_1, m_2) + \Phi(M_1, M_2) - \Phi(M_1, m_2) - \Phi(m_1, M_2)$$

is a significant quantity. It describes the individual's preference between the lotteries

$$\nu_1 = \begin{Bmatrix} (m_1, m_2) & (M_1, M_2) \\ \frac{1}{2} & \frac{1}{2} \end{Bmatrix} \quad \text{and } \nu_2 = \begin{Bmatrix} (M_1, m_2) & (m_1, M_2) \\ \frac{1}{2} & \frac{1}{2} \end{Bmatrix}.$$

If $\Delta > 0$ (i.e. $\nu_1 > \nu_2$ and so we are dealing with a 'go for broke' player), unnormalized α, β and Φ can be found so that $\Phi(\lambda, \mu) = \alpha(\lambda)\beta(\mu)$. If $\Delta = 0$,

α, β and Φ can be found so that $\Phi(\lambda, \mu) = \alpha(\lambda) + \beta(\mu)$. If $\Delta < 0$, α, β and Φ can be found so that $\Phi(\lambda, \mu) = -\alpha(\lambda)\beta(\mu)$.

Note 4. In all the above discussion we have assiduously avoided using the word 'independent' (except in the phrase 'independence of irrelevant alternatives') since this would beg too many questions in this context. Following a suggestion of John Howard, it would seem reasonable to say that a preference relation $\gtrsim$ on $\mathscr{L}$ evaluates $\mathscr{L}_1$ and $\mathscr{L}_2$ *independently* if, for any lottery $\nu \in \mathscr{L}$,

$$\nu = \begin{Bmatrix} (x_1, y_1) & (x_1, y_2) & (x_2, y_1) & \dots \\ p_{11} & p_{12} & p_{21} & \dots \end{Bmatrix},$$

the preference relation $\lesssim$ takes into account only the *marginal* probabilities

$$p_i = \sum_j p_{ij} \quad \text{and} \quad q_j = \sum_i p_{ij}.$$

Under these conditions it is clear that $\Delta = 0$ in note 3 and it follows that $\lesssim$ evaluates $\mathscr{L}_1$ and $\mathscr{L}_2$ independently if and only if we may write

$$\Phi(\lambda, \mu) = \alpha(\lambda) + \beta(\mu).$$

This form for $\Phi(\lambda, \mu)$ is, of course, usually taken for granted when a contest is to be repeated several times.

7 CONVENTION CONSISTENCY

It is fairly easy to see what is meant by saying that a convention like 'ladies first' or 'drive on the right' is used consistently. It is not quite so easy to see what it means to say that a convention is used consistently in the presence of axiom B.

We shall argue in the following way. Suppose that B is a separable bargaining contest which separates into bargaining contests B_1 and B_2. Suppose that, if B, B_1 and B_2 are each considered separately, the game theory in use assigns to each solutions which lead to ν, ν_1 and ν_2 as negotiated outcomes of the three games respectively. It need not be true that

$$\nu = (\nu_1, \nu_2)$$

but, if this is the case, then we shall say that the game theory is consistent (in its treatment of B, B_1 and B_2). In order for consistency defined in this way to be a useful concept, it is of course essential that there be no advantage to the players in linking the negotiations for the contests B_1 and B_2: hence the long discussion of separability in the previous section.

Translating this notion into language appropriate for axiom B, we obtain the following substitute for the 'independence of irrelevant alternatives' (axiom 4).

Axiom 6 (*convention consistency*)

Let $(\mathcal{X}, \mathbf{0})$ and $(\mathcal{Y}, \mathbf{0})$ be in E. If $(\mathcal{X} * \mathcal{Y}, \mathbf{0}) \in E$, then

$$f(\mathcal{X} * \mathcal{Y}, \mathbf{0}) = f(\mathcal{X}, \mathbf{0}) * f(\mathcal{Y}, \mathbf{0}).$$

Essentially, the condition is a restatement of the 'independence of irrelevant alternatives' into which an operational notion of the meaning of 'irrelevant' has been introduced.

Note 5. Requirement 2 for separability in section 6 is more stringent than perhaps it needs to be. One might drop this requirement and adapt the consistency condition so that it asserts that, *if* $\nu \in \mathcal{L}_1 \times \mathcal{L}_2$, then $\nu = (\nu_1, \nu_2)$. In axiom 6 one would then write $\mathcal{Z} = \text{conv}\,(\mathcal{X} * \mathcal{Y})$ and require that, *if* $f(\mathcal{Z}, \mathbf{0}) \in \mathcal{X} * \mathcal{Y}$, then $f(\mathcal{Z}, \mathbf{0}) = f(\mathcal{X}, \mathbf{0}) * f(\mathcal{Y}, \mathbf{0})$.

8 CONVENTION CONSISTENCY AND THE NASH BARGAINING SOLUTION

Theorem 4. Suppose that $f: D \to \mathbb{R}^2$ satisfies axioms 1, 2, 3 and 6. Then, for some $t \in (0, 1)$,

$$f = g_t.$$

We shall actually prove this theorem with D replaced by the normalized subset E introduced in section 6. Axiom 2 is needed only to deduce the theorem as stated above from the special form we prove below. The replacement of D by E means in particular, that the status quo point $\boldsymbol{\xi}$ is normalized at $\mathbf{0}$. We can therefore economize on notation by simply writing $\mathcal{X}$ or $f(\mathcal{X})$ rather than $(\mathcal{X}, \mathbf{0})$ or $f(\mathcal{X}, \mathbf{0})$. Two lemmas are required. Recall, from theorem 1, that Δ denotes the unit simplex.

Lemma 1. For any $\mathcal{X} \in E$ there exists $\mathcal{Y} \in E$ such that $\mathcal{X} * \mathcal{Y} = \Delta$.

Proof. We assume that $\mathcal{X}$ is strictly convex. The general result then follows by an approximating argument.

If $0 < \tau < 1$, we define $\mathbf{a}_\tau = (a_\tau, b_\tau) = g_\tau(\mathcal{X})$ and $\mathbf{A}_\tau = (A_\tau, B_\tau) = g_\tau(\mathcal{U}_\tau)$ where

$$\mathcal{U}_\tau = \{\mathbf{X} : \mathbf{X} \geqslant \mathbf{0} \text{ and } a_\tau X_1 + b_\tau X_2 \leqslant 1\}.$$

Then $A_\tau = \tau a_\tau^{-1}$ and $B_\tau = (1 - \tau) b_\tau^{-1}$ so that

$$\mathbf{a}_\tau * \mathbf{A}_\tau = (\tau, 1 - \tau) \tag{2.1}$$

and hence $\mathbf{a}_\tau * \mathbf{A}_\tau$ is a Pareto-efficient point of Δ.

It is easily seen that $\mathbf{A}_\tau$ is a continuous function of τ on (0, 1) and that the limits $\mathbf{A}_{0+}$ and $\mathbf{A}_{1-}$ both exist. It is natural to define $\mathbf{a}_1 = \mathbf{i} = (0, 1)$ and $\mathbf{a}_0 = \mathbf{j} = (0, 1)$. For $\tau = 0$ or $\tau = 1$, the definition given above for $\mathbf{A}_\tau$ becomes indeterminate but we take $\mathbf{A}_0 = \mathbf{A}_{0+}$ and $\mathbf{A}_1 = \mathbf{A}_{1-}$. Note that, since it is always true that $\mathbf{a}_\tau \geqslant (\tau, 1-\tau)$, we have that $0 \leqslant A_\tau \leqslant 1$ and $0 \leqslant B_\tau \leqslant 1$ for all $\tau \in [0, 1]$. Moreover, $A_1 = 1$ and $B_0 = 1$.

The set $\mathscr{Y}$ is defined as the set of all $\mathbf{X} \in \mathbf{R}^2$ for which there exists a $\tau \in [0,1]$ such that $\mathbf{0} \leqslant \mathbf{X} \leqslant \mathbf{A}_\tau$. To prove that $\mathscr{Y} \in E$ we need to show that $\mathscr{Y}$ is convex. Its extreme points are the points $\mathbf{A}_\tau (0 \leqslant \tau \leqslant 1)$ together with $\mathbf{0}$, $\mathbf{i}$ and $\mathbf{j}$. We proceed by assuming that for some value of $\tau \in (0, 1)$ there exist τ_1 and τ_2 distinct from τ such that $\mathbf{A}_\tau$ belongs to the convex hull of $\{\mathbf{0}, \mathbf{A}_{\tau_1}, \mathbf{A}_{\tau_2}\}$ and seek a contradiction. Suppose therefore that $\mathbf{A}_\tau = p_1 \mathbf{A}_{\tau_1} + p_2 \mathbf{A}_{\tau_2}$ where $p_1 \geqslant 0$, $p_2 \geqslant 0$ and $p_1 + p_2 \leqslant 1$. The set V_τ defined by $V_\tau = \{\mathbf{x} : \mathbf{x} \geqslant 0 \text{ and } A_\tau x_1 + B_\tau x_2 \leqslant 1\}$ has the property that $\mathbf{A}_\tau * V_\tau = \Delta$. Also, because $A_\tau x_1 + B_\tau x_2 = 1$ is the supporting line to X at $\mathbf{a}_\tau$, the set $V_{\tau_1} \cap V_{\tau_2}$ contains $\mathscr{X}$. Since $\mathscr{X}$ is strictly convex, we can find non-zero extreme points $\mathbf{b}_1$ and $\mathbf{b}_2$ of $V_{\tau_1} \cap V_{\tau_2}$ such that $\mathbf{a}_\tau = q_1 \mathbf{b}_1 + q_2 \mathbf{b}_2$ where $q_1 \geqslant 0$, $q_2 \geqslant 0$ and $q_1 + q_2 < 1$. (Note the strict inequality which arises from a positive weighting for zero.) Hence

$$\mathbf{a}_\tau * \mathbf{A}_\tau = \sum_i \sum_j p_i q_j \mathbf{A}_{\tau_i} * \mathbf{b}_j .$$

But the points $\mathbf{A}_{\tau_i} * \mathbf{b}_j$ all lie in Δ. Since $\Sigma\Sigma p_i q_j < 1$, it follows that $\mathbf{a}_\tau * \mathbf{A}_\tau$ cannot be a Pareto-efficient point of Δ which contradicts (2.1).

It remains to show that $\mathscr{X} * \mathscr{Y} = \Delta$. It is trivial that $\mathscr{X} * \mathscr{Y} \subseteq \operatorname{conv}(\mathscr{X} * \operatorname{ext} \mathscr{Y})$. Since $\mathscr{X} \subseteq V_\tau$ and $\mathbf{A}_\tau * V_\tau = \Delta$, it follows that $\mathbf{A}_\tau * \mathbf{x} \in \Delta$ for all $\mathbf{x} \in \mathscr{X}$. Since $\mathbf{0} * \mathbf{x} \in \Delta$, $\mathbf{i} * \mathbf{x} \in \Delta$ and $\mathbf{j} * \mathbf{x} \in \Delta$ for all $\mathbf{x} \in \mathscr{X}$, we deduce that $\mathscr{X} * \mathscr{Y} \subseteq \Delta$. But from (1) we have that $\Delta \subseteq \mathscr{X} * \mathscr{Y}$.

Lemma 2. Suppose that $\mathscr{X} \in E$, $\mathscr{Y} \in E$ and $\mathscr{X} * \mathscr{Y} \in E$. Then if $\mathbf{x} \in \mathscr{X}$ and $\mathbf{y} \in \mathscr{Y}$, we have that

$$\mathbf{x} * \mathbf{y} = g_\tau(\mathscr{X} * \mathscr{Y}) \Leftrightarrow \mathbf{x} = g_\tau(\mathscr{X}) \text{ and } \mathbf{y} = g_\tau(\mathscr{Y}).$$

Proof. Write $\mathbf{X} = g_\tau(\mathscr{X})$ and $\mathbf{Y} = g_\tau(\mathscr{Y})$.

1 We have that

$$\begin{aligned} \mathbf{X} * \mathbf{Y} &= \{\max_{\mathbf{x} \in \mathscr{X}} x_1^\tau x_2^{1-\tau}\}\{\max_{\mathbf{y} \in \mathscr{Y}} y_1^\tau y_2^{1-\tau}\} \\ &= \max_{(\mathbf{x}, \mathbf{y}) \in \mathscr{X} \times \mathscr{Y}} (x_1 y_1)^\tau (x_2 y_2)^{1-\tau} \\ &= \max_{\mathbf{z} \in \mathscr{X} * \mathscr{Y}} z_1^\tau z_2^{1-\tau} = g_\tau(\mathscr{X} * \mathscr{Y}). \end{aligned}$$

It follows that $\mathbf{x} = g_\tau(\mathcal{X})$ and $\mathbf{y} = g_\tau(\mathcal{Y}) \Rightarrow \mathbf{x} * \mathbf{y} = g_\tau(\mathcal{X} * \mathcal{Y})$.

2 If $\mathbf{x} \in \mathcal{X}$ and $\mathbf{y} \in \mathcal{Y}$ we have that $x_1^\tau x_2^{1-\tau} \leqslant X_1^\tau X_2^{1-\tau}$ and $y_1^\tau y_2^{1-\tau} \leqslant Y_1^\tau Y_2^{1-\tau}$. If $\mathbf{x} \neq g_\tau(\mathcal{X})$ or $\mathbf{y} \neq g_\tau(\mathcal{Y})$ then one of these inequalities is strict and hence $(x_1 y_1)^\tau (x_2 y_2)^{1-\tau} < (X_1 Y_1)^\tau (X_2 Y_2)^{1-\tau}$. Thus $\mathbf{x} * \mathbf{y} \neq g_\tau(\mathcal{X} * \mathcal{Y})$. It follows that $\mathbf{x} * \mathbf{y} = g_\tau(\mathcal{X} * \mathcal{Y}) \Leftrightarrow \mathbf{x} = g_\tau(\mathcal{X})$ and $\mathbf{y} = g_\tau(\mathcal{Y})$.

Proof of theorem 4. From axioms 1 and 3 we deduce the existence of a $\tau \in (0, 1)$ such that $f(\Delta) = g_\tau(\Delta)$. Now consider any $\mathcal{X} \in E$. By lemma 1 there exists a $\mathcal{Y} \in E$ such that $\mathcal{X} * \mathcal{Y} = \Delta$. By axiom 6, $f(\mathcal{X}) * f(\mathcal{Y}) = f(\Delta) = g_\tau(\Delta)$. It follows from lemma 2 that $f(\mathcal{X}) = g_\tau(\mathcal{X})$.

Note 6. Axiom 3 is used in the proof of theorem 4 only in locating $f(\Delta)$ on the boundary of Δ. We show that this fact can be deduced instead from axiom 7 below. Since axiom 7 implies axiom 6, theorem 4 remains valid with axioms 3 and 6 replaced by axiom 7.

Axiom 7. (*Strong convention consistency*)

Let $\mathcal{X}$ and $\mathcal{Y}$ be in E and let $\mathcal{Z} = \text{conv}\,(\mathcal{X} * \mathcal{Y})$. Then, if $f(\mathcal{Z}) \in \mathcal{X} * \mathcal{Y}$,

$$f(\mathcal{Z}) = f(\mathcal{X}) * f(\mathcal{Y}).$$

Suppose that $\boldsymbol{\delta} = f(\Delta)$ is an interior point of Δ. Then $\delta_1 > 0$, $\delta_2 > 0$ and $\delta_1 + \delta_2 < 1$. It follows that there exists a point $\mathbf{d}$ satisfying $0 < d_1 < 1$, $0 < d_2 < 1$ and $\delta_1 d_1^{-1} + \delta_2 d_2^{-1} = 1$. Put $\delta_1 = e_1 d_1$ and $\delta_2 = e_2 d_2$ so that $\mathbf{d} * \mathbf{e} = \boldsymbol{\delta}$. We define $\mathcal{X} = \text{conv}\ \{\mathbf{0}, \mathbf{i}, \mathbf{j}, \mathbf{d}\}$. Then $\boldsymbol{\delta} \in \mathcal{X} * \Delta$ because $\mathbf{d} \in \mathcal{X}$ and $\mathbf{e} \in \Delta$. But $\Delta = \text{conv}(\mathcal{X} * \Delta)$ and so, by axiom 7, $\boldsymbol{\delta} = f(\Delta) = f(\mathcal{X}) * f(\Delta) = f(\mathcal{X}) * \delta$ from which we deduce that $f(\mathcal{X}) = (1, 1)$. Since $(1, 1) \notin \mathcal{X}$, this is a contradiction.

REFERENCES

Harsanyi, J. C. 1977: *Rational Behaviour and Bargaining Equilibrium in Games and Social Situations.* CUP, Cambridge.

Nash, J. F. 1950: The bargaining problem. *Econometrica*, **18**, 155–62.

Nash, J. F. 1951: Non-cooperative games. *Annals of Mathematics*, **54**, 286–95.

Nash, J. F. 1953: Two-person cooperative games. *Econometrica*, **21**, 128–40.

Roth, A. E. 1979: *Axiomatic Models of Bargaining.* Lecture notes in economics and mathematical systems No. 170, Springer-Verlag, London.

3

Perfect Equilibrium in a Bargaining Model

A. Rubinstein

Two players have to reach an agreement on the partition of a pie of size 1. Each has to make, in turn, a proposal as to how it should be divided. After one player has made an offer, the other must decide either to accept it, or to reject it and continue the bargaining. Several properties which the players' preferences possess are assumed. The perfect equilibrium partitions (PEP) are characterized in all the models satisfying these assumptions.

Specially, it is proved that when every player bears a fixed bargaining cost for each period (c_1 and c_2), then: (i) if $c_1 < c_2$ the only PEP gives all the pie to 1; (ii) if $c_1 > c_2$ the only PEP gives to 1 only c_2.

In the case where each player has a fixed discounting factor (δ_1 and δ_2) the only PEP is $(1-\delta_2)/(1-\delta_1\delta_2)$.[1]

1 INTRODUCTION

When I refer in this paper to the *bargaining problem* I mean the following situation and question:

Two individuals have before them several possible contractual agreements. Both have interests in reaching agreement but their interests are not entirely identical. What 'will be' the agreed contract, assuming that both parties behave rationally?

I begin with this clarification because I would like to prevent the common confusion of the above problem with two other problems that may be asked about the bargaining situation, namely:

a the positive question - what is the agreement reached in practice;
b the normative question - what is the just agreement.

1 This research was supported by the UK Social Sciences Research Council in connection with the project 'Incentives, Consumer Uncertainty, and Public Policy', and by Rothschild Foundation. It was undertaken while I was a research fellow at Nuffield College, Oxford. I would like to thank J. Mirrlees and Y. Shiloni for their helpful comments. I owe special thanks to Ken Binmore for his encouragement and remarks.

Edgeworth (1932) presented this problem 100 years ago, considering it the most fundamental problem in economics. Since then it seems to have been the source of considerable frustration for economic theorists. Economists often talk in the following vein (beginning of Cross 1965, p. 67):

Economists traditionally have had very little to say about pure bargaining situations in which the outcome is clearly dependent upon interactions among only a few individuals.

The 'very little' referred to above is that the agreed contract is individual-rational and is Pareto optimal; i.e. it is no worse than disagreement, and there is no agreement which both would prefer. However, which of the (usually numerous) contracts satisfying these conditions will be agreed? Economists tend to answer vaguely by saying that this depends on the 'bargaining ability' of the parties.

Many attempts have been made in order to get to a clear cut answer to the bargaining problem. Two approaches may be distinguished in the published literature. The first is the strategic approach. The players' negotiating manoeuvres are moves in a non-cooperative game and the rationality assumption is expressed by investigation of the Nash equilibria. The second approach is the axiomatic method.

One states as axioms several properties that it would seem natural for the solution to have and then one discovers that the axioms actually determine the solution uniquely. (Nash 1953, p. 129)

(For a survey of the axiomatic models of bargaining, see Roth 1979.) The purpose of this approach is to bypass the difficulties inherent in the strategic approach. We make assumptions about the solution without specifying the bargaining process itself. Notice that in order to be relevant to our problem, these axioms may only either restrict the domain of the solution or be obtained from the assumption of rationality. Thus, for example, Nash's symmetry axiom can be considered as an assumption that all the differences between the players can be expressed in the set of utility pairs arising from the possible contracts and that there is no other relevant element that distinguishes between them. But, the key axiom in most axiomatizations – the 'Independence of Irrelevant Alternatives' has not received a proper defence and in fact it is more suited to the normative question (see Luce and Raiffa 1957, chapter 2).

It was Nash himself who felt the need to complement the axiomatic approach (see Nash 1950) with a non-cooperative game. (For a wider discussion, see chapter 2.) In his second paper on the solution that he proposed, Nash (1953) proved that the solution is the limit of a sequence of equilibria of bargaining games. These models, however, are highly stylized and artificial. Among the later works, I mention here three, wherein the bargaining is represented by a multi-stage game. Ståhl (1972, 1977) and Krelle (1975) assume the existence of a known finite number of bargaining periods and their solutions are based on dynamic programming. Rice (1977) uses the notion of a differential game. The bargaining period is identified with an interval, equilibrium strategies are the limits of 'step-wise' strategies, and the lengths of those steps tend to zero.

In this paper I will adopt the strategic approach. I will consider the following bargaining situation: two players have to reach an agreement on the partition of a pie of size 1. Each has to make, in turn, a proposal as to how it should be divided. After one party has made such an offer, the other must decide either to accept it or to reject it and continue with the bargaining. The players' preference relations are defined on the set of ordered pairs of the type (x, t) (where $0 \leqslant x \leqslant 1$ and t is a non-negative integer). The pair (x, t) is interpreted as '1 receives x and 2 receives $1-x$ at time t'.

This paper is limited to the investigation of a family of models in which the preferences satisfy:

(A-1) 'pie' is desirable,
(A-2) 'time' is valuable,
(A-3) continuity,
(A-4) stationarity (the preference of (x, t) over $(y, t+1)$ is independent of t),
(A-5) the larger the portion the more 'compensation' a player needs for a delay of one period to be immaterial to him.

The two elements in which the parties may differ are the negotiating order (who has 'first turn') and the preferences.

Two sub-families of models to which I will refer, are:

a *Fixed bargaining cost:* i's preference is derived from the function $y - c_i t$, i.e. every player bears a fixed cost for each period.
b *Fixed discounting factor:* i's preference is derived from the function $y \cdot \delta_i^t$, i.e. every player has a fixed discounting factor.

So my first step has been to restrict the bargaining situation to be considered. Secondly, I will give a severe interpretation to the rationality requirement by investigating *perfect* equilibria (see Selten 1965, 1975). A perfect equilibrium is one where not only the strategies chosen at the beginning of the game form an equilibrium, but also the strategies planned after all possible histories (in every subgame).

Quite surprisingly[2] this leads to the isolation of a single solution for most of the cases examined here. For example, in the fixed bargaining cost model, it turns out that if $c_1 > c_2$, 1 receives c_2 only. If $c_1 < c_2$, 1 receives all the pie.

2 Especially considering that the perfect equilibrium concept has been 'disappointing' when applied to the supergames, see Aumann and Shapley (1976), Kurz (1978), and Rubinstein (1977, 1979, 1980).

If $c_1 = c_2$, any partition of the pie from which 1 receives at least c_1 is a perfect equilibrium partition (PEP). In other words, a weaker player gets almost 'nothing'; he can at most get the loss which his opponent incurs during one bargaining round. In the fixed discounting factor model there is one PEP, 1 obtaining $(1-\delta_2)/(1-\delta_1\delta_2)$. This solution is continuous, monotonic in the discounting factors, and gives relative advantage to the player who starts the bargaining.

The work closest to that appearing here, is that of Ståhl[3] (1972, 1977). He investigates a similar bargaining situation but which has a finite and known negotiating time horizon, and in which the pie can be only partitioned discretely. Ståhl studies cases for which there exists a single PEP which is independent of who has the first move.

The discussed bargaining model may be modified in numerous ways, many being only technical modifications. However, I would like to point out one type of modification which I believe to be extremely interesting. A critical assumption in the model is that each player has complete information about the preference of the other. Assume on the other hand that 1 and 2 both know that 1 has a fixed bargaining cost. They both know that 2 has also a fixed bargaining cost, but only 2 knows its actual value. In such a situation some new aspects appear. I will try to conclude from 2's behaviour what the true bargaining cost is, and 2 may try to cheat 1 by leading him to believe that he, 2, is 'stronger' than he actually is. In such a situation one can expect that the bargaining will continue for more than one period. I hope to deal with this situation in another paper.

2 THE BARGAINING MODEL

Two players, 1 and 2, are bargaining on the partition of a pie. The pie will be partitioned only after the players reach an agreement. Each player, in turn, offers a partition and his opponent may agree to the offer 'Y' or reject it 'N'. Acceptance of the offer ends the bargaining. After rejection, the rejecting player then has to make a counter offer and so on. There are no rules which bind the players to any previous offers they have made.

Formally, let $S = [0, 1]$. A partition of the pie is identified with a number s in the unit interval by interpreting s as the proportion of the pie that 1 receives. Let s_i be the portion of the pie that player i receives in the partition s: that is $s_1 = s$ and $s_2 = 1 - s$.

Let F be the set of all sequences of functions $f = \{f^t\}_{t=1}^{\infty}$, where $f^1 \in S$, for t odd $f^t\colon S^{t-1} \to S$, and for t even $f^t\colon S^t \to \{Y, N\}$. ($S^t$ is the set of all

3 I would like to thank Professor R. Selten for referring me to Ståhl's work, after reading the first version of this paper.

sequences of length t of elements in S.) F is the set of all strategies of the player who starts the bargaining. Similarly let G be the set of all strategies of the player who in the first move has to respond to the other player's offer; that is, G is the set of all sequences of functions $g = \{g^t\}_{t=1}^{\infty}$ such that, for t odd $g^t: S^t \to \{Y, N\}$ and for t even $g^t: S^{t-1} \to S$.

The following concepts are easily defined rigorously. Let $\sigma(f, g)$ be the sequence of offers in which 1 starts the bargaining and adopts $f \in F$, and 2 adopts $g \in G$. Let $T(f, g)$ be the length of $\sigma(f, g)$ (may be ∞). Let $D(f, g)$ be the last element of $\sigma(f, g)$ (if there is such an element). $D(f, g)$ is called the *partition* induced by (f, g). The outcome function of the game is defined by

$$P(f,g) = \begin{cases} (D(f,g), T(f,g)), & T(f,g) < \infty, \\ (0, \infty), & T(f,g) = \infty. \end{cases}$$

Thus, the outcome (s, t) is interpreted as the reaching of agreement s in period t, and the symbol $(0, \infty)$ indicates a perpetual disagreement.

For the analysis of the game we will have to consider the case in which the order of bargaining is revised and player 2 is the first to move. In this case a strategy for player 2 is an element of F and a strategy for player 1 is an element of G. Let us define $\sigma(g,f)$, $T(g,f)$, $D(g,f)$ and $P(g,f)$ similarly to the above for the case where player 2 starts the bargaining and adopts $f \in F$ and player 1 adopts $g \in G$.

The last component of the model is the preference of the players on the set of outcomes. I assume that player i has a preference relation (complete, reflexive, and transitive) $\succsim_i$ on the set of $S \times N \cup \{0, \infty)\}$, where N is the set of natural numbers.

I assume that the preferences satisfy the following five assertions:
For all $r, s \in S$, $t, t_1, t_2 \in N$, and $i \in \{1,2\}$:

(A-1) if $r_i > s_i$, then $(r, t) \succ_i (s, t)$;
(A-2) if $s_i > 0$ and $t_2 > t_1$, then $(s, t_1) \succ_i (s, t_2) \succ_i (0, \infty)$; earlier settlement preferred;
(A-3) $(r, t_1) \succsim_i (s, t_1 + 1)$ iff $(r, t_2) \succsim_i (s, t_2 + 1)$;
(A-4) if $r_n \to r$ and $(r_n, t_1) \succsim_i (s, t_2)$, then $(r, t_1) \succsim_i (s, t_2)$;
if $r_n \to r$ and $(r_n, t_1) \succsim_i (0, \infty)$, then $(r, t_1) \succsim_i (0, \infty)$;
(A-5) if $(s + \epsilon, 1) \sim_i (s, 0)$, $(\bar{s} + \bar{\epsilon}, 1) \sim_i (\bar{s}, 0)$, and $s_i < \bar{s}_i$, then $\epsilon_i \leqslant \bar{\epsilon}_i$.

From (A-3) we can use the notation $(r, T) \succsim_i (s, 0)$ and $(r, T) \precsim_i (s, 0)$ for $(r, T + t) \succsim_i (s, t)$ and $(r, T + t) \precsim_i (s, t)$, respectively.

Two families of models in which the preferences satisfy the above conditions are:

a *Fixed bargaining costs.* Each player i has a number c_i such that $(s, t_1) \succsim_i (\bar{s}, t_2)$ iff $(s_i - c_i \cdot t_i) \geqslant (\bar{s}_i - c_i \cdot t_2)$.

b *Fixed discounting factors.* Each player i has a number $0 < \delta_i \leqslant 1$ such that $(s, t_1) \gtrsim_i (\bar{s}, t_2)$ iff $s_i \delta_i^{t_1} \geqslant \bar{s}_i \delta_i^{t_2}$.

We reserve $\delta_i = 0$ for the lexicographic preference: $(s, t_1) \gtrsim_i (\bar{s}, t_2)$ if $(t_1 < t_2)$ or $(t_1 = t_2$ and $s_i \geqslant \bar{s}_i)$.

Remark. In a more general framework of the model, player i would be characterized by the sequence of preferences $\{\gtrsim_i^t\}$, where $\gtrsim_i^t$ is i's preference on the outcomes assuming that the players have not reached an agreement in the first $t-1$ periods. In fact, I assume that $\gtrsim_i^t \equiv \gtrsim_i$. This assumption precludes discussion of some interesting bargaining situations such as: (1) player i has a sequence $\{c_i^t\}$ where c_i^t is the cost to i of the bargaining in period t; (2) player i has a fixed bargaining cost and his utility is not linear.

3 PERFECT EQUILIBRIUM

The ordered pair $(\hat{f}, \hat{g}) \in F \times G$ is called a *Nash equilibrium* if there is no $f \in F$ such that $P(f, \hat{g}) >_1 P(\hat{f}, \hat{g})$ and there is no $g \in G$ such that $P(\hat{f}, g) >_2 P(\hat{f}, \hat{g})$.

The following simple proposition indicates that even after the restriction of the bargaining problem to our model, the Nash equilibrium is a 'weak' concept.

Proposition. *For all $s \in S$, s is a partition induced by Nash equilibrium.*

Proof. Let us define $\hat{f} \in F$ and $\hat{g} \in G$ as follows:

$$\text{for } t \text{ odd}, \quad \hat{f}^t \equiv s, \quad \hat{g}^t(s^1 \dots s^t) = \begin{cases} Y, & s^t \leqslant s, \\ N, & s^t > s; \end{cases}$$

$$\text{for } t \text{ even}, \quad \hat{g}^t \equiv s, \quad \hat{f}^t(s^1 \dots s^t) = \begin{cases} Y, & s^t \geqslant s, \\ N, & s^t < s. \end{cases}$$

Clearly, $(\hat{f}, \hat{g})$ is a Nash equilibrium and $P(\hat{f}, \hat{g}) = (s, 1)$.

The above equilibrium highlights the inadequacy of the concept of a Nash equilibrium in the current context. Assume 1 demands $s + \epsilon (\epsilon > 0)$. At this point of the game, 2 intends to insist on the original planned contract and 1 intends to agree to this offer. But if ϵ is sufficiently small so that $(s, 1) <_2 (s + \epsilon, 0)$, 2 will prefer to agree to player 1's deviation. Thus, player 1 may carry out a manipulative manoeuvre and offer $s + \epsilon$ in the certainty that 2 will agree to it.

In order to overcome this difficulty (see also Harsanyi 1978) I will use the concept of the perfect equilibrium following the definition of Selten (1965, 1975). For this definition we need some additional notation. Let $s^1 \dots s^T \in S$. Define $f \mid s^1 \dots s^T$ and $g \mid s^1 \dots s^T$ as the strategies derived from f and g after the offers $s^1 \dots s^T$ have been announced and already rejected. (That is, for T odd and t odd,

$$(f \mid s^1 \dots s^T)^t (r^1 \dots r^{t-1}) = f^{T+t}(s^1 \dots s^T, r^1 \dots r^{t-1}),$$

$$(g \mid s^1 \dots s^T)^t (r^1 \dots r^t) = g^{T+t}(s^1 \dots s^T, r^1 \dots r^t),$$

and so on.)

Notice that if T is even it is 1's turn to propose a partition of the pie, and 2's first move is a response to 1's offer. Thus $f \mid s^1 \dots s^T \in F$ and $g \mid s^1 \dots s^T \in G$. If T is odd, it is 2's turn to make an offer and therefore $g \mid s^1 \dots s^T \in F$ and $f \mid s^1 \dots s^T \in G$. And now to the central definition which, as mentioned, follows Selten's definition of subgame perfectness (1965, 1975) (what may at first seem to be a slightly clumsy version of Selten's idea has been chosen to prevent the use of some additional notation which would be redundant in this paper):

Definition. $(\hat{f}, \hat{g})$ is *perfect equilibrium* (PE) if for all $s^1 \dots s^T$, if T is odd:

(P-1) there is no $f \in F$ such that
$P(\hat{f} \mid s^1 \dots s^T, f) >_2 P(\hat{f} \mid s^1 \dots s^T, \hat{g} \mid s^1 \dots s^T)$;

(P-2) if $\hat{g}^T(s^1 \dots s^T) = Y$, there is no $f \in F$ such that
$P(\hat{f} \mid s^1 \dots s^T, f) >_2 (s^T, 0)$;

(P-3) if $\hat{g}^T(s^1 \dots s^T) = N$, $P(\hat{f} \mid s^1 \dots s^T, \hat{g} \mid s^1 \dots s^T) \gtrsim_2 (s^T, 0)$;

and if T is even:

(P-4) there is no $f \in F$ such that
$P(f, \hat{g} \mid s^1 \dots s^T) >_1 P(\hat{f} \mid s^1 \dots s^T, \hat{g} \mid s^1 \dots s^T)$;

(P-5) if $\hat{f}^T(s^1 \dots s^T) = Y$, there is no $f \in F$ such that
$P(f, \hat{g} \mid s^1 \dots s^T) >_1 (s^T, 0)$;

(P-6) if $\hat{f}^T(s^1 \dots s^T) = N$, $P(\hat{f} \mid s^1 \dots s^T, \hat{g} \mid s^1 \dots s^T) \gtrsim_1 (s^T, 0)$.

(P-1) and (P-4) ensure that after a sequence of offers and rejections $s^1 \dots s^T$ the player who has to continue the bargaining has no better strategy other than to follow the planned strategy. (P-2) and (P-5) ensure that a player who has planned to accept the offer s^T has no better alternative than to accept it, and (P-3) and (P-6) ensure that if a player is expected to reject an offer, it is not better for him to accept the offer.

Example. To clarify the notation, let us show that the pair $(\hat{f}, \hat{g})$ (with $s = 0.5$) described in the proof of the above proposition is not a perfect equilibrium for fixed bargaining cost preferences with $c_1 = 0.1$ and $c_2 = 0.2$. Player 2 plans to reject a possible offer of 0.6 by player 1: that is, $\hat{g}^1(0.6) = N$. After such a rejection the players expect to agree on 0.5: that is, $P(\hat{f} \mid 0.6, \hat{g} \mid 0.6) = (0.5, 1)$. Player 2 prefers $(0.6, 0)$ to $(0.5, 1)$: thus, $(\hat{f}, \hat{g})$ violates condition (P-3).

Remark. Notice that a strategy has been defined in Section 2 as a sequence of functions which is interpreted as the player's plans after every history, including histories which are not consistent with his own plans. For example, $f^3(s^1, s^2)$ is required to be defined even where $f^1 \neq s^1$ and $f^2(s^1, s^2) = Y$. The reader is directed to Selten (1965, 1975), Harsanyi (1978) and Harsanyi and Selten (1972) for details on the significance of the requirement.

4 LEMMAS

In this section we have only to assume that the preferences satisfy (A-1) and (A-2). The following Lemmas establish connections between two sets: (A) the set of all PEPs in a game in which 1 starts the bargaining, that is $\{s \in S \mid$ there is a PE $(f, g) \in F \times G$ such that $s = D(f, g)\}$; and (B) the set of all PEPs in a game in which 2 starts the bargaining, that is, $\{s \in S \mid$ there is a PE $(g, f) \in G \times F$ such that $s = D(g, f)\}$.

Remark. In a generalized model in which the $\gtrsim_i^t$ are not identical the same considerations would be used to establish connections between the sets $\{A^t\}_{t=1,3,5\ldots}$ and $\{B^t\}_{t=2,4,6\ldots}$ where $A^t(B^t)$ is the set of all PEPs in a game which starts at time t, 1(2) making the first offer.

Lemma 1. *Let $a \in A$. For all $b \in S$ such that $b > a$, there is $c \in B$ such that* $(c, 1) \gtrsim_2 (b, 0)$.

Remark. Lemma 1 states that for a to be in A, it has to be 'protected' from the possibility that 1 will demand and achieve some better contract. Player 1 will certainly do so if there is $b \in S$ satisfying $b > a$ such that 2 would accept b if it were offered. Player 2 must therefore reject such an offer. In order that it be optimal for him to carry out this threat, player 2 has to expect to achieve a better partition in the future; that is, there must be a PEP $c \in B$ in the subgame that takes place after 2's rejection such that $(c, 1)$ is preferred by 2 to $(b, 0)$.

Proof. Let $(\hat{f}, \hat{g})$ be a PE such that $D(\hat{f}, \hat{g}) = a$. Let $b \in S$ and $b > a$. From (P-1), $\hat{g}^1(b) = N$ (otherwise if $f^1 = b$ then $P(f, \hat{g}) = (b, 1) \succ_1 (a, 1) \gtrsim_1 (a, T(\hat{f}, \hat{g}))$ $= P(\hat{f}, \hat{g})$ in contradiction to (P-1). From (P-3) $P(\hat{f} \mid b, \hat{g} \mid b) \gtrsim_2 (b, 0)$ thus, $(D(\hat{f} \mid b, \hat{g} \mid b), T(\hat{f} \mid b, \hat{g} \mid b)) \gtrsim_2 (b, 0)$ and by (A-2) $(D(\hat{f} \mid b, \hat{g} \mid b), 1) \gtrsim_2 (b, 0)$ and therefore $D(\hat{f} \mid b, \hat{g} \mid b)$ is the desirable c.

Similarly, it is easy to prove the following lemma.

Lemma 2. *For all $a \in B$ and for all $b \in S$ such that $b < a$, there is $c \in A$ such that* $(c, 1) \gtrsim_1 (b, 0)$.

Lemma 3. *Let $a \in A$. Then for all b such that* $(b, 1) \succ_2 (a, 0)$ *there is $c \in A$ such that* $(c, 1) \gtrsim_1 (b, 0)$.

Remark. Lemma 3 states that if a is a PEP then 1 should have a 'good reason' to reject any offer from 2 which is preferred by 2 to accepting 1's original offer. Assume that in a certain PE, player 1 plans to agree to a in the first period (case B below). Consider b such that $(b, 1) \succ_2 (a, 0)$. Then, player 2 will reject a if he thinks that 1 would agree to b. Thus player 1 must threaten to reject any such offer b. In order that this threat be credible there must be a PE in the subgame beginning with 1's offer which yields an agreement c such that $(c, 1) \succsim_1 (b, 0)$. This c must be a member of A.

Proof. Let $(\hat{f}, \hat{g})$ be a PE such that $D(\hat{f}, \hat{g}) = a$.

Case A: $\hat{g}^1(\hat{f}^1) = N$. Let $\hat{f}^1 = s$. Then $D(\hat{f} \mid s, \hat{g} \mid s) = a$ and $a \in B$. From (A-1) and (A-2), if $(b, 2) \succ_2 (a, 1)$ then $b < a$ and therefore from Lemma 2 there is $c \in A$ such that $(c, 1) \succsim_1 (b, 0)$.

Case B: $\hat{f}^1 = a, \hat{g}^1(a) = Y$. Let b satisfy $(b, 1) \succ_2 (a, 0)$, $\hat{f}^2(a, b) = N$, because otherwise, for any $f \in F$ satisfying $f^1 = b, P(\hat{f} \mid a, f) = (b, 1) \succ_2 (a, 0)$, in contradiction to (P-2). From (P-6), $P(\hat{f} \mid a, b, \hat{g} \mid a, b) \succsim_1 (b, 0)$. Thus $(D(\hat{f} \mid a, b, \hat{g} \mid a, b), 1) \succsim_1 (b, 0)$ and $D(\hat{f} \mid a, b, \hat{g} \mid a, b) \in A$.

In a similar way it is possible to prove the following lemma.

Lemma 4. *For all $a \in B$ and for all $b \in S$ such that $(b, 1) \succ_1 (a, 0)$ there is $c \in B$ such that $(c, 1) \succsim_2 (b, 0)$.*

5 THE THEOREM

Let

$$\Delta = \left\{ (x, y) \in S \times S \;\middle|\; \begin{array}{l} y \text{ is the smallest number such that } (x, 1) \precsim_1 (y, 0); \\ x \text{ is the largest number such that } (y, 1) \precsim_2 (x, 0) \end{array} \right\},$$

$$\Delta_1 = \{x \in S \mid \text{there is } y \in S \text{ such that } (x, y) \in \Delta\},$$

$$\Delta_2 = \{y \in S \mid \text{there is } x \in S \text{ such that } (x, y) \in \Delta\}.$$

Proposition 1. *If $(x, y) \in \Delta$, then $x \in A$ and $y \in B$.*

Proof. Consider the following $(\hat{f}, \hat{g})$; for t odd

$$\hat{f}^t \equiv x, \quad \hat{g}^t(s^1 \ldots s^t) = \begin{cases} N, & x < s^t, \\ Y, & s^t \leqslant x, \end{cases}$$

and for t even

$$\hat{f}^t(s^1 \ldots s^t) = \begin{cases} N, & s^t < y, \\ Y, & y \leqslant s^t, \end{cases} \quad \hat{g}^t \equiv y.$$

It is easy to check that $(\hat{f}, \hat{g})$ is a perfect equilibrium.

Proposition 2. $\Delta \neq \phi$ *(and therefore A and B are not empty).*

Proof. Let

$$d_1(x) = \begin{cases} 0 \text{ if for all } y\ (y,0) \succ_1 (x,1), \\ y \text{ if there exists } y, (y,0) \sim_1 (x,1), \end{cases}$$

and

$$d_2(y) = \begin{cases} 1 \text{ if for all } (x,0) \succ_2 (y,1), \\ x \text{ if there exists } x, (x,0) \sim_2 (y,1). \end{cases}$$

$d_1(x)$ is the smallest y such that $(y,0) \succsim_1 (x,1)$ and $d_2(y)$ is the largest x such that $(x,0) \succsim_2 (y,1)$. Therefore

$$\Delta = \{(x,y) \mid y = d_1(x) \text{ and } x = d_2(y)\}.$$

It is easy to check that d_1 (and d_2) is well defined, continuous, increasing, and strictly increasing where $d_1(x) > 0\,(d_2(y) < 1)$.

Let $D(x) = d_2(d_1(x))$. Thus, $\Delta = \{(x,y) \mid y = d_1(x) \text{ and } D(x) = x\}$. Notice that $D(1) \leqslant 1$ and $D(0) \geqslant 0$. From the continuity of D it follows that there exists x_0 such that $D(x_0) = x_0$. Thus, $(x_0, d_1(x_0)) \in \Delta$.

Proposition 3. *The graph of Δ is a closed line segment which lies parallel to the diagonal $y = x$.*

Proof. From the continuity of d_1 and d_2, the set Δ is closed. Notice that $x - d_1(x)$ is an increasing function. To see this, let x_0 satisfy $(0,0) \sim_1 (x_0, 1)$ (take $x_0 = 1$ if there is no x that satisfies $(0,0) \sim_1 (x,1)$). For $x \leqslant x_0$, $d_1(x) = 0$ and $x - d_1(x) = x$. For $x_1 > x_2 \geqslant x_0 (d_1(x_1), 0) \sim_1 (x_1, 1)$ and $(d_1(x_2), 0) \sim_2 (x_2, 1)$. The function d_1 is an increasing function. Thus, (A-5) implies $x_1 - d_1(x_1) \geqslant x_2 - d_1(x_2)$. Similarly, $d_2(y) - y$ is a decreasing function. We have to show that $x - y$ is constant for all $(x,y) \in \Delta$. Suppose that $x_2 < x_1$ and that (x_1, y_1)

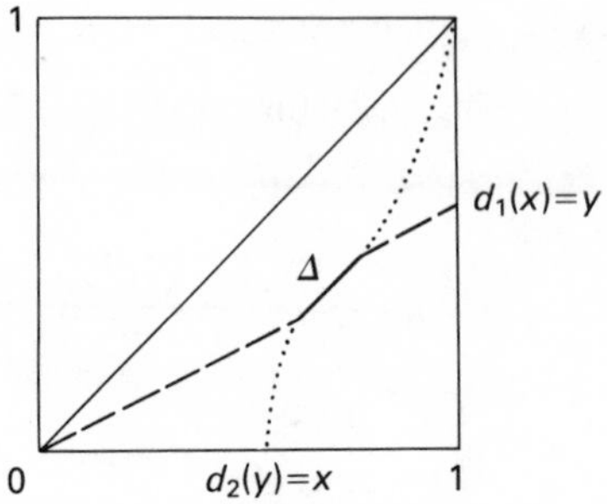

Figure 3.1

and (x_2, y_2) are both in Δ. Then $x_1 - d_1(x_1) \geqslant x_2 - d_1(x_2)$ and $x_1 - y_1 \geqslant x_2 - y_2$. Also $d_2(y_1) - y_1 \leqslant d_2(y_2) - y_2$ and $x_1 - y_1 \leqslant x_2 - y_2$. Thus $x_1 - y_1 = x_2 - y_2$.

Proposition 4. *If $a \in A$, then $a \in \Delta_1$, and if $b \in B$, then $b \in \Delta_2$.*

Proof. Suppose $\Delta_1 = [x_1, x_2]$ and $\Delta_2 = [y_1, y_2]$. Let $s = \sup\{a \in A\}$. Assume $x_2 < s$. Then $d_2(d_1(s)) < s$. Let $a \in A$ satisfy $r = d_2(d_1(s)) < a < s$. Let $b \in S$ satisfy $d_2^{-1}(a) > b > d_1(s)$. Then $a > d_2(b)$ and $(b, 1) \succ_2 (a, 0)$. From Lemma 3 there exists $c \in A$ such that $(c, 1) \succsim_1 (b, 0)$. Therefore there exists $c \in A$ satisfying $d_1(c) \geqslant b$. The facts that d_1 is an increasing function and that $d_1(c) \geqslant b > d_1(s)$ imply $c > s$ in contradiction to the definition of s.

Similarly, using Lemma 4 it is possible to show that $y_1 = \inf\{b \in B\}$. Using Lemmas 1 and 2 we get $x_1 = \inf\{a \in A\}$ and $y_2 = \sup\{b \in B\}$.

To summarize:

Theorem. *$A = \Delta_1 \neq \phi$, $B = \Delta_2 \neq \phi$. A and B are closed intervals and there exists $\epsilon \geqslant 0$ such that $B = A - \epsilon$.*

6 CONCLUSIONS

The following are applications of the theorem to the fixed bargaining cost and the fixed discounting factor models.

Conclusion 1. In the case where both the players have fixed bargaining costs, c_1 and c_2 (case I in the introduction):

1 If $c_1 > c_2$, c_2 is the only PEP.
2 If $c_1 = c_2$, every $c_1 \leqslant x \leqslant 1$ is a PEP.
3 If $c_1 < c_2$, 1 is the only PEP.

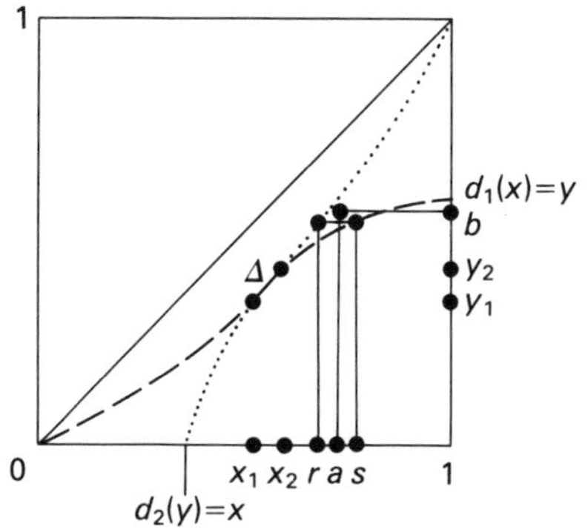

Figure 3.2

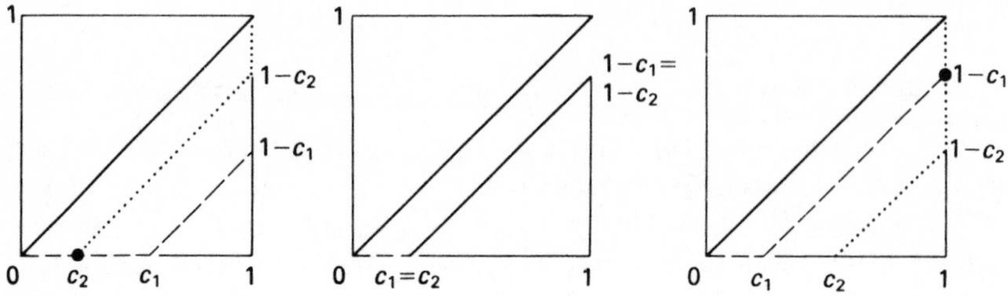

Figure 3.3

Proof. $d_1(x) = \max\{x - c_1, 0\}$ and $d_2(y) = \min\{y + c_2, 1\}$. Thus Δ is the set of all solutions to the set of equations $y = \max\{x - c_1, 0\}$ and $x = \min\{y + c_2, 1\}$. The conclusion is implied by the three diagrams of figure 3.3 related to the cases (1) $c_1 > c_2$, (2) $c_1 = c_2$, and (3) $c_1 < c_2$.

Remark. Given a particular PE, is agreement reached immediately after the very first offer? A positive answer results when $A \cap B = \phi$. If $(\hat{f}, \hat{g})$ is a PE and $T(\hat{f}, \hat{g}) > 1$, then $D(\hat{f}, \hat{g})$ is a member not only of A but also of B. Thus in almost all cases covered by Conclusions 1 and 2 (the possible exception being the case $c_1 = c_2$ in Conclusion 1) the bargaining indeed ends in the first period.

The following pair of strategies $(\hat{f}, \hat{g})$ is an example of a PE in the game where the players have fixed bargaining costs $c_1 = c_2 = c$. The pair $(\hat{f}, \hat{g})$ has the property that the negotiation ends at the second period: Let $\epsilon(x)$ be a non-negative function defined in the unit interval such that $\epsilon(x) \leqslant \max\{0, x - c\}$. Assume $\epsilon(x)$ attains its maximum at x_0 where $x_0 > 2c$ and $\epsilon(x_0) > 2c$. Let $(\hat{f}, \hat{g})$ satisfy

$$\hat{f}^1 = x_0,$$

$$\hat{g}^1(s^1) = \begin{cases} N, & c < s^1, \\ Y, & s^1 \leqslant c, \end{cases}$$

and 'after', the strategies are identical to the strategies described in Proposition 1 for the partition $\epsilon(s^1)$. The partition of this PE is $\epsilon(x_0)$ (figure 3.4).

In this example the first move by player 1 serves as a signal to player 2. Player 2 interprets 1's signal s^1 as an agreement to continue with a pair of strategies that yields the partition $\epsilon(s^1)$. Not every s^1 may serve as such a signal, since 2 will agree to every partition that gives 2 more than $1 - c$. The partition $\epsilon(s^1)$ must give 2 at least $1 - s^1 + c$; therefore $\epsilon(s^1) \geqslant s^1 - c$. A final restriction on ϵ is that $x_0 \geqslant 2c$. Otherwise 1 would prefer to offer a partition that 2 'could not refuse' (some offer between c and $x_0 - c$). This also shows

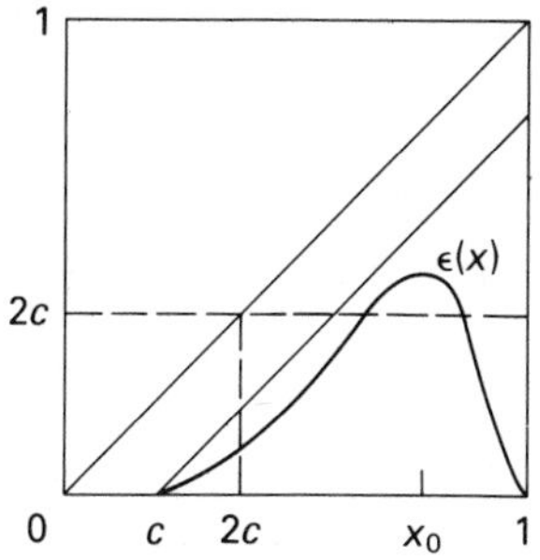

Figure 3.4

that the PE outcome may not be Pareto optimal; both players prefer to agree to $1-\epsilon(x_0)$ at the beginning of the bargaining.

Conclusion 2. In the case where the players have fixed discounting factors - δ_1 and δ_2 (Case II in the introduction) - if at least one of the δ_i is strictly less than 1 and at least one of them is strictly positive, then the only PEP is $M=(1-\delta_2)/(1-\delta_1\delta_2)$.

Remark. Notice that when $\delta_2=0$, player 2 has no threat because the pie has no worth for him after the first period. Player 1 can exploit this to get all the pie ($M=1$). When $\delta_1=0$, 1 can gain $1-\delta_2$ only, that is, the proportion of the pie that 2 may lose if he refuses 1's offer and gets 1 in the second period. When $0<\delta_1=\delta_2=\delta<1$, 1 gets $1/1+\delta>1/2$. As one would expect, 1's gain from the fact that he starts the bargaining decreases as δ tends to 1.

Proof. $d_1(x)=x\cdot\delta_1$ and $d_2(y)=1-\delta_2+\delta_2\cdot y$. The conclusion follows from figure 3.5 (the intersection of d_1 and d_2 is where $1-\delta_2+\delta_2 x\delta_1=x$, that is where $x=M$).

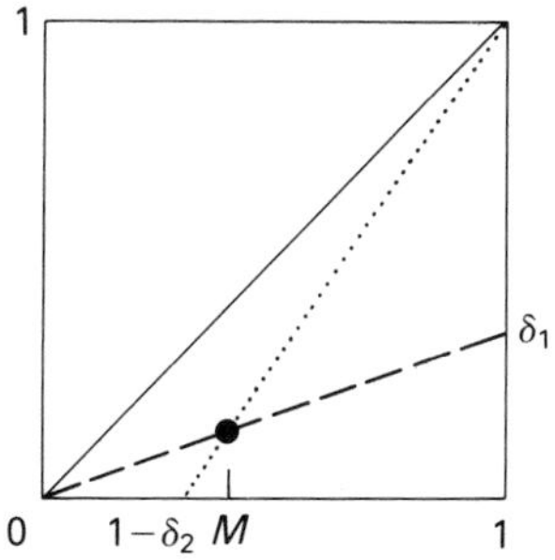

Figure 3.5

REFERENCES

Aumann, R. J. and Shapley, L. 1976: Long term competition: a game theoretic analysis. Unpublished manuscript, Stanford University.

Cross, J. G. 1965: A theory of the bargaining process. *American Economic Review*, **55**, 67–94.

Edgeworth, F. Y. 1932: *Mathematical Psychics: An Essay on the Applications of Mathematics to the Moral Sciences*. LSE Series of Reprints of Scarce Tracts in Economics and Political Sciences No. 10.

Harsanyi, J. C. 1978: A solution theory for non-cooperative games and its implications for cooperative games. Center for Research in Management Science, University of California. Berkeley, CP-401.

Harsanyi, J. C. and Selten, R. 1972: A generalized Nash solution for two person bargaining games with incomplete information. *Management Science*, **18**, 80–106.

Krelle, W. 1975: A new theory of bargaining. Unpublished manuscript.

Kurz, M. 1978: Altruism as an outcome of social interaction. *American Economic Review*, **68**, 216–222.

Luce, R. P. and Raiffa, H. 1957: *Games and Decisions*. John Wiley and Sons, New York.

Nash, J. F. 1950: The bargaining problem. *Econometrica*, **18**, 155–62.

Nash, J. F. 1953: Two-person cooperative games. *Econometrica*, **21**, 128–40.

Rice, P. 1977: A note on the Hicks theory of strike bargaining. *Zeitschrift fur die Gesamte Staatswissenschaft*, **133**, 236–44.

Roth, A. E. 1979: *Axiomatic models of bargaining*. Lecture Notes in Economics and Mathematical Systems No. 170. Springer-Verlag, Berlin.

Rubinstein, A. 1977: *Equilibrium in supergames,* Center for Research in Math. Economics and Game Theory, the Hebrew University, Jerusalem, Israel, RM 25.

Rubinstein, A. 1979: Equilibrium in supergames with the overtaking criterion. *Journal of Economic Theory*, **21**, 1–9.

Rubinstein, A. 1980: Strong perfect equilibrium. *International Journal of Game Theory*, **9**, 1–12.

Selten, R. 1965: Spieltheoretische Behandlung eines Oligopolmodels mit Nachfragetragheit. *Zeitschrift fur die Gesamte Staatswissenschaft*, **12**, 301–24 and 667–89.

Selten, R. 1975: Re-examination of the perfectness concept for equilibrium points in extensive games. *International Journal of Game Theory*, **4**, 25–55.

Ståhl, I. 1972: *Bargaining Theory*. Stockholm School of Economics.

Ståhl, I. 1977: An *N*-person bargaining game in the extensive form. In Henn, R. and Moeschlin, O. (eds) *Mathematical Economics and Game Theory*. Lecture Notes in Economics and Mathematical Systems No. 141. Springer-Verlag, Berlin.

4

Nash Bargaining Theory II

K. Binmore

1 INTRODUCTION

In three remarkable papers written in the early 1950s, Nash outlined an approach to the theory of games in general and to the theory of bargaining in particular which remains of the greatest significance.[1] This is the second of several chapters in which it is intended to discuss and to extend the general approach to bargaining outlined in those papers.

Nash attacked the bargaining problem from two different directions. The first attack employed a set of axioms which he showed determine a unique outcome for any bargaining game. This unique outcome is called the 'Nash bargaining solution' of the game. The second attack involved a specific bargaining model to be analysed as a 'non-cooperative game'. The analysis proposed by Nash leads of course to the same solution outcome as the axiomatic treatment. Nash's axiomatic method was discussed in chapter 2. A brief summary of this discussion is given in section 2. In the current chapter we shall concentrate on his second line of attack. We begin by re-iterating two fundamental points made in chapter 2 since they are equally important here.

Although this chapter is devoted to a discussion of the classical 'Nash bargaining solution', it should be emphasized that we do *not* regard this as being universally applicable to all two-person bargaining problems. In so far as we seek to justify the 'Nash bargaining solution', it will be in a context where the players communicate strictly in terms of utilities and make no use of alternative descriptions of the possible outcomes which may be available. In a later chapter it is intended to discuss some bargaining problems with a less simplified informational structure. In the case of bartering over commodities, for example, it will be argued that, with the natural informational structure in the context, the application of Nash's principles leads *not* to the 'Nash bargaining solution' but to a competitive equilibrium.

1 Large parts of sections 1 and 2 reiterate points made in chapter 2. The substance of this chapter begins with section 3.

The second fundamental point which needs to be made is that Nash's theory was intended to be understood as a strictly *game-theoretic* approach – i.e. Nash is concerned with optimal play between rational players with individual goals. Like Harsanyi (1977) and others, we feel that it is important to distinguish carefully between such an approach and *behavioural* theories (i.e. theories which seek to describe what happens in practice) and *ethical* theories (i.e. theories which prescribe outcomes which are optimal for society 'as a whole'). In particular, it is common for authors to refer to the 'Nash bargaining solution' as an arbitration scheme. Such a description invites one to consider whether or not Nash's principles are 'fair' and this is an ethical question rather than a 'game-theoretic' one. While this ethical question is doubtless meaningful, it was not a question which Nash was trying to answer. In any case, this chapter will be concerned strictly with game-theoretic questions.

Nash felt that the most fundamental type of game is what we shall call a *contest*. By this is to be understood a formal two-person game in extensive form which is to be analysed on the assumption that no pre-play communication is permitted. Other two-person games were to be reduced to contests in the following way. Suppose that G is a formal game for which some pre-play negotiation is possible. Then one should imagine the various possible steps in the pre-play negotiation procedure as moves in a larger 'negotiation game' N. The negotiation game N is then analysed as a contest and its solution leads to an outcome of G which we shall call a *negotiated outcome* of G. The determination of the appropriate negotiated outcome of G will therefore depend on a knowledge of the negotiation procedure to be employed and also on a knowledge of the game theory which the players will use in resolving the negotiation contest N.

In this chapter we restrict our attention to bargaining. Following Nash, we define a *bargaining contest* B associated with a game to be a negotiation contest in which any pair of negotiation strategies (one for each player) may be classified either as compatible or else as incompatible. If compatible, the choice of these negotiation strategies results in a binding agreement (contract) being signed which determines the (possibly mixed) strategies to be used by the players in G. If incompatible, the choice of these negotiation strategies results in a pre-determined outcome of G called the status quo.

In the context of the formal framework outlined above, Nash's bargaining axioms (see chapter 2) are assumptions about the solution outcome of certain bargaining contests. Since they determine a unique outcome (in terms of utilities), it follows that, if they are applicable in a certain situation, then they must implicitly embody all that is immediately relevant about the bargaining procedure in use and the game theory subscribed to by the players. The question then arises: do the Nash bargaining axioms actually have this property for a reasonably large class of bargaining contests?

Abstract answers to this question have some value but it is more satisfying to look at specific bargaining contests and to see whether their solution outcomes actually do satisfy the Nash axioms. In this way one can hope to gain some practical insight into the circumstances under which the 'Nash bargaining solution' is applicable (and equally into those circumstances when it is not). In considering appropriate bargaining models one has, of course, to focus on those factors which have, or appear to have, a genuine strategic relevance to the situation. This certainly does not apply to the bulk of manoeuvres common during real-life negotiations. Under this heading for example come flattery, abuse, the inducement of boredom and other more subtle attempts to put the opponent at a psychological disadvantage. These factors would certainly be of the greatest importance in a behavioural analysis but they have no place in a game-theoretic analysis. A rational player will ignore such irrelevances (assuming that his concern is only with the final contract and not with the manner in which it was reached) and address himself exclusively to the strategic issues.

The bargaining contest given by Nash as typical of the class to which his bargaining axioms apply must be viewed in the light of the remarks made in the previous paragraph. Although very simple (the bargaining is reduced to the players making simultaneous demands), it nevertheless clearly captures the strategic essence of a wide variety of bargaining situations. Nash's claim that the solution outcome of this bargaining contest coincides with the symmetric Nash bargaining solution is therefore significant. Unfortunately, Nash's analysis of this game is not quite so convincing as one would wish. As explained in chapter 2, the identification of a solution of a contest with many Nash equilibria is not necessarily straightforward.

In this chapter, we propose to discuss Nash's specific bargaining model and some related models. In particular, we wish to point out the relevance of some recent work by Rubinstein (chapter 3). Our object is to clarify the circumstances under which the result of rational bargaining will be the Nash bargaining solution or an approximation to this.

2 THE NASH BARGAINING SOLUTION

In this section, we briefly summarize the discussion of Nash's axiomatic approach given in chapter 2.

Nash takes for granted that all the bargaining contests to be considered have a solution and that the solution payoffs depend only on the set $\mathcal{X}$ (assumed compact and convex) of feasible payoff pairs and the status quo payoff pair $\boldsymbol{\xi}$. The solution payoff pair $\boldsymbol{\sigma}$ may therefore be written as $\boldsymbol{\sigma} = f(\mathcal{X}, \boldsymbol{\xi})$.

Various axioms are then given for the function f, from which it follows that $\boldsymbol{\sigma}$ is the value of $\mathbf{x}$ which maximizes

$$(x_1 - \xi_1)(x_2 - \xi_2)$$

subject to the constraints $\mathbf{x} \geqslant \boldsymbol{\xi}$ and $\mathbf{x} \in \mathcal{X}$. The resulting value of $\boldsymbol{\sigma}$ is called the (symmetric) Nash bargaining solution of the game.

If Nash's symmetry axiom is abandoned, one obtains instead that there exists τ satisfying $0 \leqslant \tau \leqslant 1$ such that $\boldsymbol{\sigma}^{(\tau)}$ is the value of $\mathbf{x}$ which maximizes

$$(x_1 - \xi_1)^{\tau}(x_2 - \xi_2)^{1-\tau}$$

subject to the constraints $\mathbf{x} \geqslant \boldsymbol{\xi}$ and $\mathbf{x} \in \mathcal{X}$. We refer to $\boldsymbol{\sigma}^{(\tau)}$ as an asymmetric Nash bargaining solution. One may think of τ as measuring the 'bargaining power' with which the bargaining procedure endows the first player. Selten, among others, has argued that the asymmetric case is of little practical interest since if, for example, the bargaining procedure involves one player moving first and this is advantageous to the player in question, than the formal bargaining will be preceded by further negotiations about who is to obtain the advantageous role. One purpose of this chapter will be to give an example of a bargaining contest in which the solution outcome is an asymmetric Nash bargaining solution but in which any pre-negotiations are pointless. (See Section 5 and Note 5.)

As explained in section 5 of chapter 2, once the symmetry axiom has been deleted, the full weight of Nash's result falls on the axiom called the 'independence of irrelevant alternatives'. As a general proposition, this axiom is not easily defended on game-theoretic grounds. (Defences offered in the literature are usually based on arbitration considerations.) An attempt is made in section 5 of chapter 2 but the argument offered is not very convincing. This fact is used in chapter 2 to introduce a more easily defended alternative axiom called 'convention consistency' which is then shown to be an adequate substitute for the 'independence of irrelevant alternatives'.

In this chapter, we are no longer interested in offering an abstract defence of the Nash axioms. Instead, we propose to exhibit some bargaining contests whose (non-cooperative) solution yields the same, or nearly the same, outcome as that characterized by the Nash axioms.

3 THE NASH DEMAND GAME

In what follows, we assume that $\mathcal{X}$ is not only compact and convex, but also satisfies

a $\mathcal{X} \subseteq \{\mathbf{x} : \mathbf{x} \geqslant \boldsymbol{\xi}\}$

b $\mathcal{X} \cap \{\mathbf{x} : \mathbf{x} > \boldsymbol{\xi}\} \neq \emptyset$

c $(\boldsymbol{\xi} \leqslant \mathbf{x} \leqslant \mathbf{y}$ and $\mathbf{y} \in \mathcal{X}) \Rightarrow \mathbf{x} \in \mathcal{X}$

Without loss of generality, we normalize so that $\boldsymbol{\xi} = \mathbf{0}$ and

$$\max\{x:(x,y)\in\mathscr{X}\} = \max\{y:(x,y)\in\mathscr{X}\} = 1.$$

In Nash's demand game, the two players simultaneously announce demands x and y. If $(x,y)\in\mathscr{X}$, each player receives his demand. Otherwise the status quo $\mathbf{0}$ results. To exclude trivial equilibria, we shall restrict demands to the ranges $0 \leqslant x \leqslant 1$ and $0 \leqslant y \leqslant 1$. In spite of this restriction, there remain too many equilibria. An admissible pair (x,y) is a Nash equilibrium if and only if it is a Pareto-efficient point of $\mathscr{X}$. To determine which of these many Nash equilibria is to be regarded as the (non-cooperative) solution, a convention is required. We shall discuss the question of what convention is appropriate very briefly in terms of the analysis given in section 6 and section 7 of chapter 2.

If B is a separable bargaining contest which separates into n Nash demand games $B_1, B_2, \ldots, B_n$, one may ask: which Nash equilibria, played separately in $B_1, B_2, \ldots, B_n$, yield a Pareto-efficient outcome for B? The answer is that the pairs of demands made in $B_1, B_2, \ldots, B_n$ must all be asymmetric Nash bargaining solutions for the n separate games, all of which correspond to the *same* value of τ.

The above analysis is, of course, too close to that given in chapter 2 to shed any new light on the issue. What is required for this purpose are some bargaining contests in which a solution can be identified without the need to appeal to the use of a convention which, of necessity, will appear arbitrary if attention is confined to a single game. Such contests can be obtained from the Nash demand game by introducing further structure. Such extra structure, however, takes the contest outside the simple class of models which Nash's axioms are designed to typify. We can therefore only hope to show that, when the distorting effect of the extra structure is sufficiently small, then the solution outcome will be *approximately* equal to the Nash bargaining solution. We begin with an elaboration of an example given by Nash (1953).

4 MODIFIED NASH DEMAND GAME I

As in the simple demand game, both players simultaneously announce demands x and y. There then follows a chance move as a result of which the players receive their demands with probability $p(x,y)$ or else receive their status quo payoffs of 0 with probability $1 - p(x,y)$. A possible interpretation is given shortly.

It will always be assumed that p is a continuous function and that $p(x,y) = 0$ when $(x,y) \notin \mathscr{X}$. Although this is not necessary for our results, we shall also assume that $p(x,y) = 1$ for $(x,y)\in\rho\,\mathscr{X}$, where $0 < \rho < 1$. Note that, unless $\rho = 1$, this game is not a bargaining contest in the sense of section

1 because a pair of strategies cannot simply be classified as compatible or incompatible. We shall, however, be interested in the case when ρ is close to 1. The game can then be regarded as 'approximately' a bargaining contest.

The expected payoffs π_1 and π_2 resulting from the demand pair (x, y) are given by $\pi_1 = xp(x, y)$ and $\pi_2 = yp(x, y)$. For $(\tilde{x}, \tilde{y})$ to be a Nash equilibrium, we require that

$$xp(x, \tilde{y}) \leqslant \tilde{x}p(\tilde{x}, \tilde{y})$$

$$yp(\tilde{x}, y) \leqslant \tilde{y}p(\tilde{x}, \tilde{y})$$

whenever $0 \leqslant x \leqslant 1$ and $0 \leqslant y \leqslant 1$. At least one Nash equilibrium always exists. A point (x^*, y^*) at which the function P defined by

$$P(x, y) = xyp(x, y)$$

achieves its maximum value is such a Nash equilibrium. Nash pointed out that (x^*, y^*) lies in the shaded region in figure 4.1. But, of course, there may be other Nash equilibria outside this region.

As $\rho \to 1-$, it is evident that (x^*, y^*) must approach the symmetric Nash bargaining solution of the game. Nash (1953) refers to the game described in this section as a 'smoothed version' of his simple demand game. He then remarks that the (symmetric) Nash bargaining solution should be regarded as the solution outcome of the simple demand game because it is the 'only necessary limit of the equilibrium points of smoothed games'.

We shall not seek to defend Nash's remark. It does not seem to be at all defensible as it stands (see Luce and Raiffa 1957, p. 142). Instead we shall demonstrate that the symmetric Nash bargaining solution is an approximation

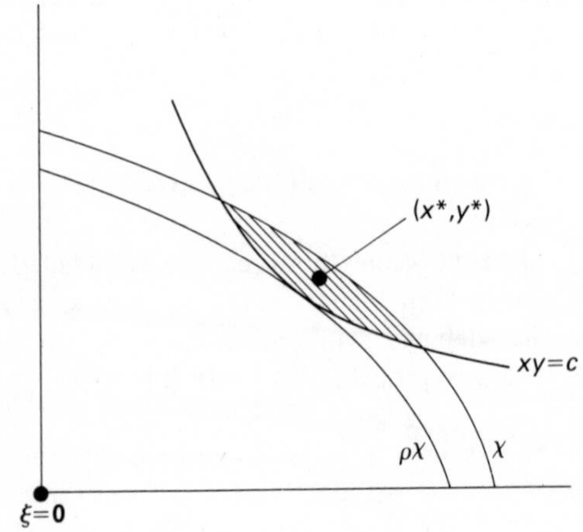

Figure 4.1

to *all* Nash equilibria of the smoothed demand game for a reasonably interesting class of functions p.

We shall suppose that the Pareto boundary of $\mathscr{X}$ is the graph of the bijective, strictly decreasing, concave function $\phi : [0,1] \to [0,1]$. The function $\theta : \mathscr{X} \to [0,1]$ will be defined by $\theta(x,y) = \alpha$, where

$$\frac{y}{\alpha} = \phi\left(\frac{x}{\alpha}\right).$$

The contour $\theta(x,y) = \alpha$ is then the Pareto boundary of the set $\alpha\mathscr{X}$.

In addition to the assumptions made so far, we shall also require that $p : \mathbb{R}^2 \to [0,1]$ be of the form

$$p(x,y) = g(\theta(x,y))$$

for $(x,y) \in \mathscr{X}$, where $g : [0,1] \to [0,1]$ is continuous and decreasing on $[0,1]$ with $g(0)=1$ and $g(0)=0$. Recall that our aim is to show that the symmetric Nash bargaining solution is an approximation to all Nash equilibria provided ρ is sufficiently close to 1. We prove this under the assumption that g is differentiable, although a weaker condition on g will suffice for this purpose (see note 1).

A feasible but not exclusive interpretation for the set-up described above is the following. The two players place their demands in sealed envelopes which are then passed to a referee. After time t, the referee opens the envelopes and determines whether the demands are compatible. Some such mechanism would seem necessary to realize in practice the requirement that the demands be 'simultaneous' (although this is not meant to imply that a real-life referee is necessary for this purpose). The players discount time using a common discount factor δ $(0 < \delta < 1)$. Thus, at time t, the set of feasible payoff pairs is $\delta^t\mathscr{X}$ (which has Pareto boundary $\theta(x,y) = \delta^t$). Assuming that each player is equipped with the same probability distribution function $F : [0,\infty) \to [0,1]$ for the random variable t, then the probability that x and y will prove compatible is assessed by both players as

$$\begin{aligned} p(x,y) &= \text{Prob}(t \leqslant T, \text{ where } \theta(x,y) = \delta^T) \\ &= \text{Prob}\left(t \leqslant \frac{\log\theta(x,y)}{\log\delta}\right) \\ &= F\left(\frac{\log\theta(x,y)}{\log\delta}\right) \\ &= g(\theta(x,y)). \end{aligned}$$

Note that a function g obtained in this way will be differentiable if the players are given a continuous probability *density* for t.

Lemma 1. Let B_y denote the set of values of α at which the maximum

$$\max_{y \leqslant \alpha \leqslant 1} \left\{ \alpha\phi^{-1}\left(\frac{y}{\alpha}\right) g(\alpha) \right\} = \max_x \{xp(x,y)\}$$

is attained. Then B_y increases in the sense that

$$y_1 < y_2 \Rightarrow \sup B_{y_1} \leqslant \inf B_{y_2}.$$

If g is differentiable, then B_y is strictly increasing in the sense that

$$y_1 < y_2 \Rightarrow \sup B_{y_1} < \inf B_{y_2}.$$

Proof. The first part of the lemma is proved by showing that, if $\alpha < \beta$ and $y < z$, then

$$\alpha\phi^{-1}\left(\frac{y}{\alpha}\right) g(\alpha) \leqslant \beta\phi^{-1}\left(\frac{y}{\beta}\right) g(\beta) \Rightarrow \alpha\phi^{-1}\left(\frac{z}{\alpha}\right) g(\alpha) < \beta\phi^{-1}\left(\frac{z}{\beta}\right) g(\beta).$$

Suppose that the antecedent in this implication is true but the consequent is false. Unless

$$\alpha\phi^{-1}\left(\frac{y}{\alpha}\right) g(\alpha) = \beta\phi^{-1}\left(\frac{y}{\beta}\right) g(\beta) \tag{4.1}$$

we may then conclude that

$$\alpha g(\alpha) \left\{ \phi^{-1}\left(\frac{y}{\alpha}\right) - \phi^{-1}\left(\frac{z}{\alpha}\right) \right\} < \beta g(\beta) \left\{ \phi^{-1}\left(\frac{y}{\beta}\right) - \phi^{-1}\left(\frac{z}{\beta}\right) \right\}. \tag{4.2}$$

But g decreases on $[0, 1]$ and so $g(\alpha) \geqslant g(\beta)$. Thus

$$\frac{\phi^{-1}(y/\alpha) - \phi^{-1}(z/\alpha)}{y/\alpha - z/\alpha} > \frac{\phi^{-1}(y/\beta) - \phi^{-1}(z/\beta)}{y/\beta - z/\beta} \tag{4.3}$$

But this contradicts the assumption that ϕ (and hence ϕ^{-1}) is concave.

If (4.1) holds, then $g(\alpha) \neq g(\beta)$ because $\alpha\phi^{-1}(y/\alpha)$ is strictly increasing in α. However, only weak inequality can be asserted in (4.2). Nevertheless, (4.3) remains true because now $g(\alpha) > g(\beta)$.

For the second part of the lemma, we also have that g is differentiable. If B_y is not strictly increasing, then there exist y_1 and y_2 with $y_1 < y_2$ such that $\sup B_{y_1} = \inf B_{y_2}$. But, for all y satisfying $y_1 < y < y_2$, we have that $\sup B_{y_1} \leqslant \inf B_y \leqslant \sup B_y \leqslant \inf B_{y_2}$. It follows that there exists α such that $B_y = \{\alpha\}$ for $y_1 < y < y_2$. But, then, for $y_1 < y < y_2$,

$$D_-\left(\alpha\phi^{-1}\left(\frac{y}{\alpha}\right) g(\alpha)\right) \geqslant 0 \geqslant D^+\left(\alpha\phi^{-1}\left(\frac{y}{\alpha}\right) g(\alpha)\right) \tag{4.4}$$

and so, for $y_1/\alpha < u < y_2/\alpha$,

$$u(D_- \log\phi^{-1})(u) \leqslant 1 + \alpha g'(\alpha)/g(\alpha) \leqslant u(D^+ \log\phi^{-1})(u).$$

Since $\log\phi^{-1}$ is concave, it follows that $(D\log\phi^{-1})(u) = G/u$, where the constant $G = 1 + \alpha g'(\alpha)/g(\alpha)$ is negative because $\log\phi^{-1}$ is strictly decreasing. But then G/u is a strictly increasing function of u which contradicts the concavity of $\log\phi^{-1}$.

Lemma 2. Let A_x denote the set of values of β at which the maximum

$$\max_{x\leqslant\beta\leqslant 1}\left\{\beta\phi\left(\frac{x}{\beta}\right)g(\beta)\right\} = \max_{y}\ \{yp(x,y)\}$$

is attained. Then A_x increases. If g is differentiable, then A_x is strictly increasing.

Proof. As lemma 1.

The set of Nash equilibria is the intersection of the two 'reaction sets'

$$A = \{(x,y) : \theta(x,y) = \beta \in A_x\}$$

$$B = \{(x,y) : \theta(x,y) = \alpha \in B_y\}.$$

In particular, $(x^*, y^*) \in A \cap B$ (see figure 4.2).

In view of lemmas 1 and 2, we have that $A \subset S$ and $B \subset T$, where S and T are as illustrated in figure 4.3. Thus $A \cap B \subset S \cap T$. If g is differentiable, then all points on $\theta = \theta(x^*, y^*)$ are excluded from S and T with the exception of (x^*, y^*) itself. The set $S \cap T$ is then

$$S \cap T = \{(x,y) : (x,y) \geqslant (x^*, y^*) \ \text{ or } \ (x,y) \leqslant (x^*, y^*)\}.$$

Since all Nash equilibria lie in $S \cap T$, it follows that all Nash equilibria lie in the region shaded in figure 4.4. Hence the Nash bargaining solution approximates all these Nash equilibria, provided that ρ is sufficiently close to 1.

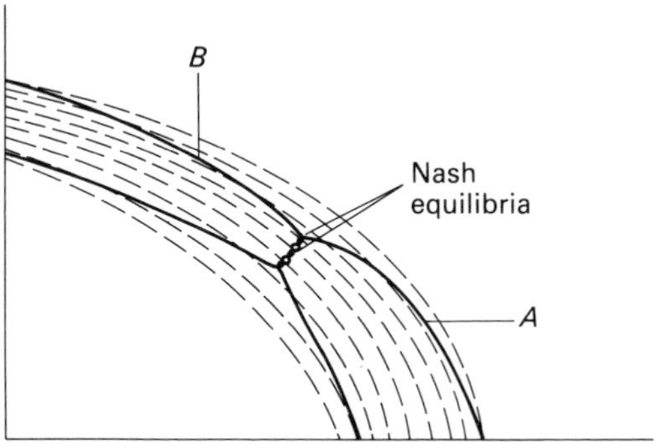

Figure 4.2

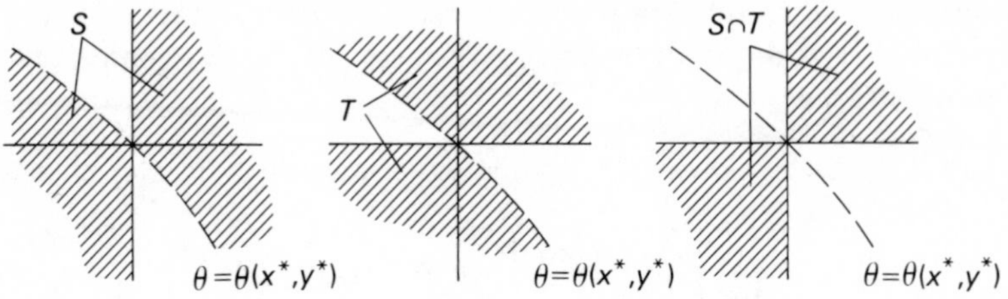

Figure 4.3

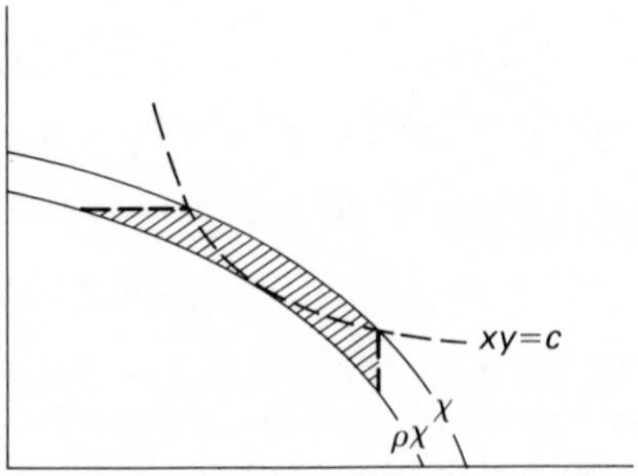

Figure 4.4

Note 1. The condition that g is differentiable is not essential for the above result. An adequate sufficient condition is

$$D_{-}g(x) = \liminf_{h \to 0-} \frac{g(x+h) - g(x)}{h} \leqslant \limsup_{h \to 0+} \frac{g(x+h) - g(x)}{h} = D^{+}g(x)$$

for each x. (This is applied in (4.4) of lemma 1.) If this condition is not satisfied, it may be that multiple Nash equilibria can be found on a contour $\theta(x, y) = \alpha$. This happens, for example, in the case when $\phi(x) = 1 - x$ and

$$g(z) = \begin{cases} W - l(z - w) & (z \leqslant w) \\ W - m(z - w) & (z \geqslant w) \end{cases}$$

where $0 < w < 1$ and $0 < l < m$. Each point on $x + y = w$ is then a Nash equilibrium.

5 MODIFIED NASH DEMAND GAME II

The requirement in the simple Nash demand game that the players make simultaneous demands and that the set of feasible payoff pairs is perfectly known to the players at the instant their demands become effective is not particularly realistic. Its usefulness therefore depends on the extent to which it retains the significant features of less idealized situations. In the previous section, we relaxed the requirement that the players are perfectly informed about the set of available payoff pairs. In this section we relax instead the requirement that the players make simultaneous demands.

Specifically, we shall assume that at time $t = 0$ the first player makes a demand which is either accepted or rejected by the other player. If accepted, the negotiations terminate with the first player receiving his demand and the second player receiving the maximum utility consistent with the first player's payoff. If the first player's demand is rejected, the negotiations continue with the second player making a demand after a time interval T which the first player can either accept or reject. And so on.

In order to inject some sense of urgency into the negotiations, we suppose that the players react negatively to the delay of an agreement. This is modelled by assuming that the set of feasible payoff pairs at time t is

$$\mathscr{X}_t = \{(\delta_1^t x, \delta_2^t y) : (x, y) \in \mathscr{X}\}$$

where $0 < \delta_1 < 1$ and $0 < \delta_2 < 1$. Our assumptions about the set $\mathscr{X}_0 = \mathscr{X}$ will be the same as in section 4. Note, however, that we depart from section 4 in allowing the players to have possibly different discount rates δ_1 and δ_2. In the case that no demand is ever accepted, we suppose that both players receive 0. The status quo is therefore $\mathbf{0}$.

Games of this kind have been considered by various authors (notably Ståhl (1972). For the purposes of this chapter, however, the significant work appears in chapter 3. Rubinstein describes the set of *perfect equilibrium* outcomes for a class of games which includes that described above (see note 2). In particular, his argument shows that, in the case of the bargaining contest described above, there exists a *unique* perfect equilibrium outcome (see note 3). The solution of the game can therefore be identified quite unequivocally. Given Rubinstein's result, it is quite easy to show that the solution outcome is an approximation to the 'asymmetric Nash bargaining solution' with 'bargaining power' τ where

$$\tau = \frac{\log \delta_2}{\log \delta_1 + \log \delta_2}$$

provided that the time interval T between successive demands is sufficiently small. (If $\delta_1 = \delta_2$, we obtain the ordinary 'Nash bargaining solution' for which $\tau = \frac{1}{2}$.) Of course, the smaller the value of T, the closer the approximation to the case in which demands are made simultaneously.

Given the existence of a unique perfect equilibrium outcome (X_1, Y_1), it is obvious that this must be as in figure 4.5a (in which we have used the notation $\Delta_i = \delta_i^T$). Figure 4.5b illustrates the unique perfect equilibrium

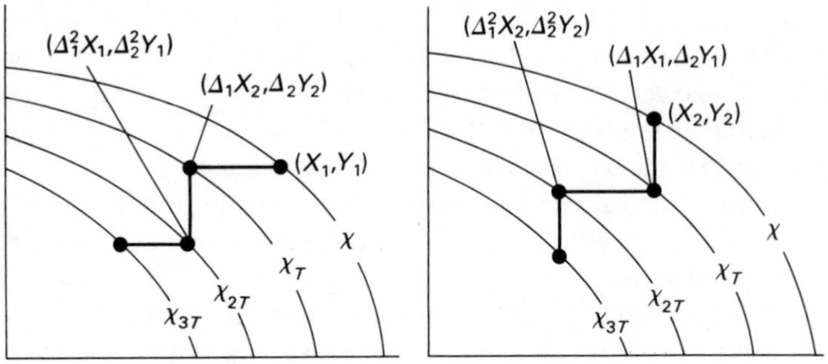

Figure 4.5a and b

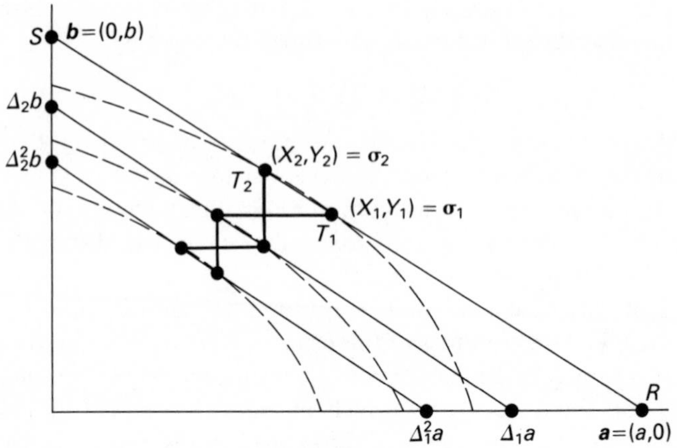

Figure 4.6

point (X_2, Y_2) which is obtained in the corresponding game in which the second player makes the opening demand.

One can think of the bargaining situation at the kth stage as a one-period bargaining game in which the player to make the demand has 'bargaining power' $\tau = 1$ and the status quo consists of the solution of the bargaining contest which ensues if the demand is refused.

In order to characterize the points (X_1, Y_1) and (X_2, Y_2), we introduce the chords illustrated in figure 4.6.

It is easily shown that

$$\boldsymbol{\sigma}_1 = \alpha \mathbf{a} + (1-\alpha)\mathbf{b}$$

$$\boldsymbol{\sigma}_2 = (1-\beta)\mathbf{a} + \beta \mathbf{b}$$

where

$$\alpha = \frac{1-\Delta_2}{1-\Delta_1\Delta_2}; \quad \beta = \frac{1-\Delta_1}{1-\Delta_1\Delta_2}.$$

The solution outcome $\boldsymbol{\sigma}_1$ (and its companion $\boldsymbol{\sigma}_2$) are therefore found by drawing the unique chord for which

$$\frac{ST_1}{SR} = \frac{1-\Delta_2}{1-\Delta_1\Delta_2}; \quad \frac{RT_2}{RS} = \frac{1-\Delta_1}{1-\Delta_1\Delta_2}.$$

Observe that $\alpha \to \tau$ as $T \to 0+$ and $\beta \to 1-\tau$ as $T \to 0+$, where

$$\tau = \frac{\log \delta_2}{\log \delta_1 + \log \delta_2}.$$

The 'asymmetric Nash bargaining solution' $\boldsymbol{\sigma}^{(\tau)}$ with 'bargaining power' τ is located at the point T in figure 4.7 where $ST/SR = \tau$.

It is evident that both $\boldsymbol{\sigma}_1$ and $\boldsymbol{\sigma}_2$ approximate $\boldsymbol{\sigma}^{(\tau)}$ for small enough values of T.

Note that, in the limit as $T \to 0+$, it does not matter which player opens the bidding. Each player receives the same whether he goes first or second. But the limiting outcome does *not* satisfy Nash's symmetry axiom unless $\delta_1 = \delta_2$.

Note 2. Rubinstein in chapter 3 gives a list of assumptions under which his theorem holds. These refer to the preferences the players hold over pairs (s, t) of possible outcomes. In Rubinstein's notation, $s \in [0, 1]$ labels the possible demands and t denotes the time at which the demand is made. In the

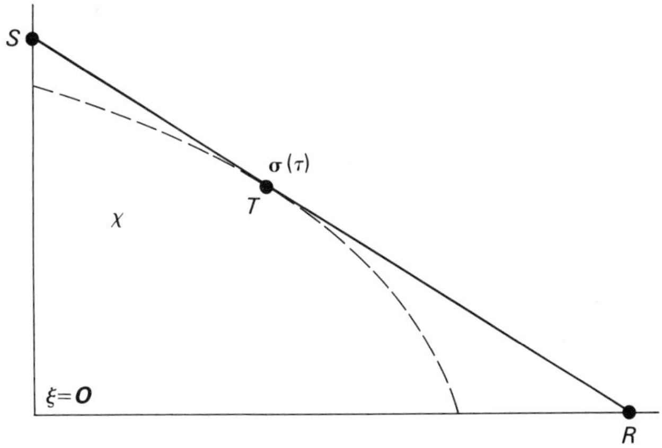

Figure 4.7

special case considered in this chapter, one may label the demands by the utility that the first player would receive if that demand had been made and accepted at time 0. The first player's preferences are then described by the utility function $\psi_1(s,t) = \Delta_1^t s$ and the second player's by $\psi_2(s,t) = \Delta_2^t \phi(s)$. This system of preferences clearly satisfies all of Rubinstein's assumptions except possibly the last (A-5) in so far as it relates to the second player's preferences. The relevant assumption is that the function ϵ defined by

$$(s + \epsilon(s), 1) \sim_2 (s, 0)$$

increases with s. In our notation we have that $\Delta_2 \phi(s + \epsilon(s)) = \phi(s)$. Since ϕ decreases, $\epsilon(s) < 0$ and $s + \epsilon(s)$ increases. From the concavity of ϕ we have that, for $s_1 < s_2$,

$$\frac{\phi(s_2) - \phi(s_1)}{s_2 - s_1} \leqslant \frac{\phi(s_2 + \epsilon(s_2)) - \phi(s_1 + \epsilon(s_1))}{(s_2 + \epsilon(s_2)) - (s_1 + \epsilon(s_1))}$$

$$= \Delta_2^{-1} \left\{ \frac{\phi(s_2) - \phi(s_1)}{(s_2 - s_1) + (\epsilon(s_2) - \epsilon(s_1))} \right\}$$

and hence

$$\epsilon(s_2) - \epsilon(s_1) \geqslant (\Delta_2^{-1} - 1)(s_2 - s_1) > 0$$

as required.

Note 3. In Rubinstein's notation, the first player's payoffs at perfect equilibria in which he opens the bidding are the x-coordinates of points in the set

$$\Delta = \{(x, y) : y = d_1(x) \text{ and } x = d_2(y)\}.$$

(The y-coordinates are the first player's payoffs when the second player opens the bidding.) The set Δ is a closed line segment parallel to the line $x = y$. In our case $d_1(x) = \Delta_1 x$ and $d_2(y) = \phi^{-1}(\Delta_2 \phi(y))$ so that Δ reduces to a single point.

Note 4. It may be worth noting that the case we consider in this chapter is not such a special instance of Rubinstein's result as it may appear. If his assumption that 'pie' and 'time' are evaluated separately is extended so that 'pie lotteries' and 'time lotteries' are evaluated separately, theorem 2 of chapter 2 restricts the forms available for the player's utility functions. If one then requires uniformity over all times, it then follows that a player's utility for (s, t) must take one of the forms $\phi(s)\delta^t$ or $\phi(s) - ct$.

Note 5. If T is not close to zero, then the solution of the bargaining contest in which the first player opens the bidding will not be an 'asymmetric Nash bargaining solution' (in the sense that different values of τ will be required for different sets $\mathscr{X}$). This may be attributed to the fact that the negotiation procedure formalized is 'incomplete'. Given the opportunity, the players will prefix the formal negotiations with further negotiations in which the question of 'who goes first' is discussed. As a result of these pre-negotiations, some compromise between the two extremes $\boldsymbol{\sigma}_1$ and $\boldsymbol{\sigma}_2$ will be reached.

This viewpoint perhaps lends some interest to providing an axiomatic characterization of the set $\Sigma(\mathscr{X}, \boldsymbol{\xi}) = \{\boldsymbol{\sigma}_1, \boldsymbol{\sigma}_2\}$.

Axiom I (invariance). For any strictly increasing affine transformation $A : \mathbb{R}^2 \to \mathbb{R}^2$,

$$\Sigma(A\mathscr{X}, A\boldsymbol{\xi}) = A\Sigma(\mathscr{X}, \boldsymbol{\xi}).$$

Axiom II (efficiency)

$$\Sigma(\mathscr{X}, \boldsymbol{\xi}) \subseteq \text{eff}(\mathscr{X}).$$

Axiom III (independence of irrelevant alternatives)

$$\Sigma(\mathscr{X}, \boldsymbol{\xi}) \subseteq \mathscr{Y} \subseteq \mathscr{X} \Rightarrow \Sigma(\mathscr{Y}, \boldsymbol{\xi}) = \Sigma(\mathscr{X}, \boldsymbol{\xi}).$$

Axiom IV (monotonicity)

$$\mathscr{X} \subseteq \mathscr{Z} \quad \text{and} \quad \Sigma(\mathscr{X}, \boldsymbol{\xi}) \subseteq \text{eff}(\mathscr{Z}) \Rightarrow \Sigma(\mathscr{Z}, \boldsymbol{\xi}) = \Sigma(\mathscr{X}, \boldsymbol{\xi}).$$

With these axioms one may demonstrate the existence of constants α and β in $[0, 1]$ such that $\Sigma(\mathscr{X}, \boldsymbol{\xi}) = \{T_1, T_2\}$ where

$$\frac{ST_1}{SR} = \alpha; \quad \frac{RT_2}{RS} = \beta.$$

(See figure 4.8.)

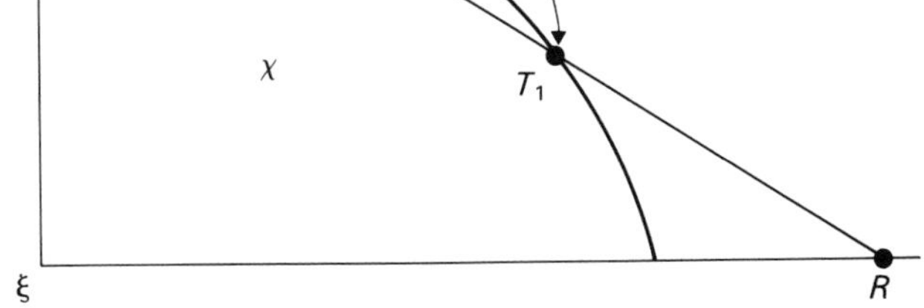

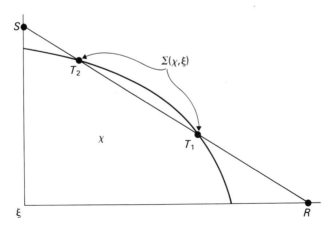

Figure 4.8

REFERENCES

Harsanyi, J. C. 1977: *Rational Behaviour and Bargaining Equilibria in Games and Social Situations*. CUP, Cambridge.

Luce, R. D. and Raiffa, H. 1957: *Games and Decisions*. Wiley, New York.

Nash, J. F. 1950: The bargaining problem. *Econometrica*, **18**, 155-62.

Nash, J. F. 1951: Non-cooperative games. *Annals of Mathematics*, **54**, 286-95.

Nash, J. F. 1953: Two-person cooperative games. *Econometrica*, **21**, 128-40.

Ståhl, I. 1972: *Bargaining Theory*. Economics Research Institute, Stockholm.

5

Perfect Equilibria in Bargaining Models

K. Binmore

1 INTRODUCTION

In chapters 2, 4, 8 and 11 and in Binmore (1981) I explore some consequences of the Nash approach to cooperative game theory for two-person bargaining games. In a later paper with the title 'Bargaining and coalitions' some of these ideas will be extended to the case of multi-person bargaining games with particular reference to the question of coalition formation. The results are obtained by examining bargaining models based on that introduced by Rubinstein in chapter 3.

We find it convenient to employ a somewhat different technique to that used by Rubinstein and the purpose of the current chapter is to illustrate this technique in the two-person case. The chapter is therefore very much a postscript to Rubinstein's work. However, we do refine his results to some extent and consider, in particular, examples in which the bargaining process involves random moves or in which the 'cake' does not shrink steadily over time.

2 A RUBINSTEIN-TYPE MODEL

Two players labelled 1 and 2 alternate in having the opportunity to communicate a proposal to the other. When a proposal is made by one player, the other may accept or reject the proposal. If the proposal is accepted, then the game ends with the implementation of the proposal. Both players prefer the acceptance of any feasible proposal to the rejection of all proposals.

It is assumed that the opportunities for making proposals are restricted to times t_n $(n = 0, 1, 2, \ldots)$ where $\langle t_n \rangle$ is strictly increasing with $t_0 = 0$. A proposal at time t consists of a utility pair chosen from a given feasible set $\mathscr{X}_t \subseteq \mathbb{R}^2$. Both $\langle t_n \rangle$ and the sets $\mathscr{X}_t$ are to be understood as pre-determined and uninfluenced by the players' strategy choices.

It is supposed throughout that each set $\mathscr{X}_t$ is closed and bounded above and that its Pareto frontier $\mathscr{P}_t$ is connected. (We say $\mathbf{x}$ is Pareto-optimal in

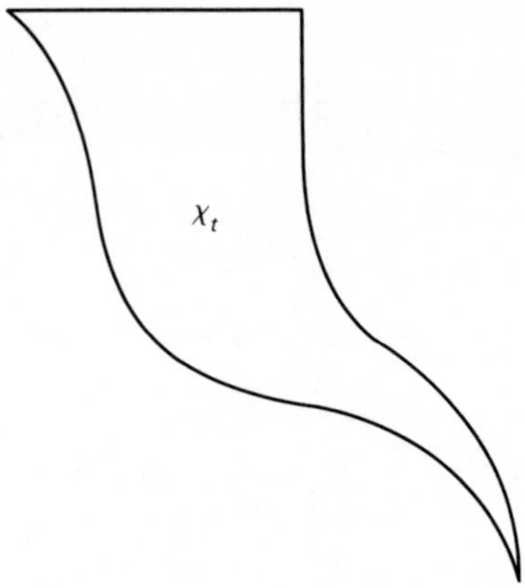

Figure 5.1

$\mathscr{X}$ if and only if $y_1 > x_1$ and $y_2 > x_2$ implies $\mathbf{y} \notin \mathscr{X}$. The Pareto frontier of $\mathscr{X}$ is the set of all its Pareto-optimal points, see figure 5.1.)

Rubinstein in chapter 3 studied the perfect equilibria (in pure strategies) of games of this general type although he found it computationally convenient to impose more stringent conditions on the sequence $\langle t_n \rangle$ and the sets $\mathscr{X}_t$. A structural rather than a technical difference between the model described above and Rubinstein's model is that the current model does not require that the players necessarily make a proposal at each time they have the opportunity to do so. This extra freedom does not affect the perfect equilibrium outcomes under a 'steadily shrinking cake' assumption as employed by Rubinstein. But its absence would seriously distort the situation when the cake is not steadily shrinking or when several players are bargaining over which coalition is to form. In these latter cases a player may well wish to avoid making a proposal at a time when he has the opportunity to do so because he anticipates having a more favourable proposal accepted in the future.

Diagrams like figure 5.2 are sometimes helpful in clarifying the reasoning which we shall employ. The figure indicates some possible perfect equilibrium choices on a section of the appropriate game tree.

3 CHARACTERIZATION OF PERFECT EQUILIBRIUM OUTCOMES

We now describe a method which provides a geometric characterization of the perfect equilibrium outcomes (in pure strategies) for the model introduced above. Note that we follow Rubinstein in using the terminology 'perfect

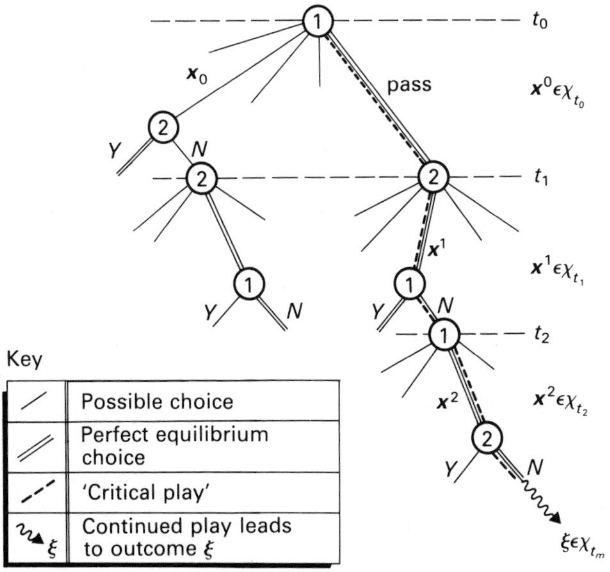

Figure 5.2

equilibrium' in Selten's original sense. Such an equilibrium is sometimes said to be 'subgame perfect'.

Proposition 1. Suppose that player 1 is the proposer at time t_{k-1} and that any perfect equilibrium outcome **z** in a subgame commencing at time t_k satisfies $m \leqslant z_2 \leqslant M$. Then any perfect equilibrium outcome **y** in a subgame commencing at time t_{k-1} satisfies $l \leqslant y_1$ and $m \leqslant y_2$ where

1 If there exists $\mathbf{x}^1 \in \mathscr{P}_{t_{k-1}}$ with $x_2^1 > M$ and $\mathbf{x}^2 \in \mathscr{P}_{t_{k-1}}$ with $x_2^2 \leqslant M$, then

$$l = \inf\{x_1 : (x_1, M) \in \mathscr{P}_{t_{k-1}}\}.$$

(See figure 5.3.)

2 If for all $\mathbf{x} \in \mathscr{P}_{t_{k-1}}$, $x_2 > M$, then

$$l = \sup\{x_1 : (x_1, x_2) \in \mathscr{P}_{t_{k-1}}\}.$$

(See figure 5.4.)

3 Otherwise, $l = -\infty$ (see figure 5.5.).

Proof. In equilibrium, player 2 will reject any proposal **y** at time t_{k-1} with $y_2 < m$. Thus either he accepts a proposal at time t_{k-1} with $y_2 \geqslant m$ or else we proceed to time t_k when all equilibrium outcomes **z** satisfy $z_2 \geqslant m$ by hypothesis. This deals with case 3 and part of case 1. The remainder of

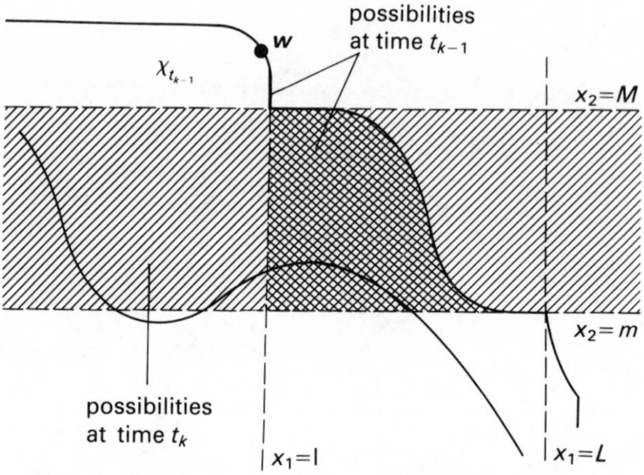

Figure 5.3 Standard case 1.

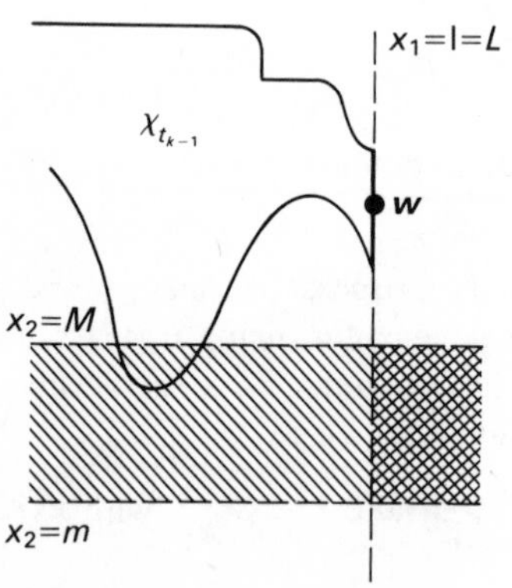

Figure 5.4 Degenerate case 2.

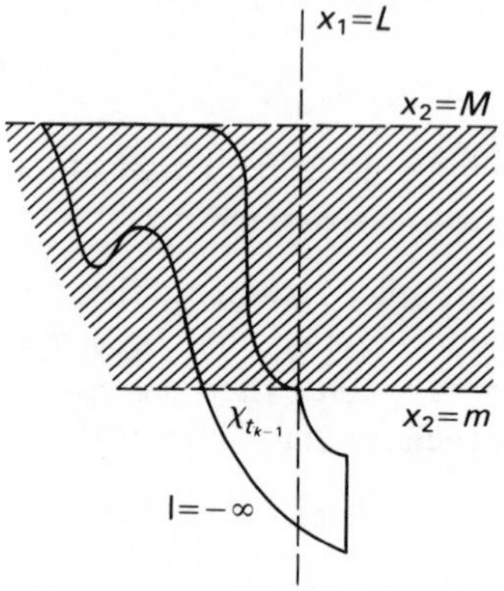

Figure 5.5 Degenerate case 3.

case 1 is dealt with by observing that any action by player 1 at time t_{k-1} which leads to an outcome y with $y_1 < l$ cannot be in equilibrium. The reason is that $\mathbf{w} \in \mathscr{P}_{t_{k-1}}$ can be found with $w_1 > y_1$ and $w_2 > M$. In equilibrium, player 2 will accept $\mathbf{w}$ at time t_{k-1} and hence player 1 would improve his payoff by proposing $\mathbf{w}$ at time t_{k-1}. Case 2 is similar in that player 2 will accept any proposal $\mathbf{w} \in \mathscr{P}_{t_{k-1}}$.

Proposition 2. If, in proposition 1, it is known that $z_1 \geqslant a$ for all $\mathbf{z}$, then l may be replaced by $\max\{a, l\}$ (see figure 5.6).

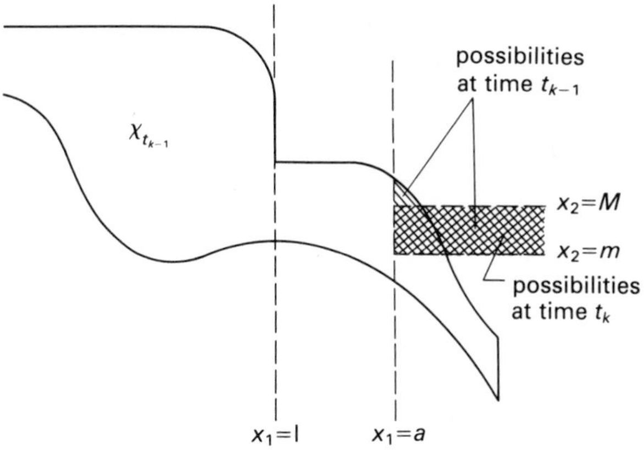

Figure 5.6

Proof. Player 1 would prefer to pass rather than to take action which leads to an outcome **y** with $y_1 < a$.

Proposition 3. If, in proposition 1, it is known that $z_1 \leqslant A$ for all **z**, then $y_1 \leqslant \max\{A, L\}$ where

1 If there exists $\mathbf{x}^1 \in \mathscr{P}_{t_{k-1}}$ with $x_2^1 \geqslant m$ and $\mathbf{x}^2 \in \mathscr{P}_{t_{k-1}}$ with $x_2^1 \leqslant m$, then

$$L = \sup\{x_1 : (x_1, m) \in \mathscr{P}_{t_{k-1}}\}.$$

2 If for all $\mathbf{x} \in \mathscr{P}_{t_{k-1}}, x_2 > m$, then

$$L = \sup\{x_1 : (x_1, x_2) \in \mathscr{P}_{t_{k-1}}\}.$$

3 Otherwise, $L = -\infty$ (see figure 5.7).

(The standard case 1 is illustrated in figures 5.3 and 5.5. The degenerate case 2 is illustrated in figure 5.4.)

Proof. This proposition merely asserts that a perfect equilibrium outcome **y** is either achieved via an acceptance at time t_{k-1} in which case $y_1 \leqslant L$ by proposition 1 or else is achieved via a rejection at time t_{k-1} in which case $y_1 \leqslant A$.

The need to deal adequately with possible degenerate cases complicates the statements of the preceding propositions but they are quite easy to use in practice. In our applications we shall have available an induction hypothesis which asserts that all perfect equilibrium outcomes **z** in a subgame which commences at time t_k satisfy an inequality of the form $\mathbf{b}_k \leqslant \mathbf{z} \leqslant \mathbf{B}_k$ where

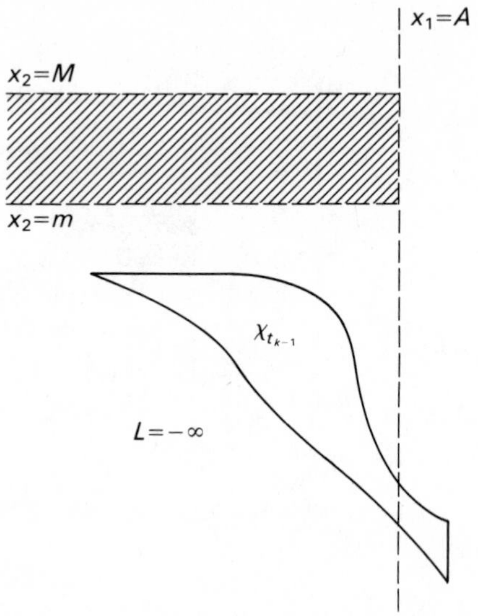

Figure 5.7 Degenerate case 3.

$\mathbf{b}_k$ and $\mathbf{B}_k$ are points in $[-\infty, +\infty)^2$. We then use propositions 2 and 3 to compute $\mathbf{b}_{k-1}$ and $\mathbf{B}_{k-1}$ with the property that

$$\mathbf{b}_{k-1} \leqslant \mathbf{y} \leqslant \mathbf{B}_{k-1}$$

for all perfect equilibrium outcomes $\mathbf{y}$ in a subgame which commences at time t_{k-1}.

4 SHRINKING CAKE ASSUMPTION

To avoid obscuring the simplicity of the technique with technicalities, we impose a further condition before continuing the discussion. This condition consists of the *shrinking cake assumption* that the Pareto frontier of $\mathcal{X}_s \cup \mathcal{X}_t$ coincides with that of $\mathcal{X}_s$ when $s \geqslant t$.

Later in the chapter we shall return to the question of how the method applies when the cake is not steadily shrinking in the context of some specific examples.

The advantage of introducing the shrinking cake assumption at this stage is that it becomes possible to describe the construction of the sets E_n which appear in the following proposition by means of a simple figure. Figure 5.8 indicates the general situation. Figures 5.9 and 5.10 indicate degenerate cases.

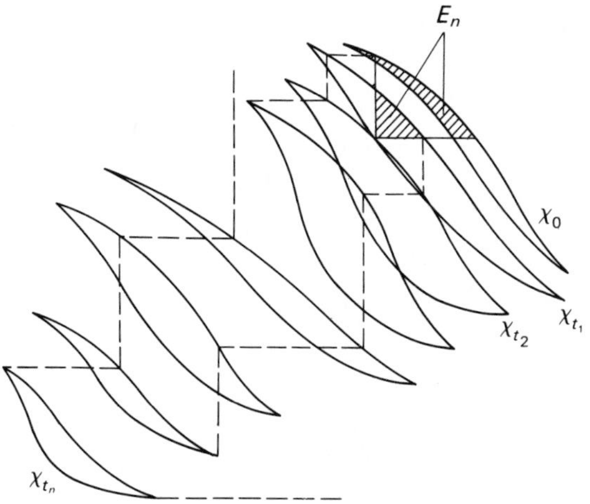

Figure 5.8 Standard case.

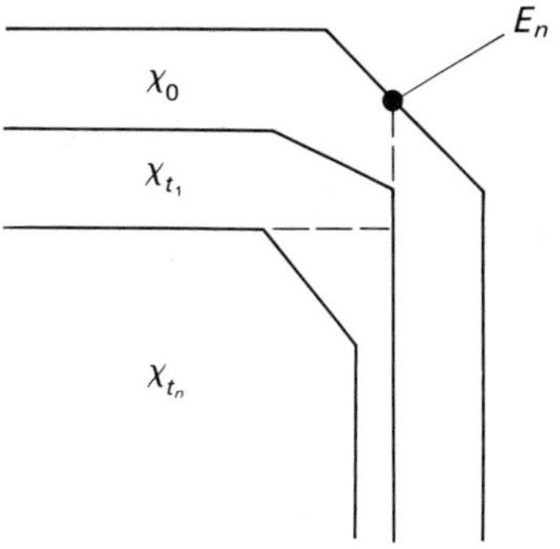

Figure 5.9 Degenerate case.

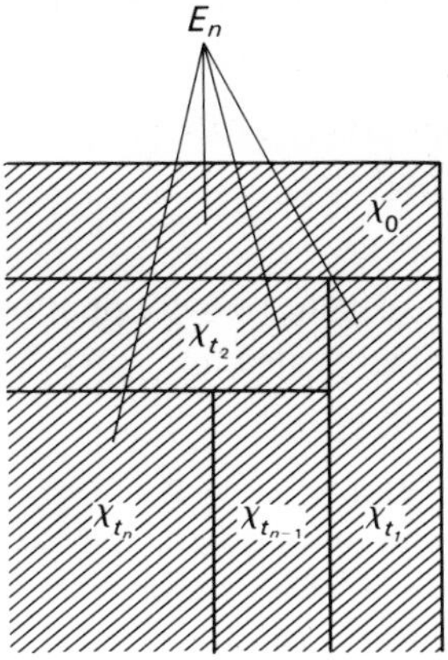

Figure 5.10 Degenerate case.

Proposition 4. All perfect equilibrium outcomes lie in the set

$$E = \bigcap_{n=1}^{\infty} E_n$$

where E_n is constructed as indicated in figures 5.8–5.10.

Proof. It is only necessary to observe that we proceed by induction beginning with the observation that all perfect equilibrium outcomes in subgames

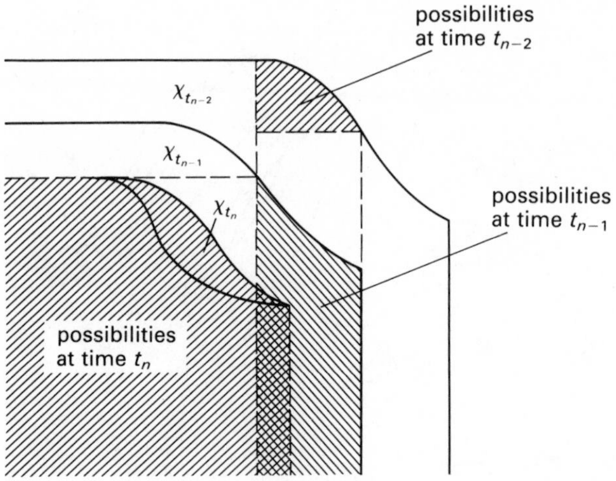

Figure 5.11

commencing at time t_n lie in the union of the sets $\mathcal{X}_{t_m}$ with $m \geqslant n$. Figure 5.11 illustrates the first two steps.

Proposition 5. Any point in the set E of proposition 4 is a perfect equilibrium outcome.

Proof. We begin by observing that any point $\mathbf{x}^0$ which is Pareto-optimal in $\mathcal{X}_0$ is a perfect equilibrium outcome provided that there exists a sequence $\langle \mathbf{x}^n \rangle$ as illustrated in figure 5.12 with $\mathbf{x}^n$ Pareto-optimal in $\mathcal{X}_{t_n}$ for each $n = 0, 1, 2, \ldots$.

The broken arrows indicate the responses planned by player 2 at time 0 in equilibrium. For example, player 2 plans to accept a proposal of $\mathbf{y}$ at time 0 but, if he finds himself at time t_1 in the position of having refused the proposal $\mathbf{y}$, to continue with the proposal $\mathbf{x}^1$.

Assuming similar history-independent plans at every time t_n $(n = 0, 1, 2, \ldots)$, we obtain a perfect equilibrium which yields the outcome $\mathbf{x}^0$. The 'critical play' (see figure 5.12b) is indicated with a broken line.

This argument justifies the assertion that every Pareto-optimal point of E is a perfect equilibrium outcome. Consider next a point $\boldsymbol{\xi}^0 \in E \cap \mathcal{X}_0$ which is not Pareto-optimal in $\mathcal{X}_0$. Figure 5.13 indicates appropriate perfect equilibrium responses for player 2 given that player 1's perfect equilibrium initial proposal is $\boldsymbol{\xi}^0$. Note that $\mathbf{X}^1$ and $\mathbf{x}^1$ in figure 5.13 may be taken to be perfect equilibrium outcomes in a subgame which commences at time t_1 by the argument given above.

There remains the question of points $\boldsymbol{\xi}^n \in E \cap \mathcal{X}_{t_n}$. Any such outcome may be achieved via a perfect equilibrium in which the first n proposals are

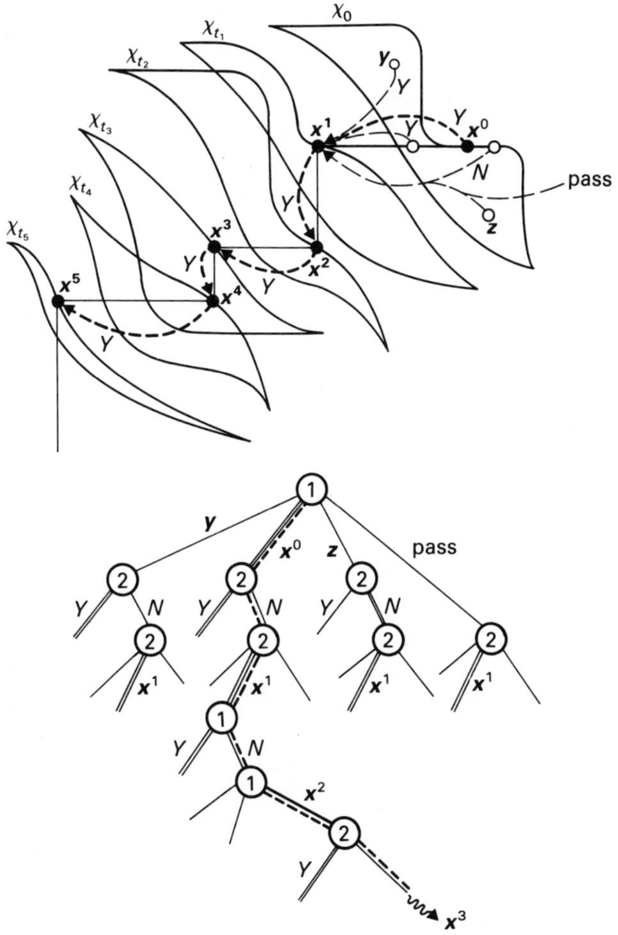

Figure 5.12a and b

rejected. Figure 5.14 illustrates an appropriate perfect equilibrium in the case $n = 2$. Note that $\mathbf{X}^1$ is a perfect equilibrium outcome in a subgame commencing at time t_1 and $\boldsymbol{\xi}^2$ and $\mathbf{x}^2$ are perfect equilibrium outcomes in a subgame commencing at time t_2.

As an example, consider the 'fixed costs' model given by Rubinstein in chapter 3. Here $t_n = nt$ $(n = 0, 1, 2, \ldots)$ for a fixed value of t and

$$\mathcal{X}_{nt} = \{(\mathbf{x}_1 - c_1 nt, \mathbf{x}_2 - c_2 nt) : \mathbf{x}_1 \geqslant 0, \mathbf{x}_2 \geqslant 0, \mathbf{x}_1 + \mathbf{x}_2 = 1\}.$$

One might think of two agents, with utilities linear in money, who bargain over how to share a dollar. Delaying agreement by time T generates a cost of $c_i T$ to player i $(i = 1, 2)$. Propositions 4 and 5 are applied geometrically in figure 5.15.

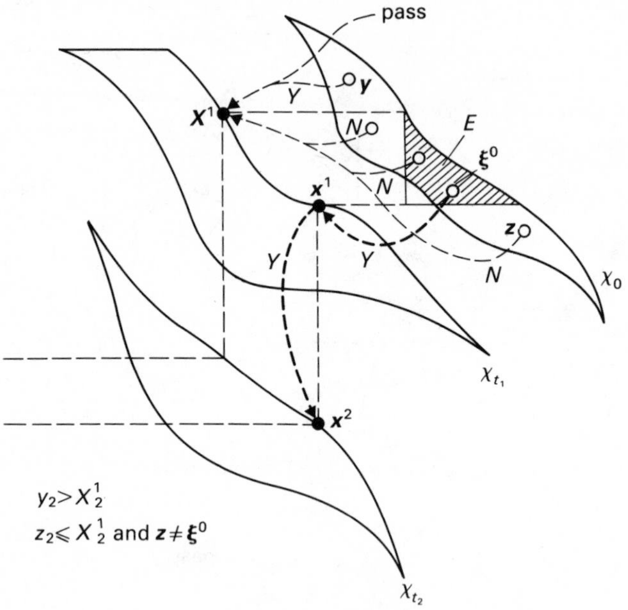

Figure 5.13

Thus, when $c_1 < c_2$, there is a unique perfect equilibrium outcome which is achieved by an acceptance of the initial proposal. Player 1 receives a payoff of 1 (the whole cake) and player 2 receives a payoff of 0. When $c_1 > c_2$, there is again a unique perfect equilibrium outcome which is achieved by an acceptance of the initial proposal. Player 1 receives $c_2 t$ and player 2 receives $1 - c_2 t$.

The case $c_1 = c_2 = c$ is more interesting in that it exhibits multiple perfect equilibria, some of which are only achieved after a rejection of the initial proposal. The situation, however, is 'unstable' in the sense that small perturbations in the parameters of the problem (i.e. the intervention of a 'trembling hand') tend to eliminate the multiplicity problem. For example, irregularities in the sequence $\langle t_n \rangle$ will typically lead to a degenerate situation similar to that obtained when unit costs are unequal. A more significant perturbation is perhaps to allow a certain measure of risk aversion by taking

$$\mathscr{X}_{nt} = \{\mathbf{x} : x_2 + nc_2 t = \phi(x_1 + nc_1 t)\}$$

where $\phi : [0, 1] \to [0, 1]$ is a continuous, decreasing, strictly concave surjection. For a point to satisfy the stationarity condition indicated in figure 5.16 it is necessary that

$$\frac{\phi(X - c_1 t) - \phi(X)}{-c_1 t} = -\frac{c_2}{c_1}$$

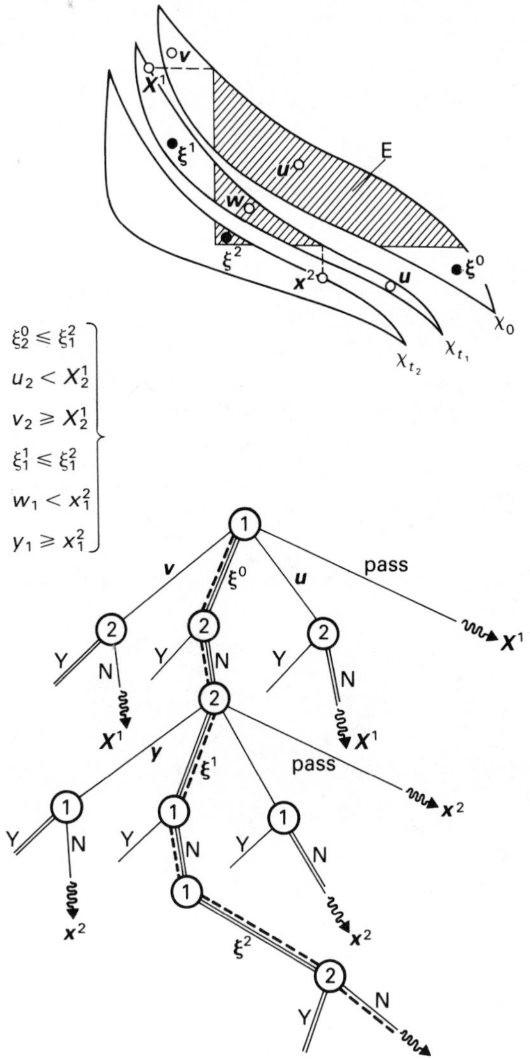

Figure 5.14

but since ϕ is strictly concave, this can hold for at most one value of X. We conclude that for all ϕ satisfying our conditions, there is a *unique* perfect equilibrium outcome.

Note 1. Continuing the discussion of the previous paragraph, it may be noted that when $t \to 0+$, we are led to the unique point on $\mathscr{X}_0$ at which a 'supporting line' has slope $-c_2/c_1$. This point may be thought of as an

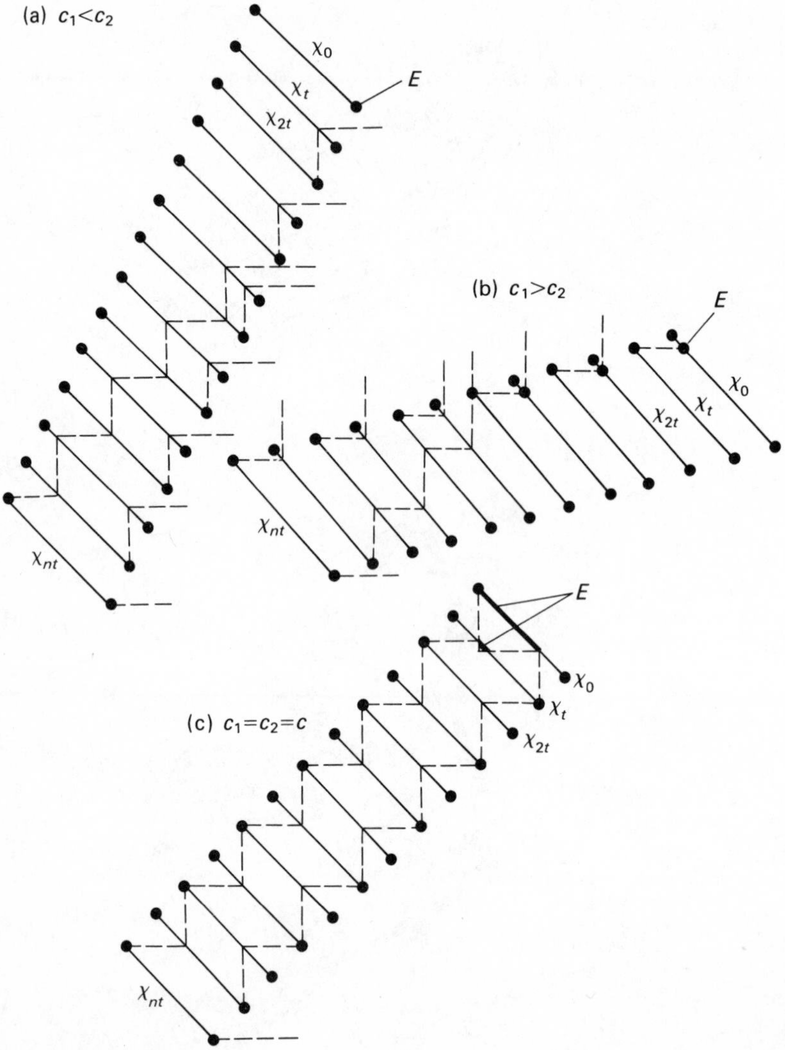

Figure 5.15

asymmetric 'Nash bargaining solution' for $\mathscr{X}_0$ (see section 7) in the case when a unit of utility for player 1 may be identified with a unit of utility for player 2 (e.g. when utility = cash) but utility levels remain uncompared. The justification is that the point in question is characterized by Nash's axioms (excluding symmetry) provided that the axiom requiring invariance under transformations of the form $(x_1, x_2) \to (Ax_1 + B, Cx_2 + D)$ where $A > 0$ and $C > 0$ is replaced by an axiom requiring invariance under transformations of

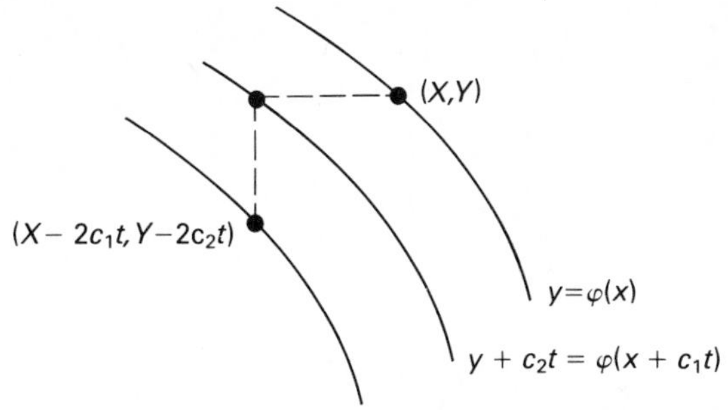

Figure 5.16

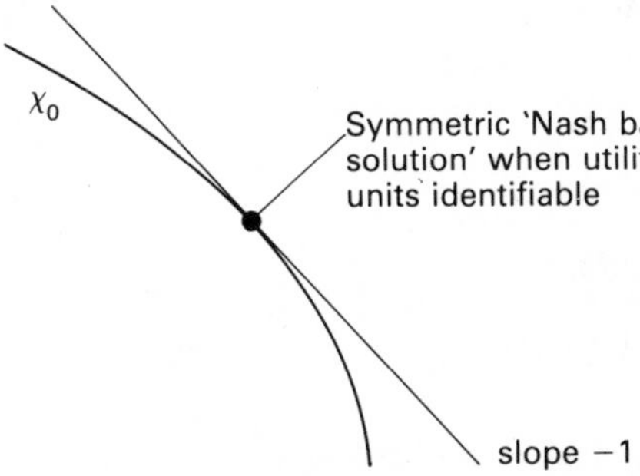

Figure 5.17

the form $(x_1, x_2) \mapsto (x_1 + B,\ x_2 + D)$ (see figure 5.17). The fact that utility units are comparable eliminates the need for a 'status quo' although this may be thought of as located at $(-\infty, -\infty)$.

Note 2. If $\mathscr{X}_{nt}$ is modified so that free disposal is admitted, then uniqueness is lost in the case $c_1 < c_2$ (although not when $c_1 > c_2$). Player 1 still receives the whole cake but player 2 may receive anything between 0 and $-(c_2 - c_1)t$. However, uniqueness is again restored by replacing the boundary of $\mathscr{X}_0$ by the graph of a strictly concave function.

Note 3. If the players have only finite cash reserves and can no longer negotiate when these are exhausted, the analysis of section 5 needs to be modified. The appropriate game is then essentially finite.

6 VANISHING CAKES

Perhaps the most interesting special case of the model of section 4 is that in which the set $\mathscr{X}_t$ shrinks to a single point as $t \to +\infty$. In studying this

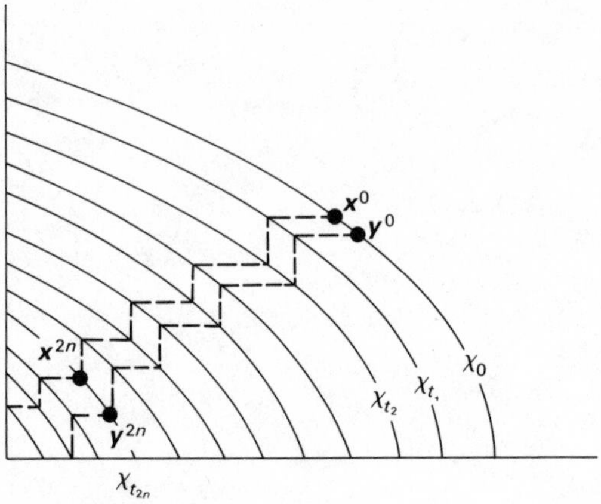

Figure 5.18

situation we shall always assume that the sets $\mathscr{X}_t$ all lie in the non-negative orthant and that the point to which they shrink is **0**. As figure 5.18 suggests, the typical result is that the set E of perfect equilibrium outcomes reduces to a single point.

However, a unique perfect equilibrium outcome is not guaranteed under these circumstances even when the sets $\mathscr{X}_t$ are line segments. To see this, we consider the case in which

$$\mathscr{X}_{t_n} = \{\mathbf{x} : x_1 \geqslant 0, x_2 \geqslant 0 \text{ and } p_n^{-1}x_1 + q_n^{-1}x_2 = 1\}$$

where $\langle p_n \rangle$ and $\langle q_n \rangle$ are decreasing sequences of positive terms. Referring to figure 5.20, we have that

$$x_1^{n+2} - y_1^{n+2} = \frac{p_{n+1}q_n}{p_n q_{n+1}}(x_1^n - y_1^n)$$

and hence

$$|x_1^0 - y_1^0| = \left(\frac{p_0}{p_1}\cdot\frac{p_2}{p_3}\cdots\frac{p_{2n}}{p_{2n+1}}\right)\left(\frac{q_1}{q_0}\cdot\frac{q_3}{q_2}\cdots\frac{q_{2n+1}}{q_{2n}}\right)|x_1^{2n+2} - y_1^{2n+2}|$$

$$\leqslant \left(\frac{p_0}{p_1}\cdot\frac{p_2}{p_3}\cdots\frac{p_{2n}}{p_{2n+1}}\right)\left(\frac{q_1}{q_0}\cdot\frac{q_3}{q_2}\cdots\frac{q_{2n+1}}{q_{2n}}\right)p_{2n+2}$$

$$= p_0\left(\frac{p_2}{p_1}\cdot\frac{p_4}{p_3}\cdots\frac{p_{2n+2}}{p_{2n+1}}\right)\left(\frac{q_1}{q_0}\cdot\frac{q_3}{q_2}\cdots\frac{q_{2n+1}}{p_{2n}}\right).$$

It follows that a sufficient condition for the existence of a unique perfect equilibrium outcome is the divergence of the series

$$\sum_{n=1}^{\infty} \log\left(\frac{p_{2n}}{p_{2n-1}} \cdot \frac{q_{2n-1}}{q_{2n-2}}\right).$$

Equally, it is clear that if this series converges then multiple perfect equilibrium outcomes exist. Convergence will occur when odd terms of the sequence $\langle p_n \rangle$ are sufficiently close to the subsequent even terms and, simultaneously, even terms of the sequence $\langle q_n \rangle$ are sufficiently close to the subsequent odd terms. Figure 5.19 indicates an appropriate configuration.

7 FIXED DISCOUNT RATES

The most important special case of a 'vanishing cake' is that in which the players discount future consumption at fixed rates δ_1 and δ_2 (where $0<\delta_1<1$ and $0<\delta_2<1$). Thus

$$\mathscr{X}_{t_n} = \{(x_1\delta_1^{t_n}, x_2\delta_2^{t_n}) : \mathbf{x} \in \mathscr{X}_0\}.$$

Rubinstein in chapter 3 showed the existence of a unique perfect equilibrium outcome when $\mathscr{X}_0$ is convex and $t_n = nt$ $(n = 0, 1, \ldots)$.

A simple modification of the argument of section 6 shows that the same result holds in the case when

$$\mathscr{X}_0 = \{\mathbf{x} : 0 \leqslant x_1 \leqslant \psi(x_2)\}$$

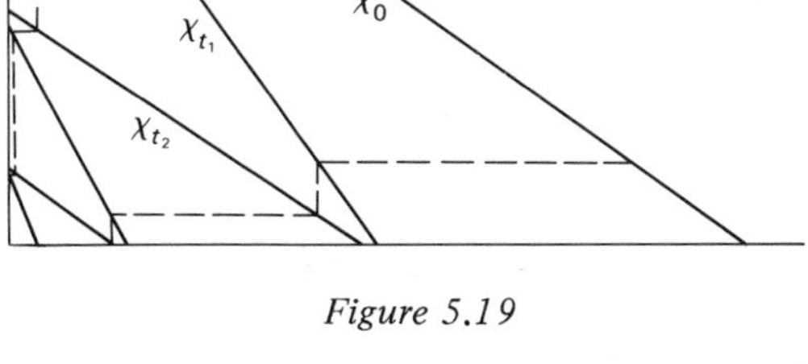

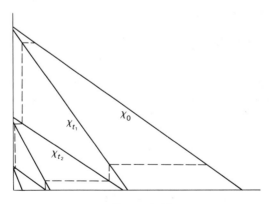

Figure 5.19

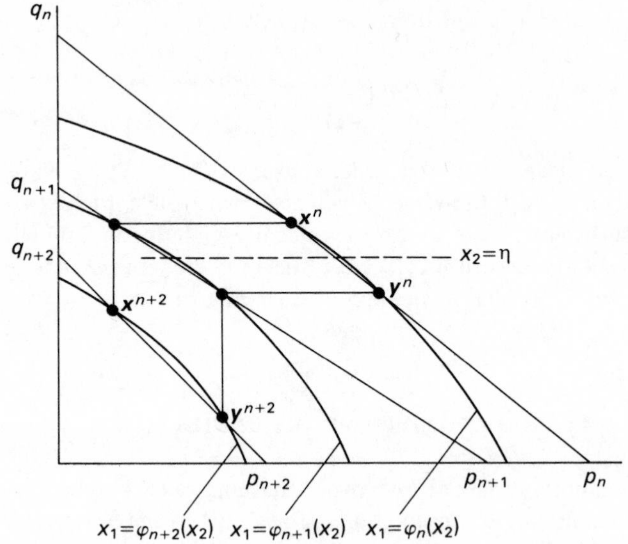

Figure 5.20

where $\psi : [0, 1] \to [0, 1]$ as a continuous, decreasing strictly concave surjection which is differentiable on $(0, 1)$ and $\langle t_n \rangle$ satisfies $t_n \to +\infty$ as $n \to \infty$. We write $\phi_n(x_2) = \delta_1^{t_n}\psi(x_2\delta_2^{-t_n})$ and apply the Cauchy mean value theorem in figure 5.20.

We obtain that

$$\frac{p_{n+1}}{q_{n+1}} \Big/ \frac{p_n}{q_n} = \frac{\phi_{n+1}(x_2^n) - \phi_{n+1}(y_2^n)}{\phi_n(x_2^n) - \phi_n(y_2^n)} = \frac{\phi'_{n+1}(\eta)}{\phi'_n(\eta)}$$

where η lies between x_2^n and y_2^n. But

$$\left|\frac{\phi'_{n+1}(\eta)}{\phi'_n(\eta)}\right| = \left(\frac{\delta_1}{\delta_2}\right)^{t_{n+1}-t_n} \left|\frac{\psi'(\eta\delta_2^{-t_{n+1}})}{\psi'(\eta\delta_2^{-t_n})}\right| \geqslant \left(\frac{\delta_1}{\delta_2}\right)^{t_{n+1}-t_n}$$

because ψ' decreases. (Note that differentiability can be dispensed with in the argument at the expense of some extra steps.) Proceeding as in section 6 but with the estimate $|x_1^{2n+2} - y_1^{2n+2}| \leqslant \delta_1^{t_{2n+2}}$, we are led to the inequality

$$|x_1^0 - y_1^0| \leqslant \left(\frac{\delta_2}{\delta_1}\right)^{t_1-t_0} \left(\frac{\delta_2}{\delta_1}\right)^{t_3-t_2} \cdots \left(\frac{\delta_2}{\delta_1}\right)^{t_{2n+1}-t_{2n}} \delta_1^{t_{2n+2}}$$

$$= \exp\left\{\log\delta_1 \sum_{k=0}^{n} (t_{2k+2} - t_{2k+1}) + \log\delta_2 \sum_{k=0}^{n} (t_{2k+1} - t_{2k})\right\}$$

$$\leqslant \exp\left\{\Delta \sum_{k=0}^{2n+1} (t_{k+1} - t_k)\right\}$$

$$= \exp\{\Delta t_{2n+2}\} \to 0 \text{ as } n \to \infty,$$

where $\Delta = \max\{\log \delta_1, \log \delta_2\} < 0$.

Note 4. It is a simple matter to obtain a formula for the first coordinate x_1^0 of the unique perfect equilibrium outcome in the example of section 6 (assuming that the appropriate series diverges). We have that

$$x_1^0 = \frac{p_0}{q_0}(q_0 - q_1) + \frac{p_0}{p_1} \cdot \frac{q_1}{q_0} \cdot \frac{p_2}{q_2}(q_2 - q_3) + \frac{p_0}{p_1} \cdot \frac{q_1}{q_0} \cdot \frac{p_2}{p_3} \cdot \frac{q_3}{q_2} \cdot \frac{p_4}{q_4}(q_4 - q_5) + \ldots.$$

In the case when $p_2 = \delta_1^n$, $q_n = \delta_2^n$ and $t_n = n$, this reduces to a result calculated by Rubinstein in chapter 3: namely

$$x_1^0 = \frac{1 - \delta_2}{1 - \delta_1 \delta_2}.$$

8 SMALL TIME INTERVALS

Various criticisms are possible of Rubinstein-type models even in those cases in which they produce a unique perfect equilibrium outcome. One criticism is that the models favour the player with the opportunity to make the first proposal and the model leaves the mechanism by means of which player 1 gains this advantage unexplained. A second criticism is the rigidity of the timetable for making proposals: what constrains the players to the use of this timetable?

Both criticisms lose their force when the model is applied in the case of vanishing small time periods between proposals. Beginning with the second criticism, suppose that no constraints exist on the timetabling of proposals. Then, after rejecting a proposal, a player will typically wish to introduce his counterproposal at the earliest possible moment. In support of this assertion we note that the analysis of section 4 works equally well when the players do not necessarily alternate as proposers (with the obvious modifications). In particular, it is never possible in the case when a unique perfect equilibrium outcome exists that a player's perfect equilibrium strategy will require his passing the opportunity to make a bid. It therefore seems reasonable to use a Rubinstein-type model in which the intervals between successive proposals are vanishingly small as a paradigm for the case in which the players are not formally constrained by an exogenously determined timetable. The use of such a model has two subsidiary merits of which the most important is that the advantage to player 1 in being the first to make a proposal disappears in the limit as the time interval between proposals tends to zero. Both criticisms mentioned at the head of this section are therefore dealt with by the same expedient. The second advantage is that the results of Rubinstein-type bargaining in the limiting case can be described for a wide variety of cases as asymmetric Nash bargaining solutions (see Roth 1979).

An asymmetric Nash bargaining solution for a compact, convex feasible set $\mathscr{X}$ with status quo point $\mathbf{0} \in \mathscr{X}$ is obtained by finding the point $\boldsymbol{\sigma}$ at which $x_1^{\tau_1} x_2^{\tau_2}$ is maximized on $\mathscr{X}$ (see figure 5.21). We call the positive parameters τ_1 and τ_2 the 'bargaining powers' for players 1 and 2 respectively. Only their ratio is significant but we normalize by requiring that $\tau_1 + \tau_2 = 1$.

In chapter 4 we showed that, in the case when $\mathscr{X}_0$ is convex, $t_n = nt$ and the players discount future consumption using fixed discount rates δ_1 and δ_2 (see the first paragraph of section 7), then the unique perfect equilibrium outcome converges as $t \to 0+$ to the asymmetric Nash bargaining solution for $\mathscr{X}_0$ with status quo $\mathbf{0}$ and 'bargaining powers':

$$\tau_1 = \frac{\log \delta_2}{\log \delta_1 + \log \delta_2}\,; \quad \tau_2 = \frac{\log \delta_1}{\log \delta_1 + \log \delta_2}\,.$$

Note 5. The result concerning asymmetric Nash bargaining solutions has been obtained independently by MacLennan (1982) and Moulin (1982) under various hypotheses. It was also derived from ostensibly different models in chapter 2 and in Binmore (1981). The result is also strongly related to Harsanyi's rationalization of Zeuthen's Principle (Harsanyi 1977) and to the earlier work of Bishop (1963) and Foldes (1964). There is perhaps room for a paper drawing all these different strands together with an appreciation of the work of Ståhl (1972), Hicks (1953) and others. In particular, a

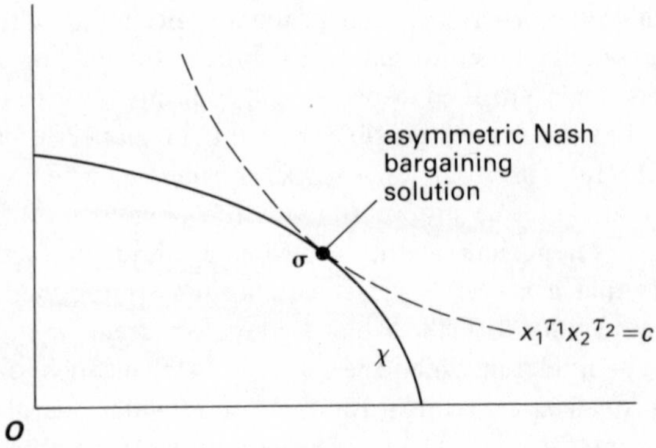

Figure 5.21

formal treatment of the idealizations required to justify the argument offered in section 8 would be useful together with a comparison of the necessary assumptions with those made by other authors who arrive at the same conclusion.

Note 6. Results similar to that mentioned in the final paragraph of section 8 may be obtained under more general conditions by resorting to the use of the calculus. As an example, we consider the case in which $\mathcal{X}_0$ is the unit simplex and the time intervals between successive intervals become vanishingly small but the length of the intervals is not uniform over time. Suppose that, at time t, the interval separating a proposal by 2 from the succeeding proposal by 1 is $u(t)$ of the interval separating successive proposals by 1. Then the proposal on the 'critical play' (see figure 5.2) at time t will assign player 2 $y(t)$ where $y(t) = (\delta_2/\delta_1)^{\frac{1}{2}t} w(t)$ and $w(t)$ is the solution of the linear differential equation

$$w'(t) + w(t)\left(\tfrac{1}{2} - u(t)\right)\log(\delta_2/\delta_1) = u(t)\,(\delta_1\delta_2)^{\frac{1}{2}t}\log \delta_1$$

which satisfies the boundary condition

$$w(t) \to 0 \quad \text{as } t \to +\infty.$$

In particular, the unique perfect equilibrium outcome assigns player 2 utility $y(0)$. The case $u(t) = \frac{1}{2}\ (t \geqslant 0)$ of uniform time intervals is particularly easily solved.

9 CHANCE MOVES IN RUBINSTEIN-TYPE MODELS

The argument of section 4 remains valid in principle although more complicated in detail when chance moves are introduced into the game tree (while retaining the perfect information assumption). Three cases of immediate interest suggest themselves:

1 The case in which the players are allowed the use of mixed strategies;
2 The case in which, at each time t_n ($n = 0, 1, 2, \ldots$), the proposer is chosen by a random device;
3 The case in which the time interval separating one proposal from the next is a random variable.

The third case is fairly easily dealt with and we shall only comment that, if in the example of section 5 in which negotiation costs are assumed linear in time, we allow all time intervals separating proposals to be independent, identically distributed random variables with expectation T, then the results are the same as if the time intervals were fixed at T. However, this result depends on the 'linearity' of the cake.

We discuss the second case in section 10 below and comment on the first case in section 12.

10 RANDOM SELECTION OF PROPOSERS

In this section we suppose that the model of section 2 is modified so that, at time t_n ($n = 0, 1, 2, \ldots$), the proposer is chosen by a chance move. For simplicity we shall continue to use the shrinking cake assumption of section 4 together with the assumption that the sets $\mathscr{X}_t$ are all convex (see section 12).

The propositions of section 3 remain valid in this new situation provided that the subgame commencing at time t_k is understood to be that which obtains *before* the chance move at time t_k has chosen a proposer while the subgame commencing at time t_{k-1} is understood to be that which obtains *after* the chance move at time t_{k-1} has chosen a proposer. The following result substitutes for proposition 4 under the new circumstances considered in this section *provided* that all chance moves are independent.

Proposition 6. All perfect equilibrium outcomes for games of the type described above lie in the set

$$E = \bigcap_{n=1}^{\infty} E_n$$

where E_n is constructed as indicated in figure 5.22 and p_k and q_k are the respective probabilities that players 1 and 2 are chosen as proposer at time $t_k (p_k + q_k = 1)$.

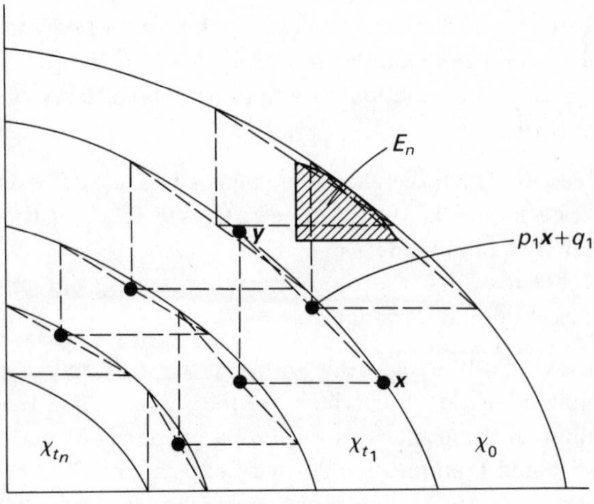

Figure 5.22

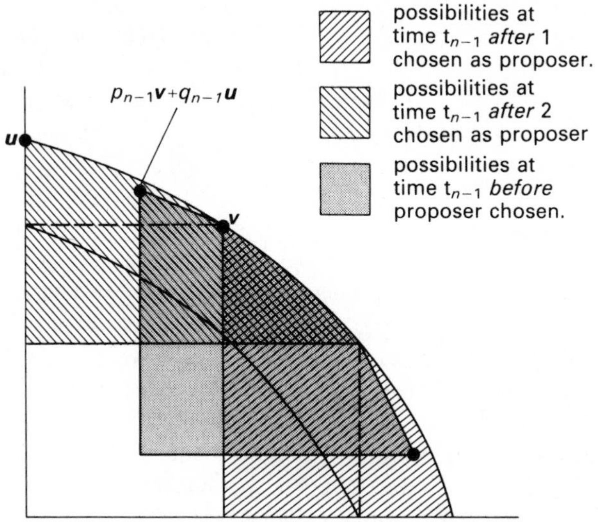

Figure 5.23

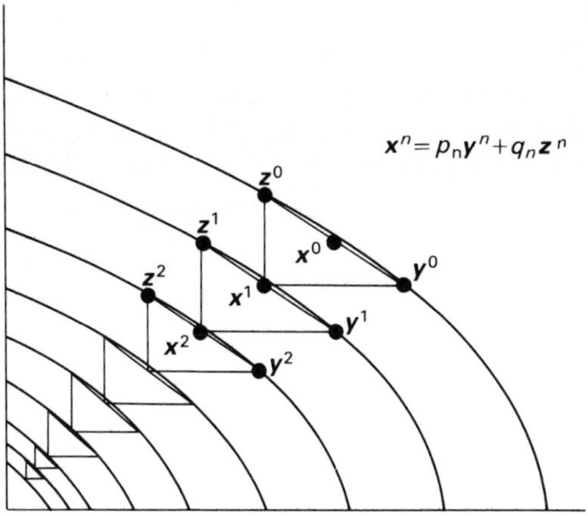

Figure 5.24

Proof. The first step of the induction process is indicated in figure 5.23.

We do not attempt to provide a substitute for proposition 5 but it will be obvious that any configuration of the type shown in figure 5.24 will support a perfect equilibrium.

11 FIXED DISCOUNT RATES WITH RANDOM PROPOSERS

As an example we return to the case in which a cake shrinks according to fixed discount rates as described in section 7. To simplify the calculations, we follow Rubinstein in considering only the case in which $t_n = nt$ $(n = 0, 1, 2, \ldots)$ and

$$\mathscr{X}_0 = \{\mathbf{x} : x_1 \geqslant 0,\ x_2 \geqslant 0 \text{ and } x_1 + x_2 = 1\}.$$

Instead of assuming that the proposer at time t_n is pre-determined as in section 7, we assume that player 1 is chosen at times t_{2n} $(n = 0, 1, 2, \ldots)$ with probability p_1 and that player 2 is chosen at times t_{2n+1} $(n = 0, 1, 2, \ldots)$ with probability $q_2 (p_1 + p_2 = 1, q_1 + q_2 = 1)$. All chance moves are assumed independent.

Convergence issues are much as in section 7 and we obtain the existence of a unique perfect equilibrium outcome which can be computed using the stationarity considerations indicated in figure 5.25.

We obtain that

$$\alpha = \frac{(1 - \delta_2^t)\{q_1(\delta_2^t p_1 + \delta_1^t p_2) + p_1\}}{1 - (\delta_2^t p_1 + \delta_1^t p_2)(\delta_2^t q_1 + \delta_1^t q_2)}$$

with $\beta = 1 - \alpha$. It may be checked that this yields the same result as obtained by Rubinstein in the case $t = 1, p_1 = 1$ and $q_2 = 1$ (see note 4).

The most interesting case is when player 1 is always selected with probability p (i.e. $p_1 = q_1 = p$). Then

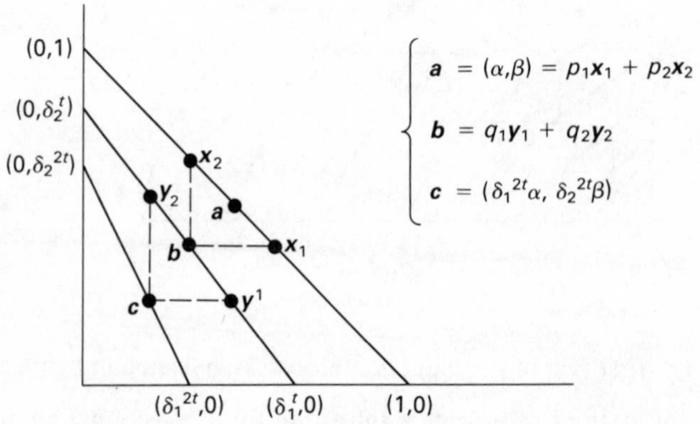

Figure 5.25

$$\alpha = \frac{p(1-\delta_2^t)}{1-(\delta_2^t p + \delta_1^t(1-p))}.$$

In particular, when $p = \frac{1}{2}$ the game is symmetric except in so far as the discount rates are concerned, and we obtain

$$\alpha = \frac{1-\delta_2^t}{2-\delta_1^t-\delta_2^t}.$$

The perfect equilibrium outcome we have calculated above is Pareto-efficient in $\mathscr{X}_0$ but this is only because $\mathscr{X}_0$ is a particularly simple set. In the general case, for example if $\mathscr{X}_0$ is strictly convex, the unique perfect equilibrium outcome will not be Pareto-efficient in $\mathscr{X}_0$. This deficiency disappears when we allow $t \to 0+$ and provides an extra reason to those considered in section 8 for considering this limiting case. What is more, the above calculations are adequate to locate the unique perfect equilibrium outcome in the limiting case for any convex set $\mathscr{X}_0$ (subject to the conditions of section 2) since the method used in chapter 4 works here also.

We obtain that, if the proposer at time t_n is player 1 with probability p and the time interval t is sufficiently small, then the unique perfect equilibrium outcome is approximately the asymmetric Nash bargaining solution for $\mathscr{X}_0$ with status quo $\mathbf{0}$ calculated with 'bargaining powers'

$$\tau_1 = \frac{p \log \delta_2}{p \log \delta_2 + (1-p) \log \delta_1}; \quad \tau_2 = \frac{p \log \delta_1}{p \log \delta_2 + (1-p) \log \delta_2}.$$

Note 7. It is natural to ask what happens when the intervals between successive proposals are vanishingly small and the probability of player 1 being selected at time t is $p(t)$. The number α above is then replaced by $\alpha(0)$ where $\alpha(t)$ satisfies the differential equation

$$\alpha'(t) + p(t)\,\alpha(t) \log\left(\frac{\delta_2}{\delta_1}\right) = \delta_1^t p(t) \log \delta_2.$$

12 MIXED STRATEGIES

The propositions of section 3 remain valid when mixed strategies are admitted except that in proposition 1 it is necessary to replace $\mathscr{P}_{t_{k-1}}$ by the Pareto frontier of the convex hull of $\mathscr{X}_{t_{k-1}}$. Thus if all the sets $\mathscr{X}_t$ are *convex*, the characterization of section 4 remains intact even if mixed strategies are permitted.

13 NON-SHRINKING CAKES

In this section we no longer admit the possibility of chance moves in order to concentrate on the situation when the shrinking cake assumption of section 4 is abandoned.

None of the propositions of section 3 depend on the shrinking cake hypothesis and it follows that proposition 4 remains valid in general provided that the sets E_n are suitably constructed. We shall not attempt an exhaustive account of the appropriate construction of the sets E_n since the degenerate cases would confuse the issue. Instead we indicate in figure 5.26 the appropriate inductive step in the four situations which we deem non-degenerate.

14 OSCILLATING CAKES WITH FIXED DISCOUNT RATES

A natural first example of the case of a non-shrinking cake is obtained by taking

$$\mathcal{X}_{t_n} = \{\mathbf{x} : x_1 \geqslant 0,\ x_2 \geqslant 0 \text{ and } p_n^{-1}x + q_n^{-1}x_2 \leqslant 1\}$$

as in section 6 but omitting the hypothesis that $\langle p_n \rangle$ and $\langle q_n \rangle$ are decreasing. There is nothing essentially new beyond that mentioned in section 6 on convergence questions and we therefore specialize by taking $t_n = nt$ ($n = 0, 1, 2, \ldots$) and

$$p_{2n} = a\delta_1^{2nt},\ q_{2n} = \delta_2^{2nt}$$

$$p_{2n+1} = \delta_1^{(2n+1)t},\ q_{2n+1} = b\delta_2^{(2n+1)t}.$$

One may imagine that there are two cakes available (or one oscillating cake), one and only one of which can be divided if the players can agree on a division. The availability of the cakes alternates with the proposers who discount future consumption at fixed discount rates δ_1 and δ_2. Figure 5.27 illustrates the two cakes at time 0.

The most interesting case is that in which $0 < a < 1$ and $0 < b < 1$ as illustrated. The advantage that a player enjoys through being proposer is then balanced by the fact that the available cake is the least favourable for him. Figure 5.28 illustrates the application of the technique described in section 13 when $b < \delta_2^t$. It will be noted that case I of figure 5.26 applies to begin with but eventually case IV takes over (since the point P will ultimately lie above the line $x_2 = B$). We are thus reduced to the situation studied in sections 6 and 7 but with a rather faster rate of convergence. Stationarity considerations allow the computation of the unique perfect equilibrium outcome. This

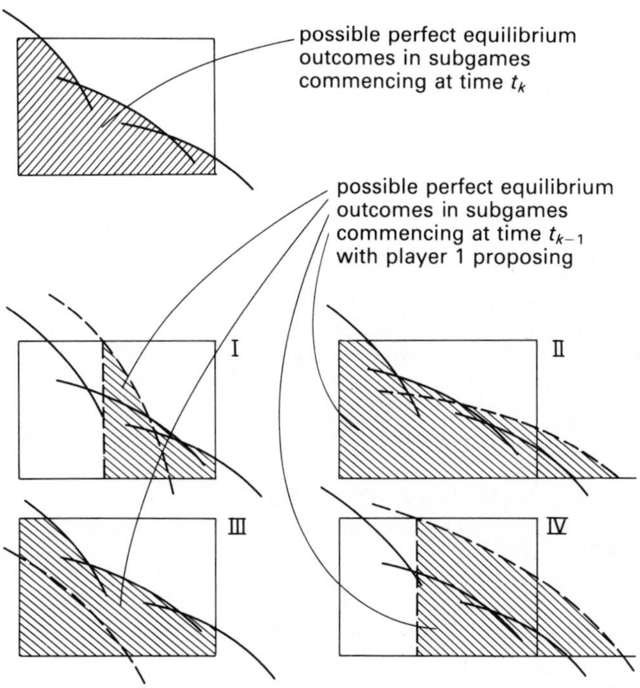
possible perfect equilibrium outcomes in subgames commencing at time t_k
possible perfect equilibrium outcomes in subgames commencing at time t_{k-1} with player 1 proposing
I
II
III
IV

Figure 5.26

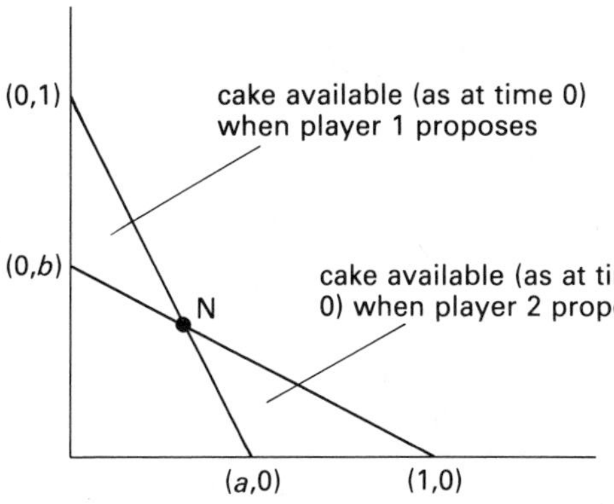
(0,1)
cake available (as at time 0) when player 1 proposes
(0,b)
N
cake available (as at time 0) when player 2 proposes
(a,0)
(1,0)

Figure 5.27

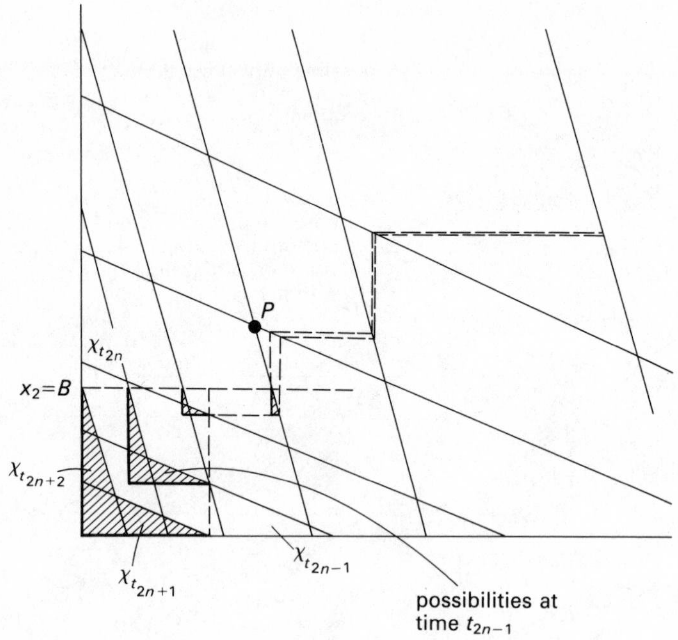

Figure 5.28

is achieved via an acceptance of player 1's initial proposal and assigns him a payoff

$$a\left(\frac{1-b\delta_2^t}{1-ab\delta_1^t\delta_2^t}\right)$$

while player 2 receives

$$b\delta_2^t\left(\frac{1-a\delta_1^t}{1-ab\delta_1^t\delta_2^t}\right).$$

This reduces to the result of Rubinstein mentioned in note 4 when $a=b=1$. As explained in section 8, it is of interest to study the case when $t\to 0+$. In the limit, the unique perfect equilibrium outcome is then the point N of figure 5.27 independently of the values of the discount rates δ_1 and δ_2.

This last observation provides some grounds for regarding N as the 'Nash bargaining solution' for the two-cake problem. Some additional grounds are mentioned in chapter 8.

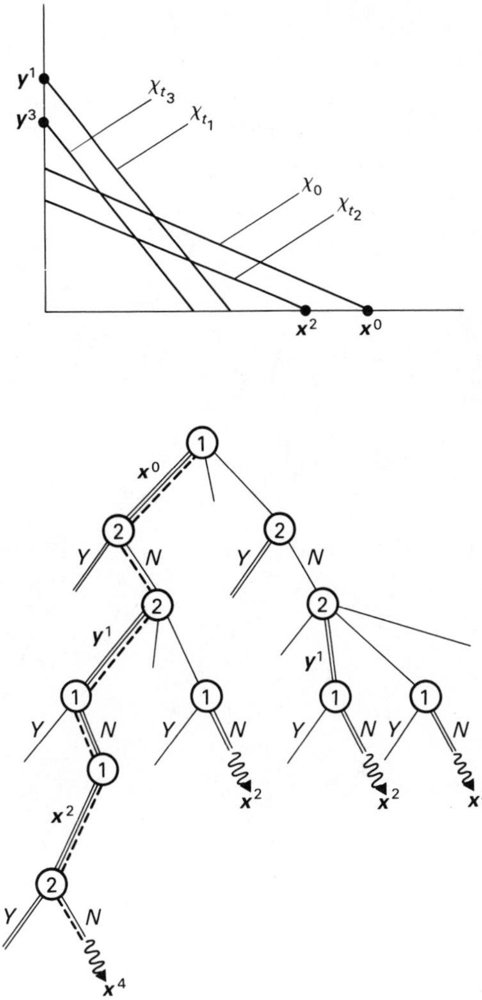

Figure 5.29

We now turn to the case when $a > 1$, $b > 1$ and $a > \delta_1^{-t}$. Under these circumstances case II of figure 5.26 applies throughout and the technique does not limit the possibilities for perfect equilibria outcomes at all. Figures 5.29 and 5.30 indicate why x^0, y^1 and x^2 of figure 5.29 are all achievable as perfect equilibrium outcomes in this case.

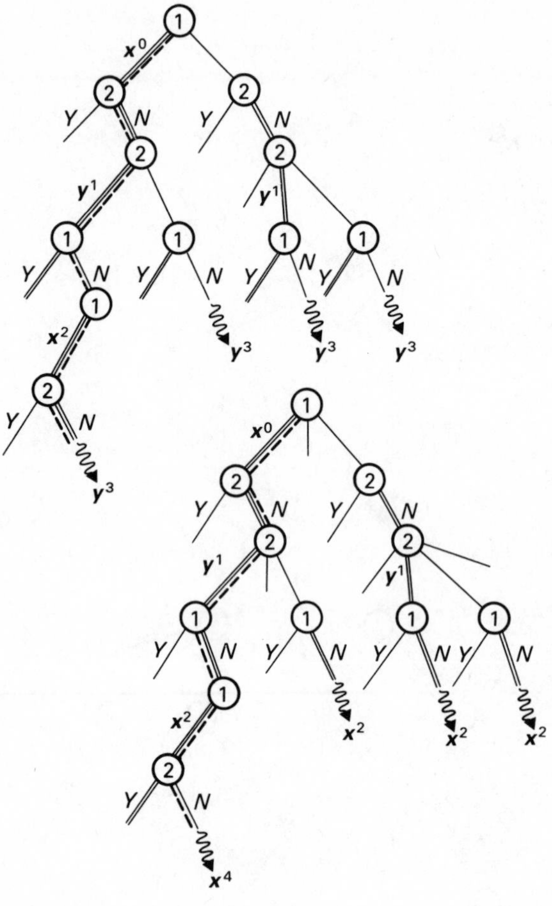

Figure 5.30

REFERENCES

Binmore, K. G. 1981: Bargaining, Proceedings of the Conference of the Association of University Teachers of Economics.

Bishop, R. L. 1963: Game theoretic analyses of bargaining. *Quarterly Journal of Economics*, **77**, 559–602.

Foldes, L. 1964: A determinate model of bilateral monopoly. *Economica*, **31**, 117–31.

Harsanyi, J. 1977: *Rational Behaviour and Bargaining Equilibrium in Games and Social Situations*. CUP, Cambridge.

Hicks, J. 1953: *The Theory of Wages*. MacMillan, London.

MacLennan, A. 1982: A general noncooperative theory of bargaining. University of Toronto mimeo.

Moulin, H. (1982: Bargaining and non-cooperative implementation. Working paper, École polytechnique, Laboratoire D'Économétrie, Paris.

Roth, A. 1979: *Axiomatic Models of Bargaining*. Lecture Notes in Economics and Mathematical Systems 170. Springer-Verlag, New York.

Rubinstein, A. 1980: Perfect equilibrium in a bargaining model. ICERD discussion paper, London School of Economics.

Ståhl, I. 1972: *Bargaining Theory*. Economic Research Institute, Stockholm.

6

Involuntary Unemployment as a Perfect Equilibrium in a Bargaining Model

A. Shaked and J. Sutton

This chapter presents an analysis of a two-person non-cooperative bargaining game in which one party is free, subject to certain frictions, to switch between rival partners. This permits us to capture the notion of an asymmetry between 'insiders' and 'outsiders' in the context of a firm bargaining with its workers, in the presence of unemployment.[1]

1 INTRODUCTION

No concept in economics is at the same time of such central importance, and so elusive of a satisfactory theoretical formulation, as the notion of an 'unemployment equilibrium'. The central idea which must be captured is that certain 'unemployed' workers would *strictly* prefer to be employed, at the prevailing wage rate. In other words, the equilibrium wage lies above the Walrasian level, and jobs are rationed ('unequal treatment').

In order to motivate such a non-Walrasian equilibrium, we need to introduce some 'imperfection' or 'friction' into the competitive model.[2]

1 Our thanks are due to the International Centre for Economics and Related Disciplines at the London School of Economics for financial support.

2 Most of the recent literature in this field has tended either (a) to invoke imperfect information, in one form or another, as an explanation (implicit contract theories, adverse selection mechanisms, search models), (b) to treat aggregate unemployment as a 'disequilibrium' phenomenon, associated with some slowness of adjustment of 'lengthy' wage contracts or (c) to replace the usual non-cooperative equilibrium concepts by a cooperative one – an appeal to 'unionization'.

The implicit contract literature is reviewed by Azariadis (1981) and Hart (1982). For the 'adverse selection' mechanism see Weiss (1980); the early search literature treating unemployment as a disequilibrium phenomenon is discussed in Phelps (1970). The 'unionization' approach is based on positing some objective function for workers which includes the level of unemployment as an argument (for an elementary introduction, see Carrter (1959, chapter 8). A further approach involves an appeal to 'fairness' (Hicks 1974; Akerlof 1980).

The competitive model may be represented in terms of an auction, in which all workers are treated in a symmetric matter. The firm can *simultaneously* announce offers to various workers, 'paying off one against the other' to establish the Walrasian wage.

In the present chapter, we depart from this 'competitive' story by introducing a minimal friction which characterizes actual labour markets: the firm in practice operates with a certain workforce at any point in time. It cannot line these up against a 'reserve workforce' of the unemployed, and play off one against the other in the manner of a Walrasian auction. It can, of course, replace its current workers - but such changes are in practice not instantaneous (or costless).

The key feature which must be captured is as follows: the firm's current workforce enjoy a bargaining advantage insofar as it takes some time to replace them; *but if the firm does replace them, then it will find itself at the same disadvantage in due course vis-à-vis the new workforce*.

Once this point is appreciated, it is not clear at first glance that the outsiders afford the firm any credible threat. We shall see below, however, that the firm can indeed gain by using this threat.

Now in trying to capture the problem involved here, we will find it convenient to work throughout in a simple *bargaining* framework. No production takes place; we merely investigate a firm engaged in negotiating a wage with some potential worker (and we assume it needs exactly one worker). We capture the 'friction' alluded to above, by supposing simply that the firm can bargain only with one individual at a time: we label that individual the 'insider'. As bargaining proceeds, however, it is free, after some specified time, to switch over to some other worker, who thereby becomes the new 'insider'. If the frequency with which it can make a switch is very low, then we approach a bilateral monopoly between the firm and its 'insider'. Here, the unemployed constitute 'no threat' to the firm's employees, and so their presence plays no part in wage determination. If, on the other hand, the firm can switch instantaneously, then the threat to do so will suffice to establish a Walrasian outcome. Our central aim in the present chapter is to present a suitable equilibrium concept, which can span the range of possibilities lying between the Walrasian pole, on the one hand, and bilateral monopoly on the other.

In proceeding along these lines, we are relying on an *analogy* between the delay (and consequent loss of production) involved in practice, in changing the firm's workforce, and the delay which we introduce in our bargaining process, in respect of the firm's ability to switch to a new partner.

In modelling this bargaining process, the crucial importance of the role of threats, and the question of whether they will be carried out, suggests that the appropriate equilibrium concept to explore is that of a perfect equilibrium (Selten 1975).

Thus, we do *not* take the approach of contract theory, where an agent undertakes 'now' to carry out some action 'later', which may not be in fact optimal. Here, we allow agents to freely revise their plans at each instant – and we investigate what sort of agreement they will reach in the light of this.

Now it is of course a commonplace that the 'bilateral monopoly' problem in itself poses serious difficulties. Rubinstein (chapter 3) has presented a solution to this problem using the notion of a perfect equilibrium in a bargaining process. Our present analysis applies the same kind of approach within a more general context, so that the Rubinstein solution to the bilateral monopoly problem will emerge as a special case.[3]

For expository reasons, however, we begin by presenting a new and very simple method of solving Rubinstein's 'bilateral monopoly' problem. The same method will then be used in analysing the general model (see figure 6.1).

2 THE 'BILATERAL MONOPOLY' CASE

A single firm requires the services of exactly one worker to produce a gross profit of one unit. In the present section, we assume that only one worker is available, with a reservation wage of zero.

The firm bargains with the worker in the following manner. At time zero, the firm makes an offer w to the worker. If he accepts, then the game ends, the payoffs received by the worker, and by the firm, being w and $1-w$ respectively.

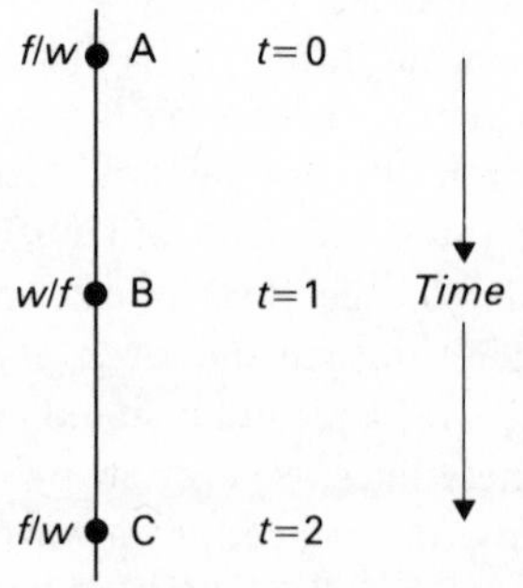

Figure 6.1 The bilateral monopoly game. The notation a/b reads: a makes a proposal to b.

3 We are concerned with the 'discount rate' case in Rubinstein's chapter. He also considers a 'fixed cost' scheme, in which equilibrium may not be unique. For an analysis of more complicated cases, see chapter 5.

If the worker rejects the offer, he may formulate a counteroffer; this takes a finite time interval, and we choose this interval as our unit of time in what follows. Thus the worker makes a counteroffer w' at time 1. If the firm accepts, the game ends. The payoffs to the worker, and the firm, are now equal to $\delta w'$ and $\delta(1-w')$, where $\delta < 1$ represents their common discount factor (the extension to the case of different discount factors is quite simple).

On the other hand, if the firm rejects the offer, it may in turn make an offer at time 2. Thereafter, firm and worker take turns to make proposals.

The notion that the time taken to formulate successive proposals is 'negligibly small' may now be captured by considering the limit as $\delta \to 1$; we will be particularly interested in the properties of the model in this limit, in what follows. We remark that, in this limit, the outcome is independent of 'who calls first'.

Rubinstein has shown that there a *unique* perfect equilibrium partition in this game.[4] We now develop a (much simpler) analysis of the game, which re-establishes this result.

The strategies of firm and worker are said to constitute a *perfect equilibrium* if, in every subgame, the strategies relating to that subgame form a Nash equilibrium. In a perfect equilibrium, a player will agree to a proposal if it offers at least as much as he will obtain in the future, given the strategies of all players. (See chapter 3 for a precise definition.)

We express all payoffs in terms of their values discounted to period zero.

Consider point C in figure 6.1. The discounted sum of the payoffs at that point is δ^2. Consider the game which begins at this point with a call by the firm. We define M as the supremum of the payoffs which the firm can obtain in any perfect equilibrium of this game. Discounted to $t = 0$, this becomes $\delta^2 \cdot M$.

Now consider a call made by the worker in the preceding period (point B in figure 6.1). Any call by the worker which gives the firm more than $\delta^2 \cdot M$ will be accepted by the firm, so there is no perfect equilibrium in which the firm receives more than $\delta^2 \cdot M$; and since the discounted value of the total payoff at time $t = 1$ is δ, it follows that the worker will get at least $\delta - \delta^2 \cdot M$ in any perfect equilibrium of the subgame beginning from that point. In fact, $\delta - \delta^2 \cdot M$ is the infimum of the payoff received by the worker in this subgame.

Now consider the offer made by the firm in the preceding period (point A in figure 6.2). In the subgame beginning from this point, the worker will not accept anything less than the infimum of what he will receive in the game beginning next period - the present value of which is $\delta - \delta^2 \cdot M$. Hence the firm will obtain at most $1 - \delta + \delta^2 \cdot M$. In fact, as before, this is the supremum of what the firm will receive here.

4 That is, this is a unique division of the cake which can be supported as a perfect equilibrium.

But the game at point C is identical to the game at point A, apart from a shrinkage of all payoffs by a factor of δ^2. Hence it follows that the supremum of the firm's payoff here must equal M. Hence

$$M = 1 - \delta + \delta^2 \cdot M$$

whence

$$M = \frac{1}{1+\delta}.$$

But the above argument may be repeated exactly, if we instead begin by letting M represent the infimum of the payoff to f in any perfect equilibrium of G, and the interchange throughout the pairs of words: more/less, most/least, supremum/infimum and accept/reject.

Hence the above equation also defines M as the infimum of the payoff to f. Thus the payoffs in any perfect equilibrium partition are uniquely defined: the firm receives $1/(1+\delta)$ and the worker receives $w = 1 - 1/(1+\delta) = \delta/(1+\delta)$.

It is easy to show that this solution is indeed supported by a pair of strategies, and so there exists a unique perfect equilibrium partition. (These strategies are such that the offers made at any point correspond to the perfect equilibrium partition, and players agree to an offer of at least that amount.)

Note that as $\delta \to 1$, we have $w \to \frac{1}{2}$.

This completes our derivation of the Rubinstein result.

Three remarks are in order here, regarding this approach to resolving the 'bilateral monopoly' problem.

Remark 1. Consider a variant of the above game in which one of the players (say the 'worker') is replaced by a succession of short-lived agents; each of these lives for exactly two periods. Thus the firm faces a succession of workers; it bargains with each for one 'round' and then faces a new rival.

It is immediately clear from the preceding proof that the equilibrium payoffs are the same as before. Even though the 'first' worker will not be involved in later rounds, his payoff is undiminished - he must offer to the firm only the amount which the firm can achieve in the appropriate lower subgame. The identity of the firm's subsequent rivals is immaterial. This is a point to which we return later.

Remark 2. An interesting interpretation of the solution is as follows: at the start of each even numbered period $2n$, the firm makes an offer. If the worker fails to accept this, then the 'cake shrinks' from size δ^{2n} to δ^{2n-1}, i.e. by $(1-\delta)\delta^{2n}$. The firm's payoff coincides with the sum of the shrinkages which occur during these time periods. To see this note that the firm's payoff can be expanded as follows:

$$\frac{1}{1+\delta} = (1-\delta)(1+\delta^2+\delta^4+\ldots).$$

This principle continues to hold good even when the discount rates are not equal, or where the time intervals between successive calls are irregular.

Remark 3. The key idea involved here is to untangle two elements in the bargaining problem: (a) the technical framework within which bargaining occurs (the 'rules of the game'); and (b) the preferences of the bargaining agents.

The Rubinstein approach imposes *symmetry* in respect of the technical framework, or the 'timetable' of offers, while allowing, in general, that the preferences of agents may differ. Now, once we move to a situation in which we have one firm, but many workers, it is natural to capture this difference in their respective positions by introducing an asymmetry into the technical framework of bargaining moves – the fact that the firm faces many workers enhancing the variety of moves open to it at any point in the game.

3 THE MODEL

Again we consider a firm which requires the services of one worker to produce a gross profit of one unit. Now, however, we assume there are $n > 1$ workers available, each with a reservation wage of zero. Thus the Walrasian 'market clearing' wage is zero.

The simplest story which leads to the Walrasian outcome here is that of the familiar 'auction' market in which the firm announces offers to two (or more) workers *simultaneously*. In this case the only equilibrium is clearly the Walrasian.

The central idea of our present approach concerns our attempt to capture the fact, alluded to in the introduction, that the typical firm in practice operates with a certain workforce at any point in time. It cannot line these up against a 'reserve workforce' of the unemployed, and play one off against the other in the manner of the Walrasian auction.

We capture this idea by constructing a bargaining framework, or timetable of moves, which has the key property that the firm can never make simultaneous offers to two different workers. At each point, one worker is identified as the current 'insider'. Bargaining between the firm and this worker proceeds as in the bilateral monopoly model just described. However, the firm can plan to 'switch' to an outsider, thereby making him the new 'insider' henceforward, subject to two restrictions.

The first is a minimal restriction designed to give the current 'insider' an advantage vis-à-vis rival 'outsiders', and to avoid the use of strategies by the firm which would be equivalent in effect to its making 'simultaneous offers'. We require that, following any offer by the firm, the 'insider' can always reply with a counteroffer before the firm switches over to negotiate with an outsider. Within our present structure, then, we require the restriction:

Condition (a). If the firm makes an offer to the insider at any time, then it must wait for at least a time of 1 unit before switching.

It is easy to show that this is a *necessary* condition for obtaining a non-Walrasian outcome. For, if the firm can make successive offers to worker 1, and to worker 2 (without any counteroffer by 1 intervening), then we have a situation analogous to that in which the firm makes simultaneous offers.

(The precise role of restriction (a) in the analysis is explained in section 5 below.)

The object of our second 'restriction' is to introduce into the model some parameter which can capture the degree to which the 'frictions' we embody here carry us away from the Walrasian solution. With this in mind we assume:

Condition (b). If the firm begins bargaining with a given worker at some point in time, then it cannot switch until some minimum time, which we label T, has elapsed. Once this time has elapsed, it is free to switch at any time subject only to (a).

The parameter T measures the degree to which the outsiders represent a threat to the insiders. We will assume $T > 1$; as $T \rightarrow 1$ it can be shown that we converge to the Walrasian solution. Here the firm can, having made an offer to worker 1, proceed to make an offer to 2 'almost' as quickly as worker 1 can make his counteroffer.

On the other hand, as $T \rightarrow \infty$, it will be shown that we obtain the case of bilateral monopoly.

We may now proceed to a description of the game.

At time 0, the firm makes an offer of w to some worker (say, worker '1'). The process of bargaining with '1' now proceeds until time T has elapsed, the firm, and worker 1, taking it in turns, period by period, to make proposals.

Once time T has elapsed, the firm faces a twofold choice: either continue bargaining in this way with 1, or else switch to some other worker (say '2'), and begin bargaining with 2, following the same pattern as before. Such a switch can be made at any time, subject to restriction (a). As of the moment the firm switches to worker 2, he becomes the new 'insider'; and time T must elapse before it can switch back to 1, or move to a third worker.

The possibility arises that the firm makes an offer to the outsider simultaneously with receiving an offer from the insider; if so, the latter gets priority in the sense that it is considered first, and only if it is rejected, is the former proposal considered.

Thus the firm may be thought of as 'lining up' an alternative worker, while continuing to bargain with the 'insider'.

The essence of the firm's strategy involves the threat of making such a switch, and thereby increasing its bargaining power vis-à-vis the current 'insider'.

While it will turn out below that agreement will be instantaneous, so that such 'switches of workforce' will not be observed, nonetheless the threat

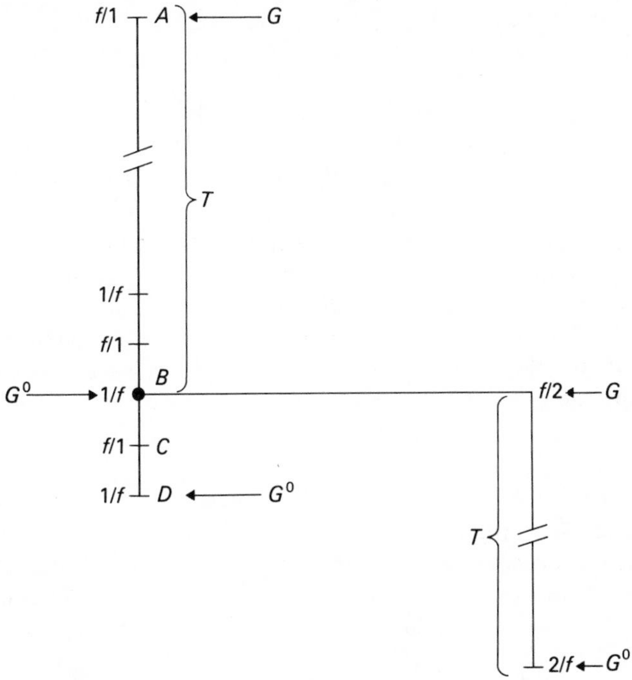

Figure 6.2 The notation a/b reads: a makes an offer to b.

involved would indeed be carried out, were the 'insider' to make certain proposals, and replies. This idea is central to the *perfect equilibrium* concept.

In working through the proofs which follow, the reader may find it helpful to imagine that each worker has complete information on all moves made in preceding negotiation, including those in which he did not take part. The essence of the method developed above, however, is that it establishes upper and lower bounds to equilibrium payoffs which are independent of the past history of the game. Thus the equilibrium we calculate below is the unique equilibrium of the game, irrespective of workers' information concerning past negotiations.

4 EXISTENCE AND UNIQUENESS OF EQUILIBRIUM

We will simplify the exposition by confining ourselves to the case where T is an integer.[5] The results for the general case are stated in section 5 below.

5 We are here following the advice of a referee. The general case is solved in a similar manner.

The game is illustrated in figure 6.2. The firm negotiates with some worker, labelled 1, for time T; it is then free to switch subject only to condition (a), i.e. the worker is allowed to first make a counteroffer in response to the firm's last offer.

Now suppose T is odd. Then the firm is first free to switch immediately after receiving a counteroffer from the insider. It is this case which we will focus on in the present section. If T is even, the firm has a choice, once T has just elapsed, between making an immediate offer to the insider *or* to the outsider (but not to both!). A simple argument shows that, in this case, it will remain with the insider for one more period – so that this case is equivalent to that in which the delay is equal to $T + 1$. (See section 5 below.)

In figure 6.2, then, the firm is free to switch to worker 2 at point B, or after (subject to restriction (a)). We will in fact show that it will choose to switch immediately, at point B. (See proposition below.) At point B, the firm receives a counteroffer from 1. If at that point, it chooses to switch, it makes an immediate offer to 2.

We define G^0 as the game which begins immediately following a call by the 'insider', and where the firm is free at this time to switch to an outsider. Clearly the homogeneity of the game permits us to define G^0 independently of the time elapsed since the beginning of the game. Let M^0 denote the supremum of the payoff to the firm taken over all equilibrium partitions of this game.

We note that the game immediately following a switch by the firm is the same as our initial game; we label this game G. Again we label the supremum of the payoff to the firm, taken over all equilibrium partitions of this game, as M.

We begin by establishing the following:

Lemma. *Let M, M^0 be the suprema (infima) of the payoffs to the firm in any perfect equilibrium of G, G^0, respectively. Then:*

1 $M^0 = \max[\delta(1 - \delta + \delta M^0); M]$,

2 $$M = \frac{1 - \delta^{T+1}}{1 - \delta} + \delta^T \cdot M^0.$$

Proof. 1 In figure 6.2, consider the game beginning from point B, at which the firm is free to switch from 1 to 2. Represent all payoffs in terms of their values discounted to this point.

At B, the firm chooses either to switch, or to remain with 1, according to which yields the higher payoff. We now note the amount which f can at most receive along each branch.

By switching to 2 it receives at most M. Suppose it remains with 1. Then (repeating the argument of section 2) at point D the firm receives at most $\delta^2 M^0$. Hence 1 receives at least $\delta^2 - \delta^2 M^0$. Hence at point C, the firm receives at most $\delta - \delta^2 + \delta^2 M^0$. Hence, by not switching, the firm can receive at most $\delta(1 - \delta + \delta M^0)$.

2 In figure 6.2, express all payoffs in values discounted to point A. Beginning from point B, note that the firm can at most obtain $\delta^T \cdot M^0$ here. Proceeding to work backwards (repeating our above argument T times) to point A, we see that the firm receives at most

$$1 - \delta + \delta^2 - \delta^3 - \ldots - \delta^T + \delta^T M^0 = \frac{1 - \delta^{T+1}}{1 - \delta} + \delta^T M^0$$

which equals M. *QED*

We now characterize the solution to the game G.

Proposition. *The game G has a unique perfect equilibrium partition in which the firm receives payoff*

$$M = \frac{1 - \delta^{T+1}}{(1 + \delta)(1 - \delta^T)} .$$

Proof. We proceed by showing that the two equations stated in the preceding lemma have a unique solution.

Two cases arise, corresponding to the possibilities that (a) the firm does not switch, and (b) the firm switches, i.e. either:

a $M^0 = \delta(1 - \delta + \delta M^0)$

or:

b $M^0 = M$.

We show that (a) is impossible, i.e. the firm will carry out its threat to switch. To see this, assume it does not switch, i.e.

$$M^0 = \delta(1 - \delta + \delta M^0).$$

Solving, we obtain

$$M^0 = \frac{\delta}{1 + \delta}, \quad \text{whence } M = \frac{1}{1 + \delta}$$

(where the last step follows from part (2) of the lemma). But equation (1) of the lemma implies that $M^0 \geqslant M$, which implies a contradiction.

It therefore follows that (b) holds, i.e. the firm does switch. Hence $M^0 = M$, and substituting this into equation (2) of the lemma, we obtain the result, that equations (1) and (2) have a unique solution. Thus the infimum, and the supremum, of the firm's payoff in any perfect equilibrium partition, coincide.

It is straightforward to show that this solution is indeed supported by a pair of strategies, as noted in section 2 above, so that there exists a unique perfect equilibrium partition. *QED*

We now turn to the interpretation of this result. The firm's payoff equals

$$M = \frac{1 - \delta^{T+1}}{(1 + \delta)(1 - \delta^T)}.$$

Now if $T = 1$, we have the Walrasian solution, $M = 1$. Notice that this is not a case of 'simultaneous' calls by the firm, but rather an equivalent 'limiting' case: the firm must call workers one at a time, but as soon as each replies with a counteroffer, the firm can instantaneously switch to another worker, and this is enough to ensure a Walrasian outcome.

We now turn to the other pole, where $T \to \infty$. Here we have

$$M = \frac{1}{1 + \delta}.$$

This corresponds to Rubinstein's solution to the bilateral monopoly problem. In this case, then, the threat of the outsiders has vanished, and the asymmetry between the relatively advantaged 'insiders' and the competing outsiders has become so large, that the outsiders do not impinge on the outcome at all.

We now wish to show that the qualitative features of these results are preserved in the limit where the delays incurred in bargaining become negligibly small. We allow δ to converge to unity. This can be interpreted in two ways, (a) the agents are 'far sighted', so that the losses incurred by delaying agreement for one bargaining round are negligible, or (b) the length of a bargaining round becomes negligibly small (reinterpreting δ in the obvious manner).

Now in the limit $\delta \to 1$, it is readily shown that

$$N = \frac{1}{2} \cdot \frac{T + 1}{T}.$$

At $T = 1$, we obtain the Walrasian solution as before, while as $T \to \infty$, we have $M = \frac{1}{2}$, corresponding to bilateral monopoly.

5 SOME FURTHER COMMENTS

We here state the general results for the case where T takes any (integer or non-integer) value. For proofs, the reader is referred to Shaked and Sutton (1980).

First suppose T is an even integer. Then, at the moment when T has just elapsed, the firm can choose whether to make an immediate offer to 1, *or* to make an immediate offer to 2. In the latter case it faces a constraint, in that it must stay with 2 for time T thereafter. In the former case, it does not (it can switch after one period). Thus, clearly, the firm will stay with the insider for one more round before switching. Thus the case where T is an even integer is equivalent to the case where the delay equals $T + 1$.

Now suppose that $T = k + \epsilon$ where k is an odd integer and $0 \leqslant \epsilon < 1$. Then two cases arise, according to whether ϵ lies below, or above, some critical value $\epsilon^*(k)$. For $\epsilon < \epsilon^*(k)$, the firm chooses to switch immediately time $T = k + \epsilon$ has elapsed. On the other hand, if $\epsilon > \epsilon^*(k)$, then the firm does not switch at once; rather, it waits to make one more offer to the insider, and on receiving the insider's counteroffer, immediately switches to an outsider. Hence its payoff will not depend on ϵ in the range $\epsilon^*(k) \leqslant \epsilon < 1$.

Finally, suppose $T = K + \epsilon$ where k is even, and $0 < \epsilon < 1$. Then restriction 1 requires the firm to wait until at least $T + 1$, and the payoff to the firm equals that which obtains when T equals $k + 1$.

The solution is:

$$Q = \begin{cases} \dfrac{1 - \delta^{k+1}}{1 - \delta^{k+\epsilon}} \cdot \dfrac{1}{1+\delta}, & \epsilon \leqslant \epsilon^*(k), \\[2ex] \dfrac{1 - \delta^{k+3}}{1 - \delta^{k+2}} \cdot \dfrac{1}{1+\delta}, & \epsilon \geqslant \epsilon^*(k), \end{cases}$$

where $\epsilon^*(k)$ is defined as the value where these two expressions coincide. In the limit $\delta \to 1$, this becomes

$$Q = \begin{cases} \dfrac{1}{2} \cdot \dfrac{k+1}{k+\epsilon}, & \epsilon \leqslant \epsilon^*(k) = \dfrac{2}{k+3}, \\[2ex] \dfrac{1}{2} \cdot \dfrac{k+3}{k+2}, & \epsilon \geqslant \epsilon^*(k) = \dfrac{2}{k+3}. \end{cases}$$

This solution is illustrated in figure 6.3.

A central feature of our result, which we wish to emphasize, concerns the behaviour of the model in the limit $\delta \to 1$. The point we wish to make is best seen by re-examining the role of restriction 1 in the analysis.

Restriction 1 forbids the firm to switch to an outsider immediately following an unsuccessful offer to the current insider; i.e. it gives the insider a 'right to reply'. Suppose we dropped this restriction, while continuing to require the firm to remain with the insider for at least time T before making a switch. The firm could then simply wait for time T to elapse, and in due course make a proposal to 1, and then immediately make a proposal to 2. Thus it could in effect make simultaneous proposals to 1 and 2, and so establish a Walrasian solution after a certain delay.

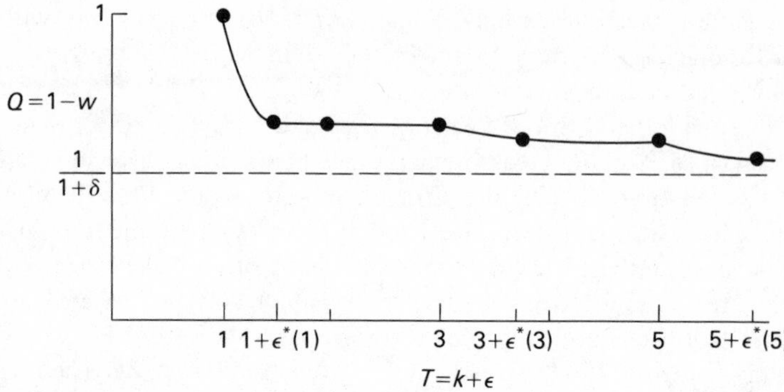

Figure 6.3 The payoff to the firm as a function of T. Note that the wage W equals $(1 - Q)$.

The outcome would, however, still be non-Walrasian. The insider could extract a positive wage at time 0 which reflected the fact that the delay involved in waiting for a Walrasian outcome imposes a loss on the firm.

However, in the limit $\delta \to 1$, the solution would now converge to the Walrasian; here, the departure from the Walrasian solution just reflects the agents' valuation of a transient delay involved in achieving a Walrasian outcome.

The effect of our use of restriction 1 in the analysis is to ensure that the firm can never make simultaneous offers to two workers; and it is this which leads to our much more dramatic departure from the Walrasian model.[6]

6 SUMMARY AND CONCLUSIONS

We have been concerned throughout with an attempt to formalize the notion that the firm in practice has an 'existing workforce' at any point in time. It cannot instantly and costlessly switch them for a rival workforce drawn from the unemployed; and this drives a wedge between the labour market we describe, and that of the Walrasian auction in which the firm can play one worker off against another by making simultaneous offers to each.

6 The following question might be asked: since the worker can extract a surplus w, might it be attractive to change the 'rules of the game' to allow the firm to charge an 'entry fee' to the worker for beginning negotiation with him, thus extracting the surplus? It is intrinsic to the 'perfect equilibrium' idea, that the firm could, having received the fee, keep switching to successive outsiders, extracting a similar fee from each – thus making the payment of such a fee unacceptable to the worker.

To capture this idea, we have followed Rubinstein's model of bilateral monopoly in describing a firm and a worker, who take turns to announce offer and counteroffer according to some timetable. It is natural to take this timetable, or structure of bargaining moves, to be symmetric, for the case of a bilateral monopoly.

Now once we move to the case where the firm faces a number of potential employees, an asymmetry exists between the bargaining position of the firm, and that of the worker. It is this asymmetry which we are trying to capture here. We have done so by describing the firm, at any point in time, as being involved in a negotiation of the 'bilateral monopoly' type vis-à-vis some 'insider' – the 'existing workforce'. Subject to the insider's 'right of reply', it is free, however, to switch after time T to an 'outsider' who thereafter becomes the new 'insider'.

The effect of this is to allow the firm to make certain choices as the game proceeds, as to whether it will switch or not. Since in each case it chooses the path advantageous to itself, this enhances its equilibrium payoff.

Now since the firm, at each time, has some particular worker as its 'insider', the resulting game can in fact be seen as equivalent to one in which the firm negotiates with a single worker – but at certain times it has a choice as to setting (or re-setting) the timetable of offers and counteroffers. As we noted in remark 1 on the Rubinstein model, the identity of the firm's rival at each point does not matter in itself here.

To illustrate this point, consider the limiting 'Walrasian' case where $T = 1$. Here, as soon as the insider makes his first counteroffer the firm can immediately switch to a new worker. Reinterpret the game as one between a firm and a single worker: the firm calls at $t = 0$; at $t = 1$ the worker replies and then the firm calls again; at $t = 2$ the worker replies and then the firm calls again – and so on. Applying the argument developed in remark 2 on the Rubinstein model, the firm clearly receives the entire 'cake' here – in fact the arrangement is analogous to that in which the firm announces a take-it-or-leave-it offer in a one-shot game.

What matters, then, is not the *identity* of the worker with whom the firm negotiates. The role played by the availability of a substitute workforce is that it allows the firm to 'change the timetable', choosing from between alternative sequences the one most advantageous to itself. More generally, *the advantage for the firm of being in this asymmetric position vis-à-vis workers is simply that it widens the range of options open to the firm in bargaining.*

This asymmetry in their respective positions is captured in the present framework, which allows the firm to 'dominate' in the bargaining process – increasing the fraction of the bargaining period in which the firm 'has an offer on the table'.

We finally remark that our model can be extended to the case where the parties have different discount factors (for the 'bilateral monopoly' case, see chapter 3). Now the discount rate here represents the loss in utility incurred as a result of delays in reaching agreement. It is through this channel for example, that the factors traditionally identified as the 'costs of a strike' to workers will appear.

The limitations of the present exercise, on the other hand, are self-evident. We have confined ourselves to the relatively tractable case in which the firm has a single worker. Progress in extending the analysis to the n-worker case requires some advance in the (notoriously difficult) n-person bargaining problem. Thus, in the present chapter, many issues which arise as to the interplay between insiders themselves (individual bargaining versus union bargaining say) are avoided. The extension of the model, to a theory of unemployment, requires, moreover, a full specification of the demand side (and so of the determinants of the level of employment).

The main contribution of the present chapter is that it allows us to characterize equilibrium in a situation where an 'asymmetry' exists between 'insiders' and 'outsiders', which may be more or less pronounced. Our central theme is that, once such an asymmetry is present, a non-Walrasian outcome is the general rule, and a Walrasian equilibrium is merely an extreme, limiting, case.

REFERENCES

Akerlof, G. 1980: A theory of social custom, of which unemployment may be one consequence. *Quarterly Journal of Economics*, **94**, 749–76.

Azariadis, C. 1981: Implicit contracts and related topics: a survey. In: Hornstein, Z., Grice, J. and Webb, A. (eds.) *The Economics of the Labour Market*. HMSO, London.

Binmore, K. 1982: *Perfect Equilibria in Bargaining Models*. ICERD Discussion Paper, No. 58, London School of Economics.

Carter, A. M. 1959: *Theory of Wages and Employment*. Irwin, Illinois.

Hart, O. 1982: *Optimal Labor Contracts under Asymmetric Information*. ICERD Discussion Paper No. 42, London School of Economics.

Hicks, J. 1974: *The Crisis in Keynesian Economics*. Basil Blackwell, Oxford.

Phelps, E. 1970: *Microeconomic Foundations of Employment and Inflation Theory*. Norton, New York.

Rubinstein, A. 1981: Perfect equilibrium in a bargaining model. *Econometrica*, **50**, 97–110.

Selten, R. 1975: Re-examination of the perfectness concept for equilibrium points in extensive games. *International Journal of Game Theory*, **4**, 25–55.

Shaked, A. and Sutton, J. 1980: *Involuntary Unemployment as a Perfect Equilibrium in a Bargaining Model*. ICERD Discussion Paper No. 63, London School of Economics.

Weiss, A. 1980: Job queues and layoffs in labor markets with flexible wages. *Journal of Political Economy*, **88**, 526–38.

7

A Theory of Disagreement in Bargaining

V. P. Crawford

This chapter proposes a simple theory to explain bargaining impasses, which is based on Schelling's view of the bargaining process as a struggle between bargainers to commit themselves to favourable bargaining positions. Because bargaining impasses are generally Pareto-inefficient, anything involving a positive probability of impasse is Pareto-inefficient as well. It is demonstrated that in spite of this avoidable inefficiency, when successful commitment is uncertain and irreversible it can still be rational for individuals to attempt commitment and thereby risk an impasse; in a leading special case, the model reduces to a Prisoner's Dilemma game, in which only strategic-dominance arguments are needed to establish the conclusion. Further, making commitment more difficult, or changing the costs of disagreement in a way that makes available a wider range of settlements that are better for both bargainers than disagreement, need not always lower the probability of impasse, in spite of the conventional wisdom to the contrary.[1]

It is also a good rule not to put overmuch confidence in the observation results that are put forward *until they are confirmed by theory.* – Sir Arthur Stanley Eddington.

1 INTRODUCTION

Bargaining, broadly construed, is a pervasive phenomenon in modern economies, ranging from labour negotiations to trade agreements to strategic arms limitation talks. One need only consider these examples in the light of past experience to realize that the potential welfare gains from improving the efficiency of bargaining outcomes are enormous, perhaps even greater than those

1 This chapter incorporates material from University of California, San Diego Discussion Papers 79-3, written in November 1978 and bearing the same title, and 80-18, 'A model of the commitment process in bargaining', written in August 1980. Support from NSF Grant SES 79-05550 is gratefully acknowledged. I also owe thanks for helpful suggestions to Peter Berck, Norman Clifford, Clifford Donn, David Lilien, Mark Machina, William Samuelson, Joel Sobel, Hugo Sonnenschein, Ichiro Takahashi, William Thomson and Allan Young. Participants in seminar presentations at Caltech, UCSD, Michigan, Cornell, MIT, VPI, Harvard and Berkeley also made useful comments.

that would result from a better understanding of the effects of macroeconomic policy. Yet the problem of designing environments to yield improved bargaining outcomes has been all but ignored by economists.

A major part of this design problem is ensuring that impasses are avoided as often as possible. Because such disagreements, whether they take the form of strikes, trade restrictions or arms races, tend to be very costly, reducing their likelihood is of great welfare importance. But, before this aspect of the problem can even be approached, a theory that relates the likelihood of disagreement to the bargaining environment is needed. Such a theory would serve an important purpose in guiding attempts to determine this relationship empirically or experimentally, even if it did not yield strong theoretical conclusions.

Almost all microeconomic and game-theoretic models of bargaining beg the question of what determines the probability of disagreements by *assuming* that an efficient settlement is always reached.[2] This is probably due to the simple and elegant theoretical results often available under the efficiency assumption and to the common belief that inefficient outcomes are inconsistent with rational behaviour by well-informed bargainers. But plainly, any theory of bargaining that assumes away the possibility of disagreement must fail to capture an aspect of bargaining that is of central importance in the design problem mentioned above.

This chapter proposes a simple theory that explains the probability of disagreement in bargaining, and, it is hoped, will therefore prove more useful in studies of the design problem than existing theories. The theory develops Schelling's (1956) view of the bargaining process as a struggle between bargainers to commit themselves to – that is, to convince their opponents that they will not retreat from – advantageous bargaining positions. The potential

2 See, for example, Harsanyi (1961) and the references therein, Kalai and Smorodinsky (1975), Nash (1950, 1953) and Roth (1979) and the references therein. Notable exceptions are Cross (1969) and Ashenfelter and Johnson (1969). But the models developed there are somewhat ad hoc, in that bargainers' motivations for behaving as they are assumed to do are weak. Chatterjee and Samuelson (1981) have developed an interesting model, discussed in footnote 4 below, that explains the occurrence of disagreement by focusing on bargainers' uncertainty about each other's preferences. And Harsanyi (1968, pp. 329–34) presents an example to show that under uncertainty, the fact that bargainers cannot make binding agreements before they have all relevant information about preferences and feasible outcomes may prevent them from reaching an agreement that is fully Pareto-efficient relative to the information that is collectively available. But Harsanyi makes no attempt to explain the occurrence of impasses, and the theory of bargaining outlined there and in Harsanyi and Selten (1972), unlike the one developed here, would predict fully efficient outcomes in the absence of uncertainty about preferences and feasible outcomes. An interesting recent development, which explains the occurrence of impasses by exploring bargainers' incentives to maintain reputations for 'toughness', is the work of Rosenthal and Landau (1979).

benefits of commitment are clear, since once one's opponent is convinced, his best strategy is to yield if he can.

It is shown that if the outcome of the commitment process is both uncertain and irreversible, it can be rational for bargainers to take actions that imply a positive probability of disagreement, an outcome ex ante inferior for both to outcomes feasible through negotiation. The theory, which determines the probability or frequency of impasse endogenously, permits an evaluation of the assumption, common in the industrial relations and law and economics literature, that enlarging the set of feasible settlements that are at least as good for both bargainers as disagreement – commonly called the *contract zone* in this literature – makes a negotiated settlement more likely. It turns out that this need not be true: in quite 'well-behaved' bargaining situations, enlarging the contract zone by changing the disagreement outcome may actually increase the probability of an impasse.[3] It is also shown that attempts (like the common requirement in labour law to bargain 'in good faith') to make commitment more difficult, with the goal of reducing the probability of impasse, may have perverse effects.

The chapter is organized as follows. Section 2 discusses modelling issues and presents the bargaining model, describing the commitment process and the rules that determine bargaining outcomes. Section 3 analyses the model under the assumption of full non-cooperative game-theoretic rationality; the solution concept employed is Harsanyi's (1968) Bayesian Nash equilibrium, with an additional requirement of perfectness (see Selten 1975). Section 4 analyses the model under the alternative assumption that bargainers use simple heuristics (rather than the assumptions of perfectness and full rationality) to evaluate the uncertain future consequences of current attempted commitments. It is shown that for a leading special case of these heuristics, the bargaining game can be reduced to a Prisoner's Dilemma, providing a simple 'textbook' explanation of the fact that bargainers do not always manage to avoid disagreement and showing that the implied inefficiency can arise even if one is willing to accept only strategic-dominance arguments about bargainers' rational strategy choices. For a more general class of heuristics, a simple condition which guarantees that both bargainers will attempt commitment to incompatible positions, thereby risking impasse, is provided. The analysis of sections 3 and 4 confirms Schelling's (1966, chapter 3) suggestion that uncertainty may enhance the strategic usefulness of attempting

3 While the rationale for this assumption is rarely made explicit, it often appears to stem from an analogy between bargaining and individual behaviour, where large costs (taking uncertainty and costs of decision making into account) are more likely to be avoided than small costs. Although this analogy is superficially plausible, adopting it as an *assumption* is quite risky. There is little reason to suppose that bargaining, one of the most interactive of economic situations, is behaviourally analogous to individual decision making in all respects.

commitment, shows that Harsanyi's (1961, p. 182, 1977, p. 187) claim that attempting commitment is irrational because it creates a risk of impasse is not valid if commitment is uncertain, and yields new insight into the properties of the 'demand game' proposed by Nash (1953) as a non-cooperative model of bargaining to provide an alternative justification for the cooperative bargaining solution he axiomatized (1950).

Section 5 studies some of the comparative statics properties of the model, showing that commonly held beliefs about the effects of enlarging the contract zone and of making commitment more difficult may be invalid and are, at any rate, not justified on a priori grounds. Section 6 concludes by discussing some possible directions for future research in this area.

2 THE BARGAINING MODEL

This section outlines a model of the bargaining process that is simple, but rich enough to explain the occurrence of disagreements under reasonable behavioural assumptions. The theory follows Schelling's classic paper (1956) in focusing on the commitment aspects of bargaining. Schelling defines commitment impressionistically and by way of examples, but the essential idea seems to involve making a demand and 'burning one's bridges', or taking actions during the negotiation process that increase the future cost of backing down from one's demand. The potential benefits from such a strategy arise from the possibility that one's opponent will thereby become convinced that one will in fact not retreat, and that he will therefore decide to yield to one's demand. Schelling (1956, pp. 295–6) suggests that the possibility of mutually incompatible commitments may explain the occurrence of impasses:

> In threat situations, as in ordinary bargaining, commitments are not altogether clear: each party cannot exactly estimate the costs and values to the other side of the two related actions involved in the threats; the process of commitment may be a progressive one, the commitments acquiring their firmness by a sequence of actions. Communication is often neither entirely impossible nor entirely reliable; while certain evidence of one's commitment can be communicated directly, other evidence must travel by newspaper or hearsay, or be demonstrated by actions. In these cases the unhappy possibility of both acts occurring, as a result of simultaneous commitment, is increased. Furthermore, the recognition of this possibility of simultaneous commitment becomes itself a deterrent to the taking of commitments.

(In the present context, Schelling's 'threats' correspond to relying on the disagreement outcome rather than agreeing on a settlement.)

But Schelling (1956) does not further the question of why bargainers might attempt commitment. The puzzle is simple: attempting commitment creates a risk of impasse, which is generally Pareto-inefficient ex post.[4] And any distribution of outcomes that puts positive probability on an inefficient outcome is not ex ante Pareto-efficient either. This chapter proposes a plausible explanation of why bargainers attempt commitment when there are feasible negotiated settlements that are better for both ex ante than attempting commitment.

Schelling (1956) argues convincingly that commitment is an important component of real bargaining and that it typically involves significant elements of uncertainty and irreversibility. The uncertainty is intrinsic to the process, which is primarily psychological (and uncertain even to psychologists). The irreversibility arises primarily because attempting commitment involves making statements about one's relative evaluations of disagreement and agreement on one's position, and linking one's reputation to the maintenance of one's position. A union leader who, in the hope that management is listening, has told his members they should replace him if he fails to get them a given wage increase, cannot comfortably back down from that position. If it later turns out that he cannot obtain that wage increase in negotiations, he may actually prefer a strike. (Of course his original preferences, which presumably better reflected those of his membership, are still the relevant ones for judging outcomes.) A management representative who has stated publicly that his company cannot grant the wage increase sought by the union without going out of business is in much the same position.

Schelling (1966, chapter 3) suggests that if attempting commitment is not *certain* to cause an impasse, it might be a viable bargaining strategy even when disagreement is extremely costly. The rest of this section discusses the issues that arise in building a model to confirm that uncertainty can provide a resolution of the puzzle posed above, and outlines the model.

First, it is important to note that an element of irreversibility is an essential component of a sensible model of the commitment process. As in all models with uncertainty, unless actions whose effects cannot be undone are taken before the uncertainty is resolved, uncertainty can have no lasting effect. In the present context, bargainers could reconsider their decisions whenever an impasse seemed imminent, and it would generally be irrational for them to do otherwise.

Uncertainty is equally essential. With irreversibility alone, questions of timing take on primary importance, as Schelling (1963, chapter 2 and appendix B) pointed out. A bargainer who knew he could be the first to communicate an irrevocable demand to his opponent would find it to his advantage to do so, ending negotiations on the spot. And if bargainers must

4 Here and in what follows, 'ex ante' and 'ex post' refer to before and after the uncertainty inherent in the bargaining process is resolved.

communicate their demands simultaneously, there is great *strategic* uncertainty, in the form of multiple Nash equilibria: under certainty, the best strategy is to demand a lot if one's opponent demands a little, and vice versa. Aside from the fact that in games with multiple equilibria and payoff functions that allow inefficient outcomes it is difficult to justify the assumption that players will coordinate their strategy choices so that a Nash equilibrium arises (see, however, Nash 1953, pp. 131-6); the resulting predictions are sensitive to the theory used to predict which, if any, equilibrium will arise. Since no such theory has yet been widely accepted (see, however, Harsanyi 1977, chapter 7; Schelling 1963, chapter 4), it seems a better research strategy to avoid uncertainty about which theory is appropriate by incorporating uncertainty into the game.

At the most general level, commitment, and the entire bargaining process is a complex multistage game with incomplete information, in which bargainers make demands, take actions to increase the difficulty of retreating from their demands, learn from their opponents' actions and periodically reconsider their strategy choices. In a detailed model of the process, bargainers' freedom to undo the effects of their decisions would be eroded only gradually over time, and uncertainty would be resolved more or less continually. While much of interest might be learned by building such a detailed model, I shall adopt the alternative strategy of building the simplest possible model in which attempting and succeeding at commitment are the outcomes of bargainers' rational decisions, and in which uncertainty and irreversibility can have an effect on the outcome. This model, which has two stages and abstracts from all uncertainty except that in the commitment process, is likely to share many of the important features of more realistic models. I believe that the approach taken here will best serve to elucidate the possibilities inherent in more general models, and in real bargaining.

A useful starting point for expositing the model is the 'demand game' presented by Nash (1953) as an alternative justification for his 'fixed-threats' solution (developed in 1950). (In Nash 1953, the demand game is later combined with a 'threat game' to provide a model of what is now commonly called 'variable-threats' bargaining. But the demand game also stands on its own as a model of fixed-threats bargaining.) In Nash's demand game, there is only one stage, in which bargainers simultaneously make demands, in utility terms. If the demands are compatible, in the sense of being collectively feasible, each bargainer gets the utility level he demanded; if the demands are incompatible, disagreement is the outcome.

Nash (1953) observes that in his demand game, any Pareto-efficient pair of demands in the contract zone is in Nash (non-cooperative) equilibrium; this is the basic source of the strategic uncertainty referred to above. He then characterizes his fixed-threats cooperative solution as the only Nash equilibrium of the demand game that is the limit of equilibria of 'smoothed' games as the amount of smoothing goes to zero. (In the smoothed games, bargainers assume they are certain to get their demands if they are compatible, but that the probability of getting them, or equivalently, of their compatibility, falls off rapidly towards zero as their distance from the set of feasible settlements increases. Nash suggests that these probabilities can be thought of as reflecting uncertainty about preferences and the information structure of the game.)

The model used here differs from Nash's in three respects. First, bargainers are not *required* to make demands (that is, to attempt commitment); one of their options is to bargain cooperatively, which entails no risk of impasse and leads, if both bargainers adopt that strategy, to a known Pareto-efficient compromise settlement. Second, if both bargainers make demands that are more than compatible and allow them to stand, they share the surplus according to a rule that generalizes Nash's but need not preclude an efficient outcome. Finally, the most important difference lies in a richer specification of how demands are made and backed up. In Nash's demand game, demands are simply irrevocable; thus, one might view his model (although he apparently did not) as a model of commitment in which commitments are completely certain. I shall adopt the alternative assumption that whenever a bargainer makes a demand, he also sets in motion a process that will make it costly, to an uncertain extent, for him to later accept less than his demand, but that he may freely choose to accept less, as long as he pays the cost. It is this cost-generating process by which bargainers give meaning to their demands.[5]

More precisely, bargaining is viewed as a two-stage process, in which bargainers are perfectly informed about everything except their costs of backing down. In the first stage, bargainers simultaneously decide whether or not to attempt commitment. An attempt, if one is made, consists of the announcement of a demand (in utility terms) and a draw from a probability distribution, whose realization is the cost (again in utility terms) that must be borne if the bargainer in question later decides to accept anything less than his demand. In the second stage, each bargainer learns his own, but not his opponent's, cost of backing down (the outcome of his draw); whether or not

5 A related model of bargaining is developed by Chatterjee and Samuelson (1981). Although they do not point out the connection, their model is essentially a generalization of Nash's demand game, in which the irrevocability of demands, or what I have called commitment, is certain, but bargainers are uncertain about each other's disagreement utilities. (Minor differences between their model and Nash's are that they make assumptions that imply a linear utility–possibility frontier, and employ a different rule for compromising when more-than-compatible demands have been made.) Chatterjee and Samuelson show that in their model, Nash equilibrium demands (which are, equivalently, 'Bayesian' equilibrium demands in the game with incomplete information; see Harsanyi 1968) generally involve 'shading', or demanding more than one's disagreement utility. Given the uncertainty, shading makes it possible that with positive probability no bargain will be struck even if there is a non-empty contract zone. Thus, Chatterjee and Samuelson's model provides an alternative explanation of the occurrence of impasses.

his opponent attempted commitment, and what demand, if any, he made. He then decides, taking into account this information, whether or not to retreat from his demand; if not, he is said to have 'achieved commitment' to the position given by his first-stage demand. The second-stage part of his strategy takes the form of a rule that relates his action to the situation in which he finds himself. These second-stage decisions then determine the final outcome as in Nash's demand game, with the qualification noted above: incompatible demands from which neither bargainer retreats lead to the disagreement outcome; and compatible demands lead to a compromise that yields each bargainer at least what he demanded. In other situations, which cannot arise in Nash's game, the outcome is as follows: if exactly one bargainer has made a demand and not backed down from it, the Pareto-efficient settlement in which he gets his demand is the final outcome; and if neither bargainer has made a demand and stuck to it, the outcome is a compromise settlement. This compromise settlement is assumed to be Pareto-efficient, to avoid begging the question that lies at the heart of this chapter: whether rational behaviour in bargaining must lead to efficient outcomes.

In the above specification of the bargaining game, it is assumed that bargainers do not learn their costs of backing down until after their first-period decisions. In general multi-stage games with incomplete information, players can learn about their opponents by observing their actions in early stages and drawing inferences based on game-theoretic rationality (or other behavioural) assumptions. Players must, therefore, evaluate early-stage actions taking into account the later stage effects of the information those actions reveal to opponents. (See Kreps and Wilson (1982) for an especially clear discussion of these and other problems that arise in modelling behaviour in multi-stage games with incomplete information.) These issues do not arise in my specification of the bargaining game, because a bargainer's choice of first-stage strategy is based on precisely the same information that is assumed to be available to his opponent, who can, therefore, draw no additional inferences from it. (Information is transmitted by second-stage actions, but by then, given my assumptions, it is too late for this information to change the outcome.) This fact allows complex issues of strategic information transmission, which are conceptually separate from the issues considered here (although certainly important in real bargaining, mainly in connection with information about preferences, however), to be avoided.

The bargaining environment can be formally described as follows. It includes two bargainers, indexed $i = 1, 2$. The index j, when it appears, refers to the bargainer other than i; and to avoid needless repetition of '$i = 1, 2$', i alone will be understood to refer to each bargainer. Each bargainer i is assumed to have preferences over agreements and probability distributions of agreements that can be represented by a von Neumann-Morgenstern utility function, denoted u^i. These preferences may reflect the anticipated effects of the current agreement on future negotiations, but bargainers are assumed not to contemplate coordinating their current bargaining strategies in future negotiations. Following Nash (1950) and many others, I shall assume that the set of utility pairs, denoted U, that can be attained by negotiating an agreement is a closed, bounded, convex and non-empty subset of two-dimensional Euclidean space.[6] And I shall consider only the 'fixed-threats' case, where the *disagreement outcome* - denoted $(\underline{u}^1, \underline{u}^2)$ and defined as the pair of expected utilities bargainers associated with the (possibly uncertain) consequences of not negotiating an agreement - is independent of bargainers' actions. The disagreement outcome is assumed not to be Pareto-efficient in U. The *contract zone*, denoted V, is defined as the set of utility pairs $(u^1, u^2) \in U$ such that $u^i \geqslant \underline{u}^i$, $i = 1, 2$. For simplicity, it is assumed that the part of the boundary of the utility-possibility set that lies in V is strictly downward sloping and differentiable. And a bargainer faced with a choice between outcomes between which he is indifferent will be assumed always to choose as his opponent prefers.

Let $\bar{u}^i$ denote the largest utility for bargainer i that is compatible with $(u^1, u^2) \in V$; given the above assumptions, $\bar{u}^i > \underline{u}^i$. Write the equation of the portion of the utility-possibility frontier that lies in V, which is downward sloping by assumption, as $u^j \equiv \phi(u^i)$, and let $\psi \equiv \phi^{-1}$, so that $u^i \equiv \psi(u^j)$ also represents the utility-possibility frontier in V. (Even though the superscript i is normally understood to refer to either bargainer, this notation will be used to indicate whether the utility-possibility frontier is viewed as parameterized by u^i or u^j as they appear elsewhere in the argument; no confusion should result.) A *compatible* pair of utilities is defined as one such that, if u^1 and u^2 are the utilities, $(u^1, u^2) \in U$; *more-than-compatible* utilities lie, as a pair, in the interior of U; and *just-compatible* utilities lie on the utility-possibility frontier.

The outcome function is formally specified as follows. Without loss of generality, we may restrict bargainer i's possible commitment positions to the interval $[\underline{u}^i, \bar{u}^i]$. If neither bargainer achieves commitment, the outcome is a Pareto-efficient compromise, denoted $(\tilde{u}^1, \tilde{u}^2)$; it is assumed that $\underline{u}^i < \tilde{u}^i < \bar{u}^i$, so that $(\tilde{u}^1, \tilde{u}^2) \in V$. If both bargainers achieve commitment, to incompatible positions, the outcome is $(\underline{u}^1, \underline{u}^2)$. If bargainers achieve commitment to compatible positions $(\hat{u}^1, \hat{u}^2)$, the outcome $(u^1, u^2) \in U$ satisfies $u^1 \geqslant \hat{u}^1$ and $u^2 \geqslant \hat{u}^2$; it is not required to be Pareto-efficient. However, u^i is assumed to be a strictly increasing function of $\hat{u}^i$ in this case, as long as the positions $(\hat{u}^1, \hat{u}^2)$ remain compatible. Nash's (1953) rule, where $u^i = \hat{u}^i$, is a simple example of

6 Convexity could be dispensed with, but since bargainers are allowed to choose strategies that lead to probabilistic outcomes, it seems unnatural to prohibit them from negotiating random settlements. Without such a prohibition, U is well known to be convex.

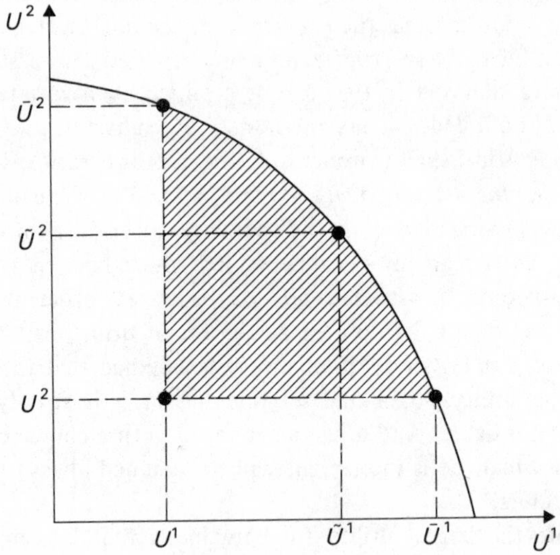

Figure 7.1

one satisfying this assumption. Finally, if bargainer i alone achieves commitment, to a position $\hat{u}^i \leqslant \bar{u}^i$, the outcome is $(\hat{u}^i, \phi(\hat{u}^i))$.

The justification of this outcome function is clear, given the above definition of achieving commitment. Figure 7.1, which illustrates a typical bargaining environment, may make it easier to remember the notation; the contract zone is the shaded portion of the utility-possibility set in the figure.

To complete the specification of the model, consider the determinants of the costs of backing down for bargainers who have attempted commitment. In principle, a bargainer's cost might be systematically related to the position he attempts commitment to; and he might, by his choice of negotiating tactics, also be able to exert an influence on the distribution of these costs that is separate from the influence of his commitment position. Introspection and casual empiricism have not made clear to me whether it should tend to be harder to back down from an extreme position or a moderate one, and I have even less intuition about how this effect should interact with bargaining tactics. For these reasons, and because these complications seem to add little to what light the model sheds on bargaining, I shall study explicitly only the leading special case where the distribution of costs, given that a bargainer has attempted commitment to some position, is independent of that position and of his other actions. The only influence he can exert on these costs is by exercising his option not to attempt commitment. This assumption does not, of course, imply that the probability of achieving commitment is necessarily independent of the commitment position as well. Additional simplicity is gained at little expense by assuming that bargainers' cost distributions are independent.

More formally, let $F^i(c^i)$ denote the distribution function of bargainer i's cost, in utility terms. F^i and F^j are assumed to represent independent and, for most of the analysis, continuous distributions, and to be common knowledge. The supports of the F^i contain only non-negative values, so there is never a 'subsidy' for backing down.

3 PERFECT BAYESIAN NASH EQUILIBRIUM

This section analyses the model developed in section 2 when the solution concept is Harsanyi's (1968) Bayesian Nash equilibrium, with an additional requirement of perfectness (see Selten 1975). These assumptions embody what has come to be known as full game-theoretic rationality in multi-stage non-cooperative games with incomplete information: they therefore provide the most stringent possible test of the model's ability to rationalize the occurrence of impasses in bargaining. The result obtained here, that impasses can occur with positive probability in equilibrium, does not stem from bargainers having irrational expectations or making sub-optimal decisions.

A Bayesian Nash equilibrium, in the context of the model, is simply a Nash equilibrium in first-stage actions and in the rules that relate bargainers' second-stage actions to the situation created by first-stage actions and to their observed costs of backing down. Each bargainer responds optimally to his opponent's strategy choice, taking into account its implications in view of his probabilistic beliefs about his opponent's cost. An additional requirement of perfectness will be imposed, which has the effect of ruling out equilibria in which a bargainer makes implausible bluffs about what he will do in the second stage that are not called in equilibrium. To put it another way, perfectness requires bargainers always to respond optimally in the second stage to any situation that might be created by their actions in the first stage. The resulting equilibrium concept is equivalent to Kreps and Wilson's (1982) notion of 'sequential rationality'.

The Bayesian Nash equilibrium is both the natural generalization of the ordinary Nash equilibrium to games with incomplete information and a natural extension of the familiar concept of rational-expectations equilibrium to situations where strategic interactions are important. It has the further, closely related advantage that rational players who have chosen strategies that are in Bayesian Nash equilibrium, and are given an opportunity to revise them before the uncertainty in the game is resolved, will never do so.

When a Bayesian Nash equilibrium exists in the model (the existence question is discussed further below), it can be characterized as follows. First, note that no equilibrium can have neither bargainer attempting commitment: in such a situation, either bargainer could attempt commitment to a position better than the compromise settlement for himself and not back down in the second stage (which is consistent with perfectness), obtaining his demand. For similar reasons equilibrium cannot, with one possible exception, involve only one bargainer attempting commitment: in such cases, that bargainer would always have an incentive to increase his demand. The exception occurs when bargainer i is already demanding $\bar{u}^i$, in which case it might be an equilibrium strategy for bargainer j to acquiesce. But in that case, j might just as well attempt commitment to $\underline{u}^j$, a convention that allows the simple classification of equilibria presented below.

It follows from the above observations that equilibrium can always be taken to involve each bargainer attempting commitment to some position; but these positions need not be incompatible. In fact, there are two significantly different types of equilibria in this model: those with compatible commitments and those with incompatible commitments. Given my assumption that the utility a bargainer gets if both bargainers achieve commitment, to compatible positions, increases strictly with his position as long as compatibility is preserved, it is immediately clear that compatible commitments can be in equilibrium only if they are just compatible. It is also clear that a bargainer can never gain by unilaterally backing down from a commitment position that is just compatible with that of his opponent, even if the cost of doing so turns out to be zero. Thus, equilibrium involves either just-compatible commitments (in which it can be assumed without loss of generality that neither bargainer backs down in the second stage, no matter what his cost) or incompatible commitments. The just-compatible commitment equilibria are the ones identified by Nash (1953), although in general, not all of those equilibria are equilibria here as well. Relaxing Nash's assumption that demands are certain to be irrevocable both destroys some compatible-commitment equilibria and allows the existence of some incompatible-commitment equilibria, as we shall see.

Suppose that bargainers have attempted commitment to the incompatible positions $(\hat{u}^i, \hat{u}^j)$ in the first stage. How do they decide when to back down in the second stage? Under my assumption of perfectness, the rules that answer this question must be in Nash equilibrium in the second-stage game created by bargainers' choices of demand in the first stage. It is clear that best-response decision rules will involve cutoff levels of costs, below which bargainers will back down, and above which they will stand firm. Suppose the F^i represent continuous distributions, so that how bargainers break ties is unimportant. The equilibrium cutoff levels are determined as follows.

Let d^i denote bargainer i's cutoff level of cots. When $c^i \leqslant d^i$, he backs down in the second stage; otherwise, he stands firm. Given the definitions and assumptions, it is easy to verify that bargainer i's expected payoff in this game, when demands are incompatible, is given by

$$w^i(d^i, d^j) \equiv F^i(d^i)\,[F^j(d^j)\,\bar{u}^i + (1 - F^j(d^j))\,\psi(\hat{u}^j)] + [1 - F^i(d^i)]\,[F^j(d^j)\,\hat{u}^i + (1 - F^j(d^j))\,\underline{u}^i] - \int_0^{d^i} c^i f^i(c^i)\,dc^i, \qquad (3.1)$$

where f^i denotes the density associated with F^i. This expression for $w^i(d^i, d^j)$ is derived by considering the four possible combinations of bargainers backing down and standing firm, weighted by their probabilities, and subtracting the expected costs incurred by backing down. An easy computation reveals that

$$w_1^i(d^i, d^j) \equiv f^i(d^i)\,[F^j(d^j)\,(\bar{u}^i - \hat{u}^i) + (1 - F^j(d^j))\,(\psi(\hat{u}^j) - \underline{u}^i) - d^i]. \qquad (3.2)$$

Suppose, for example, that the support of F^i is the interval $[0, e^i]$, where $e^i > \bar{u}^i - \underline{u}^i$. Then it is not hard to see that i's best choice of d^i must be interior, and must, therefore, satisfy $w_1^i(d^i, d^j) = 0$, which holds, given that $f^i(d^i) > 0$ whenever $d^i \in [0, e^i]$, if and only if the term in brackets on the right-hand side of (3.2) equals zero. Another easy computation reveals that whenever $w_1^i(d^i, d^j) = 0$,

$$w_{11}^i(d^i, d^j) = -f^i(d^i) < 0. \qquad (3.3)$$

Thus, it follows that the w^i are strictly quasi-concave. Since they are also clearly continuous, and the strategy spaces can be taken to be compact and convex without ruling out any 'good' strategies, Debreu's social equilibrium existence theorem (restated as Theorem 1 in Dasgupta and Maskin 1977) implies the existence of a pure-strategy Nash equilibrium in the second-stage game, given any incompatible choices of $\hat{u}^i$ and $\hat{u}^j$ in the first stage. In general, it is not possible to prove uniqueness without considerably stronger restrictions on the payoff functions, but I shall assume uniqueness to facilitate the rest of this discussion.

Given this determination of equilibrium strategies in the second stage, a perfect Bayesian Nash equilibrium in the entire game can be constructed, in dynamic programming fashion, by finding Nash equilibrium demands in the first stage, evaluating the second-stage consequences of first-stage actions using the already determined second-stage equilibrium strategy rules. Of course, that there is always an equilibrium in pure strategies in the second-stage game does not imply that there is always an equilibrium, in pure *or* mixed strategies, in the game taken as a whole. In fact, it does not appear possible to prove a general existence result under the maintained assumptions, due to the discontinuities that may arise in the payoff functions when bargainers' demands are just compatible.[7] I shall provide a partial remedy for this by exhibiting a class of examples in which equilibrium generally exists and, later, by providing restrictive, but not unreasonable, assumptions about behaviour and the bargaining environment under which existence is guaranteed.

The remainder of this section analyses in detail a class of simple examples in which pure-strategy existence can be guaranteed under easily interpretable and reasonable assumptions, and in which for some parameter configurations there are only incompatible-commitment equilibria. It is hoped that the analysis of these examples will also serve to illustrate better the workings of the model.

Normalize $\underline{u}^i = \underline{u}^j = 0$ and $\bar{u}^i = \bar{u}^j = 1$, and depart from the earlier assumptions by letting F^i and F^j be Bernoulli distributions, with probabilities q^i and q^j respectively of yielding a cost greater than unity, and probabilities $1-q^i$ and $1-q^j$ of yielding zero costs. I shall assume to avoid trivialities that q^i and q^j lie strictly between zero and one. When costs are high in this case, bargainers always stand firm in the second stage; when costs are low, they are zero. Thus, costs are never paid; they serve only to make it effectively 'impossible' to back down, with given probability. As will become clear, however, the game is far from trivial even in this simple case.

The reader can easily verify that my analysis of the necessary conditions that must be satisfied at compatible-commitment equilibria remains valid here: with inessential exceptions, any such equilibria must involve just-

7 The discontinuities prevent the use of Debreu's social equilibrium existence theorem. Dasgupta and Maskin have shown (1977, Theorems 3 and 4 and n. 16) that at least a mixed-strategy Nash equilibrium will exist, given the compactness and convexity of the strategy spaces, provided only that payoff functions are upper semi-continuous and graph continuous. (Graph continuity requires, roughly, that the graph of a player's payoff, viewed as a function of his own actions, vary continuously with changes in other players' actions.) If, in addition, the payoff functions are quasi-concave, pure-strategy existence is obtained. These more general results do not imply existence in the present model, because it is not generally true that both payoff functions are upper semi-continuous when bargainers' demands are just compatible. The intuitive reason for this is that by moving from a just-compatible to a just-incompatible demand, a bargainer may induce both his opponent and himself to back down frequently enough that the compromise settlement occurs with significant probability. (Allowing for such changes in second-stage actions is the proper way to evaluate the consequences of such a change, since it is the rule that determines the bargainer's second-stage actions, rather than the action itself, that must be held constant in viewing the perfect Nash equilibrium as a Nash equilibrium in first-stage demands, when the payoff functions are dynamic-programming value functions.) If a bargainer does this starting from a configuration of demands that is worse for himself than the compromise settlement, it can make him better off, even though there is also a risk of impasse that was not present before. This failure of upper semi-continuity appears to be intrinsic to the model rather than to the present formulation, although of course it does not *imply* non-existence of equilibrium.

compatible demands and must not be vulnerable to defections involving incompatible demands. In the remainder of this section, I shall argue first that the examples are capable of supporting compatible-commitment equilibria for some parameter configurations; second, that there are configurations where incompatible-commitment equilibria always exist; and finally, that there are configurations where only incompatible-commitment equilibria exist.

While it is difficult to characterize compatible-commitment equilibria in general, it is not hard to show that when $\tilde{u}^i \geqslant 1 - q^j$ and $\tilde{u}^j \geqslant 1 - q^i$, $(\hat{u}^i, \hat{u}^j) = (\tilde{u}^i, \tilde{u}^j)$ is such an equilibrium. Consider possible defections from the hypothesized equilibrium configuration by bargainer i, and recall the perfectness requirement, which implies that the consequences of a defection must be evaluated under the assumption that bargainer j responds optimally (in the sense of choosing an equilibrium strategy) to the second-stage situation created by the defection. To evaluate these consequences, we must first consider bargainers' choices of strategy in the second stage when they have attempted commitment to incompatible positions in the first stage. Recall that at perfect equilibria, bargainers always stand firm when costs are high, because the costs outweigh any possible gains. Bargainers' unconditional expected payoffs, as a function of the actions that they take when costs are low and taking into account their optimal actions when costs are high, are as follows, with i's expected payoff given first in each case. If bargainer i stands firm, if j also stands firm the outcome is $(\underline{u}^i, \underline{u}^j)$, and if j backs down the outcome is $(q^j \underline{u}^i + (1 - q^j)\hat{u}^i, q^j \underline{u}^j + (1 - q^j)\phi(\hat{u}^i))$. If bargainer i backs down, if j stands firm the outcome is $(q^i \underline{u}^i + (1 - q^i)\psi(\hat{u}^j), q^i \underline{u}^j + (1 - q^i)\hat{u}^j)$ and if j also backs down the outcome is $(q^i q^j \underline{u}^i + q^i(1 - q^j)\hat{u}^i + (1 - q^i)q^j \psi(\hat{u}^j) + (1 - q^i)(1 - q^j)\tilde{u}^i, q^i q^j \underline{u}^j + q^i(1 - q^j)\phi(\hat{u}^i) + (1 - q^i)q^j \hat{u}^j + (1 - q^i)(1 - q^j)\tilde{u}^j)$.

Given the definition of the outcome function when commitments are compatible, from $(\hat{u}^i, \hat{u}^j) = (\tilde{u}^i, \tilde{u}^j)$ only a defection to some incompatible commitment $\mathring{u}^i > \tilde{u}^i$ could possibly yield a better outcome for bargainer i. Inspection of the conditional payoff function given above reveals that in the case under consideration, any such defection makes backing down when costs are low a dominant strategy for bargainer j (that is, optimal for either of i's possible second-stage actions) no matter what commitment position in $(\tilde{u}^i, \bar{u}^i)$ bargainer i defects to. Thus, setting $\mathring{u}^i$ at (or near) unity is the optimal defection of this type for bargainer i, and given that $\tilde{u}^i \geqslant 1 - q^j$, backing down when costs are low is his optimal second-stage policy. This defection yields bargainer i an expected payoff of $q^i(1 - q^j) + (1 - q^i)\tilde{u}^i$, which is unprofitable when compared to his original payoff if and only if $\tilde{u}^i \geqslant 1 - q^j$. Given the symmetry of the situation across bargainers, imposing the analogous condition $\tilde{u}^j \geqslant 1 - q^i$ for bargainer j guarantees that $(\hat{u}^i, \hat{u}^j) = (\tilde{u}^i, \tilde{u}^j)$ is, in fact, a compatible-commitment equilibrium. Generally, it is not unique; I have singled this one out because it is easy to identify.

I shall now argue that when q^i and q^j are near enough to zero that $\tilde{u}^j < 1 - q^i$ and $\tilde{u}^j < 1 - q^i$, there always exist pure-strategy incompatible-commitment equilibria. This confirms Schelling's (1966, chapter 3) intuition that low probabilities of success (because they imply a low probability that one's opponent will succeed) tend to favour attempting commitment to incompatible positions. Conceivably, for some values of the parameters there are equilibria where one, or even both, of the $\hat{u}^i \leqslant 1 (= \bar{u}^i)$ constraints on bargainers' demands are binding. Since in these cases the analysis is unduly complicated by possible multiplicity of equilibria in the second-stage game. I shall confine my discussion to interior equilibria, where this multiplicity is easily dealt with. The analysis yields a simple characterization of incompatible-commitment equilibria.

The analysis of interior incompatible-commitment equilibria is greatly simplified by the observation that such equilibria must have both bargainers backing down in the second stage if (and only if) costs are low. Any other combination of second-stage strategies yields either bargainer i less than $\psi(\hat{u}^j)$ or bargainer j less than $\phi(\hat{u}^i)$; since bargainers can unilaterally guarantee themselves these amounts by defecting to just-compatible commitments in the first stage and standing firm in the second stage (which is consistent with perfectness), such strategies are incompatible with equilibrium. That bargainers are in equilibrium in the second-stage game backing down when costs are low places the following restrictions on interior incompatible-commitment equilibria, recalling the $\bar{u}^i = 0$ normalization:

$$q^i(1 - q^j)\,\hat{u}^i + (1 - q^i)\,q^j\psi(\hat{u}^j) + (1 - q^i)(1 - q^j)\,\tilde{u}^i \geqslant (1 - q^j)\,\hat{u}^i,$$

which reduces to

$$\hat{u}^i \leqslant \frac{q^j}{1 - q^j}\psi(\hat{u}^j) + \tilde{u}^i; \tag{3.4}$$

and

$$q^i(1 - q^j)\,\psi(\hat{u}^i) + (1 - q^i)\,q^j\hat{u}^j + (1 - q^i)(1 - q^j)\,\tilde{u}^j \geqslant (1 - q^i)\,\hat{u}^j,$$

which reduces to

$$\hat{u}^j \leqslant \frac{q^i}{1 - q^i}\phi(\hat{u}^i) + \tilde{u}^j. \tag{3.5}$$

An interior incompatible-commitment equilibrium must have (3.4) and (3.5) satisfied with equality, because in these examples, given the anticipated equilibrium in the second-stage game, the probabilities of backing down are independent of $\hat{u}^i$ and $\hat{u}^j$ as long as (3.4) and (3.5) are satisfied. Thus, setting $\hat{u}^i$ or $\hat{u}^j$ lower than these inequalities allow, but still incompatible, gives up something without gaining any advantage in return. In addition, as noted above, an equilibrium with incompatible commitments must yield bargainer i at least $\psi(\hat{u}^j)$ and bargainer j at least $\phi(\hat{u}^i)$; thus,

$$q^i(1-q^j)\hat{u}^i + (1-q^i)q^j\psi(\hat{u}^j) + (1-q^i)(1-q^j)\bar{u}^i \geqslant \psi(\hat{u}^j) \quad (3.6)$$

and

$$q^i(1-q^j)\phi(\hat{u}^i) + (1-q^i)q^j\hat{u}^j + (1-q^i)(1-q^j)\bar{u}^j \geqslant \phi(\hat{u}^i). \quad (3.7)$$

Substituting (3.4), with equality, into (3.6) and simplifying yields $\bar{u}^i \geqslant \psi(\hat{u}^j)$, which is equivalent, given the Pareto efficiency of $(\bar{u}^i, \bar{u}^j)$, to $\hat{u}^j \geqslant \bar{u}^j$. Similar use of (3.5), with equality, reduces (3.7) to $\hat{u}^i \geqslant \bar{u}^i$. Given the $\hat{u}^i \leqslant 1$ and $\hat{u}^j \leqslant 1$ constraints, these conditions are automatically satisfied when (3.4) and (3.5) are satisfied with equality. So the question of existence of interior incompatible-commitment equilibria reduces to the question: can (3.4) and (3.5) both be satisfied with equality for some $(\hat{u}^i, \hat{u}^j)$ in $(\bar{u}^i, 1) \times (\bar{u}^j, 1)$?

To see that they can, substitute the equality form of (3.5) into the equality form of (3.4) to obtain

$$\hat{u}^i = \frac{q^j}{1-q^j}\psi\left[\frac{q^i}{1-q^i}\phi(\hat{u}^i) + \bar{u}^j\right] + \bar{u}^i. \quad (3.8)$$

When $\hat{u}^i = \bar{u}^i$, the right-hand side of (3.8) becomes $q^j\psi[\bar{u}^j/(1-q^i)]/(1-q^j) + \bar{u}^i$, so the left-hand side is less than the right-hand side if and only if $\bar{u}^j < 1-q^i$. When $\hat{u}^i = 1$, the right-hand side simplifies to $\bar{u}^i/(1-q^j)$, so the left-hand side exceeds the right-hand side if and only if $\bar{u}^i < 1-q^j$. It therefore follows by the intermediate value theorem from the continuity of ψ and ϕ that, when $\bar{u}^j < 1-q^i$ and $\bar{u}^i < 1-q^j$, there exists a value of $\hat{u}^i$ strictly between $\bar{u}^i$ and unity that satisfies (3.8). Now consider the corresponding value of $\hat{u}^j$, as given by the equality form of (3.5). When $\hat{u}^i < 1$, $\hat{u}^j > \bar{u}^j$. And when $\hat{u}^i > \bar{u}^i$, that ϕ is strictly decreasing implies that $\hat{u}^j < \bar{u}^j/(1-q^i)$. Thus, if $\bar{u}^j < 1-q^i$ as assumed above, the corresponding value of $\hat{u}^j$ also satisfies the desired restrictions, and we have an interior incompatible-commitment equilibrium, which is completely characterized by the equality forms of (3.4) and (3.5). It does not seem possible to establish uniqueness in general, however.

This section closes with an example to show that there are parameter configurations in the region where incompatible-commitment equilibria always exist for which there are no compatible-commitment equilibria. The example will be constructed by deriving a necessary condition that compatible-commitment equilibria must satisfy, and then constructing a parameter configuration where this necessary condition cannot be satisfied; it is hoped that this will be more informative than simply pulling the parameter values in question out of a hat.

Since any equilibrium pair of compatible demands must be just compatible, either $\hat{u}^i \leqslant \bar{u}^i$ or $\hat{u}^j \leqslant \bar{u}^j$. Suppose the first for definiteness; since the example will be symmetric across bargainers, this involves no loss of generality. Given that $\hat{u}^i \leqslant \tilde{u}^i$, (3.4) holds with strict inequality for any $\hat{u}^j < 1$. Thus, by simple first-order stochastic dominance arguments, if bargainer j defects to such an incompatible commitment position, backing down (when costs are low) is a strictly dominant strategy for bargainer i in the second-stage game. This is true for *any* $\hat{u}^j < 1$, so bargainer j can guarantee himself a payoff as close as desired to $1 - q^i$ by defecting to a position near unity and standing firm in the second stage. Imposing the perfectness requirement can only raise this expected payoff further, since bargainer i has a dominant second-stage strategy in this case. It follows that a necessary condition for the demands $(\hat{u}^i, \hat{u}^j)$ to be a compatible-commitment equilibrium is that $\hat{u}^j \geqslant 1 - q^i$ if $\hat{u}^j \geqslant \tilde{u}^j$.

Consider a compatible pair of demands $(\hat{u}^i, \hat{u}^j)$ with $\hat{u}^j \geqslant 1 - q^i > \tilde{u}^j$ and $\hat{u}^i = \psi(\hat{u}^j)$. Bargainer i has the option of defecting to $\hat{u}^i + \epsilon$, where ϵ is near zero. Since $\hat{u}^i < \tilde{u}^i$, this strategy is potentially advantageous in that it may induce bargainer j to back down when costs are low in order to reduce the risk of impasse. Of course, this risk of impasse is costly for i as well, but the costs may be outweighed by the increased probability of getting the compromise settlement. For such a defection to help, two things must be true: the defection must cause bargainer j to back down in the second stage, and the resulting expected payoff for i must exceed $\psi(\hat{u}^j)$. The first condition is just inequality (3.5), which simple algebra reveals to be satisfied with strict inequality for all $\hat{u}^j \in [1-q^i, 1]$ and for $\hat{u}^i$ near $\psi(\hat{u}^j)$, provided that $\tilde{u}^j > (1 - 2q^i)/(1 - q^i)$. Impose this and the symmetric condition $\tilde{u}^i > (1 - 2q^j)/(1 - q^j)$; these are easily seen to be compatible with the $1 - q^i > \tilde{u}^j$ and $1 - q^j > \tilde{u}^i$ restrictions. They are also compatible with the fact that $\tilde{u}^i = \psi(\tilde{u}^j)$ when ψ is linear and $q^i = q^j > 1/3$, for example. The second condition, for small ϵ, requires that

$$q^i(1 - q^j)\hat{u}^i + (1 - q^i)q^j\psi(\hat{u}^j) + (1 - q^i)(1 - q^j)\tilde{u}^i > \psi(\hat{u}^j). \quad (3.9)$$

Given that $\hat{u}^i = \psi(\hat{u}^j)$, (3.9) reduces to

$$(1 - q^i)(1 - q^j)(\tilde{u}^i - \hat{u}^i) > q^i q^j \hat{u}^i, \quad (3.10)$$

which, recalling that $\hat{u}^i = \hat{u}^i - \underline{u}^i$ given the normalization, is easily interpretable in terms of weighing the benefits of the greater probability of compromise against the costs of the greater risk of impasse. For defections of the kind being analysed to rule out all possible compatible-commitment equilibria, (3.10) must hold for all $\hat{u}^i \in [0, \psi(1 - q^i)]$, given that $\hat{u}^j \geqslant 1 - q^i$ at any candidate for an equilibrium; preserving symmetry guarantees that the analogous condition for bargainer j will be satisfied.

Now suppose that ψ is linear, that $\tilde{u}^i = \tilde{u}^j = 1/2$, and that $q^i = q^j = 7/20$. This preserves symmetry, and it is easy to check that for these parameter values, the conditions for a successful defection are always met. In particular, (3.10) is satisfied for all $\hat{u}^i < 169/436$, but $\hat{u}^i \leqslant \psi(1 - q^i) = q^i = 7/20$, which is less than 169/436. Thus, there can exist *only* incompatible-commitment equilibria for these parameter values.

4 EQUILIBRIUM WITH EXPECTATIONAL HEURISTICS

Assumptions of full game-theoretic rationality such as those maintained in section 3 provide a useful discipline, whose value is obvious when the goal is to show that the occurrence of inefficient bargaining outcomes is compatible with rational behaviour by bargainers. And it is tempting simply to view perfect Bayesian Nash equilibrium as the 'right' equilibrium concept for non-cooperative multi-stage games with incomplete information, and to reject all others as being irrational or arbitrary. But in my opinion, it would be unfortunate if the search for truly descriptive behavioural assumptions in such games were to end there.

My reasons for this opinion are as follows. Suppose we give the 'fully rational' theory the greatest possible benefit of the doubt by accepting the usual iterative story about how players come to choose equilibrium strategies, in which they are free to keep revising their actions until both are satisfied. Suppose further, for the sake of argument, that the implied dynamic process converges. Even under these circumstances, in the model of this chapter, players are not relieved of the need to formulate expectations about their opponents' actions, because irreversible actions whose payoffs depend on those expectations must be taken in the first stage. Most will agree that it is reasonable to assume that what a player does in the first stage can be viewed as maximizing expected von Neumann–Morgenstern utility, where the expectation is taken over his probability distributions of his cost and of his opponent's future actions. The question is: where does the latter distribution come from?

The standard story among game theorists is that players are (except as limited by incomplete information) fully informed about the game, assume that their opponents are fully rational and actually compute self-confirming strategy rules for all players under these assumptions. Bayesian Nash equilibrium strategies (and, if best responses are unique, only such strategies) are self-confirming. Thus, unless bargainers play strategies that are in Bayesian Nash equilibrium, at least one will eventually have to revise his probabilistic expectations about opponents' actions. (Compare the concept of 'stable conformistic expectations' that characterizes Harsanyi and Selten's 1972, pp. P92–P93 'strict' equilibrium points.) This is why the Bayesian Nash equilibrium is a natural generalization of the rational-expectations equilibrium to environments where strategic interactions are important.

Given the great complexity of the simplified bargaining game analysed in section 3, it seems a reasonable conclusion that the search for a sensible solution concept to describe real bargainers should not end here. But the story just outlined, it can be argued, is a straw man. What is really being assumed is not this kind of unlimited computational skill, but rather that players who live in a population of rational players will come to learn the correct probability distributions of what their opponents will do in various strategic situations. While this view may have some justification, its theoretical underpinnings are extremely weak. And even if the learning process could be counted on to converge to the correct distributions, this view's reliance on the assumption that players have spent a long time in a stable environment (but are not repeatedly matched with the same players, for then the situation would eventually cease to be one of incomplete information) greatly limits its applicability, in a way that does not also limit the applicability of the approach about to be proposed.

In this section, I shall study the implications of an intermediate position. It will be assumed that bargainers reach a Nash equilibrium in first-stage actions, but that their expectations about what will happen in the second stage are formed more simply than fully 'rational' expectations would be. This approach has two main advantages: it is uncomplicated enough to be a reasonable candidate for a descriptive model, and it does not ignore the most immediate (first-stage) strategic aspects of bargainers' choices of demand. In evaluating this approach, the reader may find it useful to recall to what extent he used game-theoretic rationality assumptions in deciding on the best strategy to pursue in the negotiations accompanying his last purchase of a house. The analysis that follows attributes, as I hope the reader's recollections will show, a reasonable level of strategic sophistication to bargainers, and leads to conclusions that differ in some interesting respects from, but basically confirm, the results in section 3. The assumptions made here also provide a framework in which it is convenient to study the relationship of the probability of impasse to the bargaining environment, as is done in section 5.

I shall begin with the simplest possible assumption about expectations: that the probability bargainer i assigns to his deciding not to back down in the second stage is a constant, denoted p_i, and assumed to lie strictly between zero and unity. It is also assumed that bargainers view their decisions whether or not to back down as probabilistically independent events (a natural extension from section 3, where independence of the cost distributions implies conditional independence of bargainers' decisions whether or not to back down), and, for simplicity, that both share the same perception of the probabilities of these events. This case is closest in spirit to the examples of section 3, in which the equilibrium probabilities of backing down are constant, as long as bargainers' commitment positions leave them enough incentive to back down when costs are low. The present assumption fails to be fully rational because it ignores these constraints on the commitment positions. Finally, I shall maintain the assumption made in the examples of section 3, that costs are either zero or prohibitively high. This seems to allow the points of this and the next section to be made most simply, and the reader can easily check how things would change with more general assumptions about the cost distributions.

When the probabilities are constant, a striking result is obtained. A strategy is said to *dominate* another strategy if it yields the player who employs it an outcome at least as good, no matter what strategy his opponent employs. I shall now argue that in this case, successive deletion of dominated strategies reduces the bargaining game to a Prisoner's Dilemma. In it, each bargainer has two strategies: not attempting commitment, and attempting commitment to the position in the contract zone most favourable to him. This is so in spite of the fact that neither of these strategies *alone* dominates all other strategies. After the deletion of dominated strategies, the latter strategy dominates the former for each player, so that if bargainers do not play dominated strategies, bargainers attempt commitment to $(\hat{u}^i, \hat{u}^j) = (\bar{u}^i, \bar{u}^j)$ and there is a positive probability $p_1 p_2$ of impasse.[8] Since the impasse outcome $(\underline{u}^1, \underline{u}^2)$ is not Pareto-efficient, the distribution of outcomes that results is not ex ante Pareto-efficient, even though bargainers could have avoided it only by playing dominated strategies.

The argument proceeds in three steps. First, I shall show that attempting commitment to a position $\hat{u}^i \leqslant \tilde{u}^i$ is dominated by not attempting commitment. Then, I shall observe that attempting commitment to a position $\hat{u}^i > \bar{u}^i$ is dominated by attempting commitment to $\bar{u}^i$. And finally, I shall argue that after deletion of the above dominated strategies, attempting commitment to a position $\hat{u}^i$ such that $\tilde{u}^i < \hat{u}^i < \bar{u}^i$ is also dominated by attempting commitment to $\bar{u}^i$.

To see that attempting commitment to $\hat{u}^i \leqslant \tilde{u}^i$ is dominated by not attempting commitment, note that there is no difference between these strategies unless the attempt succeeds. Assuming it does, bargainer i obtains utility $\hat{u}^i$ if bargainer j does not achieve commitment; u^i, where $\hat{u}^i \leqslant u^i \leqslant \psi(\hat{u}^j)$, if j achieves commitment to a compatible position $\hat{u}^j$ (where the latter inequality follows from the requirements that $u^j \geqslant \hat{u}^j$ and $(u^i, u^j) \in U$); and $\underline{u}^i$ if j achieves commitment to an incompatible position $\hat{u}^j$ or to a position $\bar{\hat{u}}^j > \bar{u}^j$. If, on the other hand, bargainer i does not attempt commitment, he obtains, respectively, utilities $\tilde{u}^i$, $\psi(\hat{u}^j)$, and $\psi(\hat{u}^j)$ or $\underline{u}^i$. In each case, i does at least as well by not attempting commitment.

Attempting commitment to $\hat{u}^i > \bar{u}^i$ is dominated by attempting commitment to $\bar{u}^i$ simply because if commitment is achieved, the former strategy

8 Since there is no guarantee that bargainers' expectations about second-stage events are correct, $p_1 p_2$ need not be the 'true' probability of impasse, which depends on how bargainers actually behave in the second stage. To keep the discussion as simple as possible, I shall maintain throughout the practice of making ex ante judgements using bargainers' own expectations.

always yields utility $\underline{u}^i$, while the latter sometimes yields $\underline{u}^i$, but also yields $\tilde{u}^i$ and $\bar{u}^i$ with positive probabilities.

Finally, attempting commitment to $\hat{u}^i$, where $\tilde{u}^i < \hat{u}^i < \bar{u}^i$, is dominated by attempting commitment to $\bar{u}^i$. To see this, note first that p_i is the same for all $\hat{u}^i$. By the above arguments, only cases where bargainer j does not attempt commitment or attempts commitment to $\hat{u}^j$, where $\tilde{u}^j < \hat{u}^j \leqslant \bar{u}^j$, need be considered. All strategies under consideration are equivalent unless the attempted commitment succeeds. If bargainer i succeeds in committing himself to position $\hat{u}^i$, he obtains utility $\hat{u}^i$ if bargainer j does not achieve commitment and $\underline{u}^i$ if j does achieve commitment, since in this case the commitments must be incompatible because $\hat{u}^i > \tilde{u}^i, \hat{u}^j > \tilde{u}^j$, and $(\tilde{u}^i, \tilde{u}^j)$ is Pareto-efficient. Thus, since p_i is constant, it is clear that attempting commitment to $\bar{u}^i$ dominates all other strategies in this class for bargainer i.

Consider the bargaining game, reduced by deletion of dominated strategies to a game in which each bargainer i has only two strategies: not attempting commitment, and attempting commitment to the position $\bar{u}^i$. I shall now argue that the latter strategy in fact strictly dominates the former for both bargainers.

There is clearly no difference between the strategies unless commitment is successful. If bargainer i achieves commitment to $\bar{u}^i$ and bargainer j does not achieve commitment, i obtains utility $\bar{u}^i$, whereas he could have obtained only $\tilde{u}^i < \bar{u}^i$ by not attempting commitment. If, on the other hand, bargainer j achieves commitment to $\bar{u}^j$, which is necessarily incompatible with $\bar{u}^i$, bargainer i obtains utility $\underline{u}^i$ whether he attempts commitment or not, regardless of whether his attempted commitment is successful. Thus, attempting commitment to $\bar{u}^i$ stochastically dominates not attempting commitment for bargainer i.

To complete the interpretation of the game, note that it is natural to assume that $(\tilde{u}^1, \tilde{u}^2)$, the outcome when neither bargainer attempts commitment, is ex ante Pareto-superior to the outcome when both bargainers attempt commitment. It is unlikely, given the convexity of the utility-possibility set, that a bargainer would negotiate an agreement inferior to what he could obtain ex ante by attempting commitment; and the prospect of attempting commitment is at least as good as the outcome when both bargainers attempt commitment, because having an opponent attempt commitment is the worst case for a bargainer who has attempted commitment.

To summarize, in the constant-probabilities case the bargaining game can be reduced by successive deletion of dominated strategies to a classical Prisoner's Dilemma, in which attempting commitment to the most favourable position in the contract zone dominates not attempting commitment for both bargainers. There is, therefore, a unique Nash equilibrium, which results provided only that bargainers do not employ dominated strategies. Intuitively, this is so because, given the deletion of dominated strategies carried out above, attempting commitment to $\bar{u}^i$ increases bargainer i's chance of getting a favourable outcome at no cost, since the probabilities of success are constant and an impasse is no worse for i than letting bargainer j achieve commitment to $\bar{u}^j$. The Nash equilibrium of the commitment game has the usual property of the non-cooperative equilibria of Prisoner's Dilemma games – in spite of its clear individual rationality, it leads to an outcome that is collectively 'irrational' because the positive probability of impasse that results implies that its distribution of outcomes is not ex ante Pareto-efficient. While I would not wish to argue that the extreme demands that occur in equilibrium in this case are realistic, the result provides a nice 'textbook' example and a striking demonstration that, under these expectational assumptions, inefficiency can occur even if one is willing to accept only strategic-dominance arguments about bargainers' rational actions.

At this point, it is natural to consider more sophisticated heuristics relating bargainers' second-stage expectations to the positions to which they attempt commitment in the first stage. Consider what determines these expectations in the fully rational model of section 3. There, bargainers' first-stage estimated probabilities that they will not back down in the second stage depend on both bargainers' commitment positions and, in general, the entire bargaining environment. For example, bargainer i's probability in the continuous model of section 3 is $1-F^i(d^i)$, where d^i is the equilibrium cutoff level (assumed unique for the purposes of this discussion), which depends on both commitment positions and the entire bargaining environment. In section 5's comparative statics analysis, $(\underline{u}^i, \underline{u}^j)$ is the only aspect of the bargaining environment that is allowed to vary; therefore, for my purposes, $\hat{u}^i$, $\underline{u}^i$, $\hat{u}^j$ and $\underline{u}^j$ are the variables on which bargainers' estimated probabilities might be allowed to depend.

What is a reasonable form for this dependence? Since we are studying the implications of bounded rationality, there is little but intuition and common sense to go on here (although the same sort of intuition and common sense come into place in deciding what factors to take into consideration in a fully rational model). I have therefore chosen plausible hypotheses that lead to a rich, but analytically tractable model. Bargainer i's probability of successful commitment – of not backing down in the second stage – is assumed to be a twice continuously differentiable function of $\hat{u}^i$ and $\underline{u}^i$, denoted $p^i(\hat{u}^i, \underline{u}^i)$. In the continuous model of section 3, comparative statics calculations not reproduced here reveal that a fully rational $p^i(\cdot)$ should depend positively on $\underline{u}^i$, negatively on $\underline{u}^j$ and ambiguously on $\hat{u}^i$ and $\hat{u}^j$. I shall assume, however, that, denoting partial derivatives by subscripts in the usual way, $p^i_1(\cdot) < 0$, $p^i_2(\cdot) > 0$, and $p^i_{11}(\cdot) \leqslant 0$, except where the natural zero/one probability boundaries come into play. In words, the likelihood of successful commitment decreases at an increasing (algebraically decreasing) rate with increases in the position to which commitment is attempted, and increases when the bargainer's disagreement utility rises. To avoid trivialities, I shall also assume that $p^i(\bar{u}^i, \underline{u}^i) > 0$ for all $\underline{u}^i$. (Note that $\bar{u}^i$ may depend on $\underline{u}^i$.) I shall continue to assume that bargainers are well informed about everything but the outcomes of attempted commitments, and that they agree on the forms of the p^i functions.

In this model, as in the model of section 3, it does not seem possible to demonstrate without further assumptions that pure-strategy, or even mixed-strategy, Nash equilibria always exist. To see why this is so, note that the proof given in the analysis of the constant-probabilities case that not attempting commitment dominates attempting commitment to a position $\hat{u}^i \leqslant \tilde{u}^i$ is still valid in the variable-probabilities case, because the proof involved only the value of the commitment probability at a single position. Thus, such overly conservative commitment strategies can be ruled out a priori, and the notation can be simplified by assigning to $\hat{u}^i = \tilde{u}^i$ the meaning of not attempting commitment (rather than section 3's meaning of attempting commitment to $\tilde{u}^i$, which is dominated here by not attempting commitment). Given this notational convention, which will be maintained from now on, each bargainer has, in effect, a compact strategy space - the set of $\hat{u}^i$ such that $\tilde{u}^i \leqslant \hat{u}^i \leqslant \bar{u}^i$ - and the question of existence can be resolved by examining the behaviour of the payoff functions.

Let $v^i(\hat{u}^i, \hat{u}^j)$ denote bargainer i's expected payoff when bargainers' strategies are $(\hat{u}^i, \hat{u}^i)$. Given my assumptions, the payoff function can be written:

$$v^i(\hat{u}^i, \tilde{u}^j) \equiv p^i(\hat{u}^i, \underline{u}^i)\, \hat{u}^i + [1 - p^i(\hat{u}^i, \underline{u}^i)]\, \tilde{u}^i,$$
$$v^i(\tilde{u}^i, \hat{u}^j) \equiv p^j(\hat{u}^j, \underline{u}^j)\, \psi(\hat{u}^j) + [1 - p^j(\hat{u}^j, \underline{u}^j)]\, \tilde{u}^i, \qquad (4.1)$$

and

$$\begin{aligned} v^i(\hat{u}^i, \hat{u}^j) \equiv\ & p^i(\hat{u}^i, \underline{u}^i)\, p^j(\hat{u}^j, \underline{u}^j)\, \underline{u}^i + p^i(\hat{u}^i, \underline{u}^i)\, [1 - p^j(\hat{u}^j, \underline{u}^j)]\, \hat{u}^i \\ & + [1 - p^i(\hat{u}^i, \underline{u}^i)]\, p^j(\hat{u}^j, \underline{u}^j)\, \psi(\hat{u}^j) \\ & + [1 - p^i(\hat{u}^i, \underline{u}^i)]\, [1 - p^j(\hat{u}^j, \underline{u}^j)]\, \tilde{u}^i \quad \text{if } \hat{u}^i > \tilde{u}^i \text{ and } \hat{u}^j > \tilde{u}^j. \end{aligned}$$

The v^i functions are clearly continuous in $(\hat{u}^i, \hat{u}^j)$ whenever $\hat{u}^i > \tilde{u}^i$ and $\hat{u}^j > \tilde{u}^j$, but discontinuous when either $\hat{u}^i = \tilde{u}^i$ or $\hat{u}^j = \tilde{u}^j$. Thus, the standard results used to guarantee the existence of Nash equilibrium are not applicable.[9]

9 See footnote 7. In the present context, it is not hard to show that, because a bargainer is better off if his opponent does not attempt commitment than if he attempts commitment to a position that differs only slightly from the compromise settlement, and because the bargainer himself is better off not attempting commitment than attempting commitment to such a position, the v^i are everywhere upper semi-continuous. The problem is that Dasgupta and Maskin's (1977) condition of graph continuity is not satisfied here, because there is a radical change in the graph of $v^i(\hat{u}^i, \hat{u}^j)$, viewed as a function of $\hat{u}^i$, when $\hat{u}^j$ rises above $\tilde{u}^j$. This change occurs because such an increase in $\hat{u}^j$

In spite of these technical problems, the results in the constant-probabilities case indicate that pure-strategy Nash equilibria will exist much of the time, in particular when $p^i(\hat{u}^i, \underline{u}^i)$ does not fall too rapidly with increases in $\hat{u}^i$ for either bargainer i. In fact, if (but not necessarily only if)

$$p^j(\tilde{u}^j, \underline{u}^j)\, \underline{u}^i + [1 - p^j(\tilde{u}^j, \underline{u}^j)]\, \bar{u}^i > \tilde{u}^i \tag{4.2}$$

holds for $i = 1, 2$, there always exists a pure-strategy Nash equilibrium, as I shall now argue. The reason is that condition (4.2), which depends only on the parameters of the bargaining environment, guarantees that bargainer i will wish to attempt commitment to some position even if bargainer j is attempting commitment to a position near $\tilde{u}^j$. When $\hat{u}^j$ is near $\tilde{u}^j$, the relative attractiveness of attempting commitment is at its lowest, both because the associated risk of impasse is highest there and because the cost of yielding to j's demand is lowest. Thus, it is intuitively clear that if bargainer i attempts commitment in these unfavourable circumstances, he will, a fortiori, always attempt commitment. Note that $p^i(\tilde{u}^i, \underline{u}^i)$ and $p^j(\tilde{u}^j, \underline{u}^j)$, the probabilities of bargainers' successful commitment to their compromise settlements, can always be made low enough so that (4.2) is satisfied. This can be viewed as a confirmation of Schelling's (1966, chapter 3) suggestion that low success probabilities favour attempting commitment.

More formally, straightforward computations reveal that when (4.2) is satisfied, there is always some value of $\hat{u}^i$ (in fact, $\hat{u}^i = \bar{u}^i$ will do, although it is not generally a best response) such that $v^i(\hat{u}^i, \hat{u}^j) > v^i(\tilde{u}^i, \hat{u}^j)$ for *all* $\hat{u}^j$ such that $\tilde{u}^j \leqslant \hat{u}^j \leqslant \bar{u}^j$. Further, it is easy to verify that for all such $\hat{u}^j$,

$$\lim_{\hat{u}^i \to \tilde{u}^i} v_1^i(\hat{u}^i, \hat{u}^j) > 0. \tag{4.3}$$

It follows that $\hat{u}^i$ can be restricted to lie in the compact and convex interval $[\tilde{u}^i + \epsilon, \bar{u}^i]$, for some $\epsilon > 0$, without ruling out any 'good' strategies. Thus, if condition (4.2) holds, each bargainer i will always choose to attempt commitment to some position $\hat{u}^i > \tilde{u}^i$. His strategy space can therefore be taken to be compact and convex; and given (4.1), his payoff function will be continuous in the relevant range. If, in addition, the payoff function $v^i(\hat{u}^i, \hat{u}^j)$ is quasi-concave in $\hat{u}^i$, $i = 1, 2$, Debreu's social equilibrium existence theorem yields the existence of a Nash equilibrium in pure strategies.

To see that $v^i(\hat{u}^i, \hat{u}^j)$ is quasi-concave in $\hat{u}^i$, note first that when $\hat{u}^j > \tilde{u}^j$,

$$\begin{aligned} v_1^i(\hat{u}^i, \hat{u}^j) \equiv p_1^i(\hat{u}^i, \underline{u}^i)\,[p^j(\hat{u}^j, \underline{u}^j)\,[\underline{u}^i - \psi(\hat{u}^j)] + [1 - p^j(\hat{u}^j, \underline{u}^j)] \\ \times (\hat{u}^i - \tilde{u}^i)] + p^i(\hat{u}^i, \underline{u}^i)\,[1 - p^j(\hat{u}^j, \underline{u}^j)] \end{aligned} \tag{4.4}$$

---

creates a risk of impasse that was not present before. As a result, it is easy to imagine situations where bargainers' reaction correspondences, even allowing mixed strategies, are not upper semi-continuous. In such situations, there may not exist any pure-strategy, or even mixed-strategy, Nash equilibria.

and

$$v^i_{11}(\hat{u}^i, \hat{u}^j) \equiv p^i_{11}(\hat{u}^i, \underline{u}^i)\,[p^j(\hat{u}^j, \underline{u}^j)\,[u^i - \psi(\hat{u}^j)] + [1 - p^j(\hat{u}^j, \underline{u}^j)] \times (\hat{u}^i - \tilde{u}^i)] + 2p^i_1(\hat{u}^i, \underline{u}^i)\,[1 - p^j(\hat{u}^j, \underline{u}^j)]. \qquad (4.5)$$

It is easy to verify from (4.4) and (4.5) that if $v^i_1(\hat{u}^i, \hat{u}^j) = 0$, $v^i_{11}(\hat{u}^i, \hat{u}^j) < 0$; thus, $v^i(\hat{u}^i, \hat{u}^j)$, viewed as a function of the single variable $\hat{u}^i$, is strictly quasi-concave. It follows that whenever (4.2) holds for $i = 1, 2$, there is a pure-strategy Nash equilibrium, at which bargainers both attempt commitment, to incompatible positions.

In spite of this result, it is not true that in the general case where (4.2) does not hold for $i = 1, 2$, a Nash equilibrium, if one exists, necessarily involves both bargainers attempting commitment. To see this, consider the case where, for all $\underline{u}^j$, $p^j(\hat{u}^j, \underline{u}^j)$ is unity (or nearly unity) for all $\hat{u}^j$ up to and including a certain level between $\tilde{u}^j$ and $\bar{u}^j$ and falls rapidly beyond that point, so that in equilibrium $\hat{u}^j$ will generally be set at (or near) that level. This might be the case, for example, if there is some feature of the bargaining situation (for example, a Council on Wage and Price Stability guideline) that makes it especially credible that bargainer j will not accept less than some particular settlement. If bargainer i attempts commitment to some position $\hat{u}^i$ where $p^i(\hat{u}^i, \underline{u}^i) > 0$, (4.1) and the fact that $p^j(\hat{u}^j, \underline{u}^j) \approx 1$ imply that

$$v^i(\hat{u}^i, \hat{u}^j) \approx p^i(\hat{u}^i, \underline{u}^i)\,\underline{u}^i + [1 - p^i(\hat{u}^i, \underline{u}^i)]\;\psi(\hat{u}^j) < \psi(\hat{u}^j). \qquad (4.6)$$

On the other hand, if bargainer i does not attempt commitment (or, equivalently, attempts commitment to a position where $p^i(\cdot) = 0$),

$$v^i(\hat{u}^i, \hat{u}^j) \approx \psi(\hat{u}^j). \qquad (4.7)$$

Thus, in this case there is a Nash equilibrium where one bargainer attempts commitment while the other does not.

It does, however, follow from my assumption that $p^i(\tilde{u}^i, \underline{u}^i) > 0$ and (4.1) that $(\hat{u}^i, \hat{u}^j) = (\tilde{u}^i, \tilde{u}^j)$ - that is, neither bargainer attempting commitment - can never be a Nash equilibrium. For, since p^i is a continuous function, there will always exist some $\hat{u}^i > \tilde{u}^j$ such that $p^i(\hat{u}^i, \underline{u}^i) > 0$ as well; if $\hat{u}^j = \tilde{u}^j$, this $\hat{u}^i$ yields a higher value of $v^i(\hat{u}^i, \hat{u}^j)$ than $\tilde{u}^i$, by (4.1).

5 COMPARATIVE STATICS

The main interest of making the probability of impasse variable is that it provides a framework in which the relationship between the bargaining environment and the probability of impasse can be investigated. This section presents comparative statics results that indicate how the probability of impasse responds to changes in the size of the contract zone (caused by changes in the costs of disagreement) and to changes in the difficulty of commitment.

The conclusions are easy to summarize: two assumptions that are almost invariably maintained in the industrial labour relations and law and economics literature about bargaining – that the frequency of impasse is reduced by increases in the size of the contract zone (brought about by increases in the costs of disagreement) and by increases in the difficulty of commitment – cannot be supported on theoretical grounds if commitment to incompatible demands is part of the explanation for the occurrence of impasses. The comparative statics results that test these assumptions are ambiguous in the models studied here; and this ambiguity is highly robust to changes in behavioural assumptions and to special assumptions about the bargaining environment. A non-pathological example shows that strong and implausible assumptions, at the very least, would be required to resolve the ambiguities and justify the conventional wisdom.

To demonstrate the extent of this robustness, I shall adopt rather special assumptions for the remainder of this section. First, only equilibria where both bargainers attempt commitment, to incompatible positions, will be considered. Second, since the ambiguous sensitivities of the p^i functions implied by the analysis of section 3 would result immediately in ambiguous conclusions here, I shall instead maintain the assumptions about the p^i functions used in section 4; attention is further restricted to the leading special case where $p^i(\hat{u}^i, \underline{u}^i)$ can be written in the form $p^i(\hat{u}^i - \underline{u}^i)$. (No confusion should result from this abuse of notation.) Finally, it will be assumed that the $\tilde{u}^i$, viewed as differentiable functions $\tilde{u}^i(\underline{u}^i, \underline{u}^j)$ of the disagreement outcome, satisfy $0 \leqslant \tilde{u}^i_1(\cdot) \leqslant 1$ and $\tilde{u}^i_2(\cdot) \leqslant 0$ for all $(\underline{u}^i, \underline{u}^j)$; these assumptions are satisfied in most descriptive bargaining theories (see, for example, Kalai and Smorodinsky 1975 and Nash 1950), and are quite plausible.[10]

In what follows, I shall suppress the arguments of functions for notational clarity whenever this can be done without causing confusion. First, consider the effects of varying the costs of disagreement. When the p^i can be written in the form $p^i(\hat{u}^i - \underline{u}^i)$,

$$\frac{dp^i p^j}{d\underline{u}^i} \equiv p^i p^j_1 \frac{d\hat{u}^j}{d\underline{u}^i} + p^j p^i_1 \left[\frac{d\hat{u}^i}{d\underline{u}^i} - 1 \right], \tag{5.1}$$

where $\hat{u}^i$ and $\hat{u}^j$ now denote the equilibrium attempted commitments for given values of $\underline{u}^i$ and $\underline{u}^j$. If $dp^i p^j / d\underline{u}^i > 0$, shrinking the contract zone by raising $\underline{u}^i$ also raises the probability of impasse; this is the conventional wisdom. Given that $p^j_1(\cdot) < 0$ and $p^i_1(\cdot) < 0$, this will be true in general only if $d\hat{u}^j / d\underline{u}^i \leqslant 0$ and $d\hat{u}^i / d\underline{u}^i \leqslant 1$, with strict inequality holding at least once.

10 One might reasonably argue that the cooperative outcome ought to be determined instead by the actual strategic possibilities – in this case the utilities bargainers can guarantee themselves by attempting or not attempting commitment – rather than the disagreement outcome. From this point of view, that the Nash and Raiffa–Kalai–Smorodinsky solutions satisfy my assumptions that $0 \leqslant \tilde{u}^i_1(\cdot) \leqslant 1$ and $\tilde{u}^i_2(\cdot) \leqslant 0$ is less compelling. But the assumptions still appear quite plausible.

Under the maintained assumptions, straightforward but extremely tedious calculations, not reproduced here, reveal that it is not true in general that $d\hat{u}^j/d\underline{u}^i \leqslant 0$ or $d\hat{u}^i/d\underline{u}^i \leqslant 1$. Further, this ambiguity is not of the type that can be resolved by naive application of Samuelson's correspondence principle: the denominators of the expressions for $d\hat{u}^j/d\underline{u}^i$ and $d\hat{u}^i/d\underline{u}^i$ can be signed by postulating the local stability of a simple gradient adjustment process, but the numerators remain sufficiently indeterminate that the above questions cannot be resolved a priori.

In the $p^i(\hat{u}^i - \underline{u}^i)$ case, the indeterminacy can be shown to depend on interactions between bargainers' strategy choices, in the following sense. Suppose the problem is simplified (as was done in section 4, in the example used to show that there could be Nash equilibria at which one bargainer did not attempt commitment) by specifying p^j so that bargainer j's optimal commitment strategy leads to a constant probability of success for all $\hat{u}^i$. Then only bargainer i's equilibrium conditions need be considered.

The first-order condition for the problem that determines i's optimal commitment strategy requires that $v_1^i(\hat{u}^i, \hat{u}^j) = 0$, and the second-order sufficient condition requires that $v_{11}^i(\hat{u}^i, \hat{u}^j) < 0$; these derivatives are given by (4.4) and (4.5). As noted in section 4, the second-order condition is satisfied whenever the first-order condition is; it follows that $v^i(\hat{u}^i, \hat{u}^j)$ is strictly quasi-concave in $\hat{u}^i$ and has a unique maximum, which is characterized by the first-order condition, for any given value of $\hat{u}^j$.

Total differentiation of the first-order condition with respect to $\underline{u}^i$ reveals that

$$\frac{d\hat{u}^i}{d\underline{u}^i} \equiv \frac{p_{11}^i[p^j(\underline{u}^i - \psi(\hat{u}^j)) + (1-p^j)(\hat{u}^i - \tilde{u}^i)] - p_1^i[2p^j - 1 - (1-p^j)\tilde{u}_1^i]}{2p_1^i(1-p^j) + p_{11}^i[p^j(\underline{u}^i - \psi(\hat{u}^j)) + (1-p^j)(\hat{u}^i - \tilde{u}^i)]}$$
$$< 1, \tag{5.2}$$

where the inequality follows, after a simple computation, from the facts that $\tilde{u}_1^i \leqslant 1$ and the denominator, which is $v_{11}^i(\hat{u}^i, \hat{u}^j)$, is strictly negative. Similarly, total differentiation of the first-order condition with respect to $\underline{u}^j$ yields

$$\frac{d\hat{u}^i}{d\underline{u}^j} \equiv \frac{p_1^i[p^j\psi_1 + (1-p^j)\tilde{u}_2^i]}{2p_1^i(1-p^j) + p_{11}^i[p^j(\underline{u}^i - \psi(\hat{u}^j)) + (1-p^j)(\hat{u}^i - \tilde{u}^i)]} < 0, \tag{5.3}$$

where the inequality follows immediately from my assumptions and the fact that the denominator is, again, negative by the second-order condition. These results, in conjunction with (5.1), imply that when bargainers' strategy choices do not interact, the conventional wisdom relating the probability of impasse to the size of the contract zone is confirmed in the $p^i(\hat{u}^i - \underline{u}^i)$ case. Of course, these are extremely stringent requirements.

In deriving (5.3), I made use of the assumption that, from bargainer i's standpoint, $d\hat{u}^j/d\underline{u}^j \equiv 1$. This assumption follows from the assumptions in the

$p^i(\hat{u}^i - \underline{u}^i)$ case, given that bargainer i knows the form of the function p^j and really believes that $p^j(\cdot)$ will remain constant when $\underline{u}^j$ changes. But it can be argued that this is too sophisticated to be really plausible. If one instead adopts the position that bargainer i, in addition to expecting $p^j(\cdot)$ to remain constant, expects no change in $\hat{u}^j$ when $\underline{u}^j$ changes, the $p^j\psi_1$ term in the numerator of the right-hand side of (5.3) disappears, altering the magnitude, but not the sign, of $d\hat{u}^i/d\underline{u}^j$.

To make things more concrete, one would like to supplement the formal ambiguity referred to above with a simple example in which bargainers' strategy choices interact and shrinking the contract zone actually decreases the probability of an impasse. Unfortunately, I have been unable to find an analytically tractable example with this property in the $p^i(\hat{u}^i - \underline{u}^i)$ case, which I consider the most interesting; interactions between strategy choices, while necessary for the perverse result in this case, also seem to prevent explicit solutions. The following simple example, in which $p^i(\cdot)$ depends only on $\hat{u}^i$ and $p^j(\cdot)$ is constant, may help to dispel some of the reader's dissatisfaction with this state of affairs and illustrate why the perverse result can occur.

Suppose that the relevant part of the utility-possibility set lies in the non-negative quadrant and that its frontier is given by $u^i \equiv \psi(u^j) \equiv 1 - u^j$. Let $p^i(\hat{u}^i, \underline{u}^i) \equiv 1 - \hat{u}^i$ and $p^j(\hat{u}^j, \underline{u}^j) \equiv p^j$, a constant. By continuity, small amounts of dependence on $\underline{u}^i$ and $\underline{u}^j$ could be introduced without significantly altering the result. The compromise agreement that is reached if neither bargainer achieves commitment is taken to be the Nash (1950) cooperative solution (or equivalently, the Raiffa-Kalai-Smorodinsky (1975) solution, which is the same in this case), with $(\underline{u}^i, \underline{u}^j)$ as the threat point. Thus, $\tilde{u}^i \equiv (\underline{u}^i - \underline{u}^j + 1)/2$ and $\tilde{u}^j \equiv (\underline{u}^j - \underline{u}^i + 1)/2$. (One might argue as in footnote 10 that the strategic possibilities, rather than the disagreement outcome alone, ought to determine the $\tilde{u}^i$. But since the utilities bargainers can guarantee themselves, by attempting or not attempting commitment, are influenced by the $\tilde{u}^i$, this would involve a recursive definition of the $\tilde{u}^i$, adding much complexity but little insight.)

Easy calculations show that condition (4.2) guarantees that bargainer i will always attempt commitment at a Nash equilibrium provided that $p^j < 1/2$; similarly, bargainer j will always attempt commitment when $\underline{u}^j < \underline{u}^i$.[11] When these conditions are satisfied, it is clear that $\hat{u}^j = 1 - \underline{u}^i$ is bargainer j's

11 Readers of footnote 10 may be curious about whether $\tilde{u}^i$ and $\tilde{u}^j$ in this example are in fact at least as good for bargainers i and j as what they could guarantee themselves by attempting commitment. When a bargainer attempts commitment, the worst case is when the other bargainer also does. It is not always true that $\tilde{u}^i$ and $\tilde{u}^j$ are better for i and j than these worst cases, but it can be shown to be true near $(\underline{u}^i, \underline{u}^j) = (0, 0)$, for example, whenever, roughly, $1/9 < p^j < 1/3$.

optimal commitment strategy. Given this and the above assumptions, bargainer i's expected utility when he attempts commitment to $\hat{u}^i$ is given by

$$v^i(\hat{u}^i, 1-\underline{u}^i) \equiv p^j\underline{u}^i + (1-\hat{u}^i)(1-p^j)\hat{u}^i + \hat{u}^i(1-p^j)(\underline{u}^i - \underline{u}^j + 1)/2. \tag{5.4}$$

Maximizing the expression on the right-hand side of (5.4) with respect to $\hat{u}^i$ is clearly equivalent to maximizing $(1-\hat{u}^i)\hat{u}^i + \hat{u}^i(\underline{u}^i - \underline{u}^j + 1)/2$. The second-order sufficient condition is satisfied everywhere, so solving the first-order condition yields the unique optimal strategy:

$$\hat{u}^i = (\underline{u}^i - \underline{u}^j + 3)/4. \tag{5.5}$$

Now we are ready to examine the effects on the probability of impasse of changing $\underline{u}^i$ and $\underline{u}^j$. In Nash equilibrium, this probability is given by

$$p^i(\hat{u}^i, \underline{u}^i)\, p^j \equiv p^j(1-\underline{u}^i + \underline{u}^j)/4. \tag{5.6}$$

Thus, $dp^ip^j/d\underline{u}^i \equiv -p^j/4 < 0$, so increasing $\underline{u}^i$, which shrinks the contract zone, actually makes an impasse *less* likely. On the other hand, $dp^ip^j/d\underline{u}^j \equiv p^j/4 > 0$, so shrinking the contract zone by increasing $\underline{u}^j$ makes impasse more likely. (If $\underline{u}^i$ is near zero, but not exactly zero, it is possible to raise $\underline{u}^j$ while continuing to satisfy the $\underline{u}^j < \underline{u}^i$ constraint.) Thus, the conventional wisdom is valid in this example when $\underline{u}^j$ changes, but not when $\underline{u}^i$ changes. These results follow simply from the fact that changing the disagreement outcome also changes the relative costs and benefits of attempting commitment to different positions, possibly, as the example shows, in a way that induces bargainers to take more extreme positions when the contract zone shrinks, making impasse less likely.

To summarize the results obtained so far, it is impossible to establish on a priori grounds that the conventional wisdom about the effect of changing the size of the contract zone on the probability of impasse is always valid. But there is some weak theoretical evidence to support it in the leading special case where $p^i(\hat{u}^i, \underline{u}^i)$ can be written as $p^i(\hat{u}^i - \underline{u}^i)$. The expected relationship holds in this case when interactions between bargainers' strategy choices are relatively unimportant.

Certain provisions in the customs and laws governing bargaining behaviour - for example, the common requirement in labour law to bargain 'in good faith' - can be viewed as attempts to make commitment more difficult, presumably with the purpose of reducing the probability of impasse. This section concludes by asking whether such attempts that stop short of making commitment completely impossible, viewed in the present framework, appear likely to succeed.

More formally, rewrite the function p^i as $p^i(\hat{u}^i, \underline{u}^i, \alpha)$, where α is a shift parameter that shifts only $p^i(\cdot)$; without essential loss of generality, it is assumed that $p^i_3(\cdot) > 0$ everywhere. Lowering α makes commitment more difficult, in the sense of lowering the probability of successful commitment to any position. Now,

$$\frac{dp^i p^j}{d\alpha} \equiv p^i p_1^j \frac{d\hat{u}^j}{d\alpha} + p^j \left[p_1^i \frac{d\hat{u}^i}{d\alpha} + p_3^i \right], \tag{5.7}$$

where $\hat{u}^i$ and $\hat{u}^j$ denote the equilibrium attempted commitments for the given value of α. If $dp^i p^j / d\alpha > 0$, making commitment more difficult by lowering α lowers the probability of impasse, which is the conventional wisdom. Given that $p_1^j(\cdot)$ and $p_1^i(\cdot)$ are everywhere negative, this is *guaranteed* in general only if $d\hat{u}^i/d\alpha < 0$ and $d\hat{u}^j/d\alpha < 0$.

As with the above analysis of the effects of changes in the costs of disagreement on the $\hat{u}^i$, the comparative statics of Nash equilibria when α changes are complex and unilluminating, leading to formal ambiguities very similar to those encountered above. But as before, studying the response when strategy choices do not interact gives a feel for the likely effects of changing α, and can be viewed as an approximate analysis of the response of Nash equilibria when interactions between bargainers' strategy choices are not expected to be important.

Straightforward computations verify that in this case,

$$\frac{d\hat{u}^i}{d\alpha} \equiv -\frac{p_{13}^i[p^j(\underline{u}^i - \psi(\hat{u}^j)) + (1-p^j)(\hat{u}^i - \tilde{u}^i)] + p_3^i(1-p^j)}{2p_1^i(1-p^j) + p_{11}^i[p^j(\underline{u}^i - \psi(\hat{u}^j)) + (1-p^j)(\hat{u}^i - \tilde{u}^i)]}. \tag{5.8}$$

The term in brackets in the numerator of the right-hand side of (5.8) must be strictly positive by the first-order condition; the denominator must be negative by the second-order condition. Because $p_3^i(\cdot) > 0$ by assumption, it is plain that $d\hat{u}^i/d\alpha$ is strictly positive if $p_{13}^i(\cdot) > 0$ and ambiguous in sign otherwise. (A similar computation shows that $d\hat{u}^j/d\alpha$ is always ambiguous in sign.) In any event, $d\hat{u}^i/d\alpha$ can never be unambiguously negative, so that it is certainly not possible to validate the conventional wisdom theoretically in this case. The common reasoning errs in considering only the partial effect of changing α, which is obviously in the 'right' direction. But the total effect, taking into account the responses of $\hat{u}^i$ and $\hat{u}^j$ (which tend to go in the 'wrong' direction), may easily be counterintuitive.

6 CONCLUSION

This chapter develops a theory that explains why rational bargainers might take actions that could lead to an impasse.[12] By formalizing Schelling's (1956,

12 It might be argued that the rationalization of the occurrence of bargaining impasses presented here is incomplete because my assumption that bargainers can communicate only by their demands is unduly restrictive. When outcomes are always efficient, bargaining

1963, 1966) view of the bargaining process as a struggle between bargainers to commit themselves to favourable bargaining positions, the theory determines the probability of impasse endogenously and permits an investigation of its relationship to the bargaining environment. Its crucial elements, which allow an explanation of the occurrence of inefficient outcomes without assuming irrationality, are the uncertainty and irreversibility of commitment. To the extent that commitment is part of the explanation for the impasses frequently observed in real bargaining, doubt is cast on two widely held beliefs about the influence of the bargaining environment on the frequency of impasse. Contrary to what is assumed almost universally in those parts of the law and economics and industrial relations literature that consider the question, increasing the costs of disagreement and thereby enlarging the contract zone need not decrease the probability of impasse. Malouf and Roth (1981) report experimental evidence that casts further doubt on this assumption. Neither is there support for the common belief that making commitment more difficult (in the sense of lowering the probability of success) makes impasse less likely. The reason is that the 'commonsense' argument used to reach this conclusion considers only the partial effect of lowering the success probability; it ignores the resulting changes in attempted commitment positions, whose effects generally go the other way and could easily swamp the partial effect.

Two lines of research appear especially promising at this point. As mentioned above, Chatterjee and Samuelson (1981) have developed an alternative explanation of the occurrence of impasses, based on a stylized complete irreversibility of demands (certainty about commitment, in the language of this chapter) and on bargainers' uncertainty about each others' preferences. Since in their model, bargainers generally find it advantageous to demand more than their disagreement utilities, the possibility of impasse due to uncertainty about preferences is present. As it happens, their model also casts doubt on the conventional wisdom about the effects of expanding the

---

is essentially a zero-sum game, and direct communication cannot play a significant role. But when impasses are possible, bargainers have a mutual interest in avoiding them, and it is no longer clear that communication would not occur in equilibrium if it were allowed. (In the present model, bargainers might wish to communicate about their costs of backing down between the first and second stages.) Joel Sobel and I (1982) have considered the general question of strategic communication in an abstract setting, obtaining results that seem to shed some light on whether the above criticism is valid. In the paper, it is shown under reasonable assumptions (which have not, however, been verified for the present model) that perfect communication is not compatible with non-cooperative rationality unless agents' interests completely coincide, and that once their interests differ by a given, 'finite' amount, only no communication is compatible with rationality. Thus, it seems likely that direct communication would not alter the qualitative results obtained above.

contract zone. In another recent paper, Sobel and Takahashi (1980) have combined uncertainty about preferences with uncertainty about the number of stages which shares some of the effects of uncertainty about commitment, as it is conceptualized here. They obtain interesting results about how the frequency of impasse depends on the bargaining environment. Natural next steps in this direction would be to build models with endogenous numbers of stages, and to deal with the issues of strategic information transmission that are avoided in Chatterjee and Samuelson (1981), Sobel and Takahashi (1980) and the present chapter.

The second line of research involves a deeper investigation of the process by which commitment is achieved. In this chapter, bargainers communicate only by making demands and making it costly to back down from them. And the actual process of commitment is not modelled explicitly; rather, its influence on bargainers' strategy choices is summarized by the probability distributions of their costs of backing down. To explain the occurrence of impasses and to evaluate the effects of changes in the bargaining environment on their likelihood, it suffices to place certain mild, plausible restrictions on these distributions. The situation is somewhat similar to that encountered in consumer and producer theory, where many of the conclusions do not depend on detailed knowledge of technology and tastes. But to make sharper predictions, or to test the theory empirically, it would be highly desirable to know more about the nature of communication and commitment in bargaining. This would require a detailed model of the process by which bargainers back up their demands; Schelling (1956, 1963, appendix B) and Ellsberg (1975) contain fascinating discussions of the issues that such a model would have to resolve.

REFERENCES

Ashenfelter, O. and Johnson, G. 1969: Bargaining theory, trade unions, and industrial strike activity. *American Economic Review*, **59**, 35–49.

Chatterjee, K. and Samuelson, W. 1981: The simple economics of bargaining. Boston University School of Management Working Paper 2/81.

Crawford, V. P. and Sobel, J. 1982: Strategic information transmission. *Econometrica*, **50**, 1431–52.

Cross, J. 1969: *The Economics of Bargaining*. Basic Books, New York.

Dasgupta, P. and Maskin, E. 1977: The existence of economic equilibrium: continuity and mixed strategies. Stanford IMSSS Technical Report 252.

Ellsberg, D. 1975: The theory and practice of blackmail. In Young, O. R. (ed.) *Bargaining: Formal Theories of Negotiation*. University of Illinois Press, Urbana, Illinois.

Harsanyi, J. C. 1961: On the rationality postulates underlying the theory of cooperative games. *Journal of Conflict Resolution*, **5**, 179–96.

Harsanyi, J. C. 1968: Games with incomplete information played by 'Bayesian' players. II. Bayesian equilibrium points. *Management Science*, **14**, 320–34.

Harsanyi, J. C. 1977: *Rational Behavior and Bargaining Equilibrium in Games and Social Situations*. Cambridge University Press, London and New York.

Harsanyi, J. C. and Selten, R. 1972: A generalized Nash solution for two-person bargaining games with incomplete information. *Management Science*, **18**, P80–P106.

Kalai, E. and Smorodinsky, M. 1975: Other solutions to Nash's bargaining problem. *Econometrica*, **43**, 513–18.

Kreps, D. M. and Wilson, R. 1982: Sequential equilibria. *Econometrica*.

Malouf, M. and Roth, A. 1981: Disagreement in bargaining: an experimental study. *Journal of Conflict Resolution*, **25**, 329–48.

Nash, J. 1950: The bargaining problem. *Econometrica*, **18**, 155–62.

Nash, J. 1953: Two-person cooperative games. *Econometrica*, **21**, 128–40.

Rosenthal, R. and Landau, H. 1979: A game-theoretic analysis of bargaining with reputations. *Journal of Mathematical Psychology*, **20**, 1353–66.

Roth, A. E. 1979: *Axiomatic Models of Bargaining*. Springer-Verlag, Berlin and New York.

Schelling, T. C. 1956: An essay on bargaining. *American Economic Review*, **46**, 281–306; reprinted as chapter 2 of Schelling (1963).

Schelling, T. C. 1963: *The Strategy of Conflict*. Oxford University Press, London and New York.

Schelling, T. C. 1966: *Arms and Influence*. Yale University Press, New Haven and London.

Selten, R. 1975: Reexamination of the perfectness concept for equilibrium points in extensive games. *International Journal of Game Theory*, **4**, 25–55.

Sobel, J. and Takahashi, I. 1980: A multi-stage model of bargaining. University of California, San Diego, Discussion Paper 80–25.

8

Nash Bargaining and Incomplete Information

K. Binmore

In three remarkable papers written in the early 1950s, Nash outlined an approach to the theory of games and to the theory of bargaining in particular which remains of the greatest significance. Harsanyi and Selten have extended this work in various directions including that of bargaining under incomplete information. In this chapter, we elaborate on Nash's basic bargaining model and examine Harsanyi and Selten's bargaining solution in the light of the results obtained. The chapter continues the pre-occupations of chapters 2, 4 and 11 in that it is concerned with informational questions in a strictly game-theoretic[1] context but is independent of these other chapters.

1 INTRODUCTION

Nash complemented his axiomatic analysis (Nash 1950) by also describing a simple bargaining model in which the players exchange unilateral utility demands (see chapter 4). In this chapter we begin by elaborating at length on this model with the object of indicating that the model can be applied to a fairly wide range of bargaining situations. The analysis is then used as a tool in the main part of the chapter which is concerned with two-person bargaining problems in which neither negotiator is fully informed about the preferences of the other. Nash explicitly excluded such problems from his discussion (1950) and suggested that they be classified as 'haggling problems' rather than 'bargaining problems'. Nowadays, however, in the light of Harsanyi and Selten's theory of 'games of incomplete information', it is more natural to see such problems as multi-person games in which two players who actually negotiate are chosen by preliminary random moves from given populations according to given probability distributions. The fundamental paper on this subject is by Harsanyi and Selten (1972). A brief summary of the theory they develop is given below.

1 See section 2 of chapter 2 for a brief account of what is meant by saying that we adopt a 'strictly game-theoretic' viewpoint.

We begin with some notation. Suppose that the two negotiators are labelled with the Roman numerals I and II. We then require two sets, I and J, of players.[2] The players in set I may be thought of as type I players and those in set J as type II players. A random move which is unobserved by type II players selects with probability λ_i one of the players i in the set I to fill the role of negotiator I. An independent[3] random move which is unobserved by type I players selects with probability μ_j one of the players j in the set J to fill the role of negotiator II. Each player is assumed to select a negotiation strategy which he will employ if chosen to negotiate. We then have to consider a multi-person (see footnote 2) game (of complete information) which has to be analysed as a contest[4] (i.e. non-cooperatively).

Harsanyi and Selten emphasize that an important feature of the problem is that the negotiators may learn something about the identity of their opponent as a result of their observation of the negotiation strategy he has chosen. They then introduce a specific multi-stage bargaining procedure which, although rather special in some respects, nevertheless captures much of one's intuition about the relevant strategic features of such procedures. In particular, it allows for the possibility that the players may learn about each other as the negotiations proceed. Harsanyi and Selten then note that a bargaining game based on this procedure will normally have multiple equilibria. This assertion is clarified by Selten (1974). In this paper, Selten analyses a simple numerical example in detail. In his example, there are only two type I players (a strong player and a weak player) and two type II players (also strong and weak). Selten identifies two qualitatively distinct equilibria. The first and simpler of these is a 'non-revealing equilibrium'. This is an equilibrium in which all players of the same type employ the same negotiation strategy. The second of these is a '*revealing equilibrium*' in which the strong and weak players choose different negotiation strategies and hence identify themselves during the course of the negotiations. In the much more general framework of the original Harsanyi--Selten paper (1972) there will also be other more complicated equilibria.

The problem of multiple equilibria in bargaining theory is a familiar one and Harsanyi and Selten draw attention to the multiplicity of equilibria in

2 Harsanyi would say that I and II are 'players' in a 'game of incomplete information'. We prefer to use the word game in its traditional sense (which subsumes complete information) and to reserve the word player for a participant in such a game. One can then refer to I and II as roles to be filled eventually by players. Pursuing the theoretical analogy, one might perhaps refer to a 'game of incomplete information' as a script.

3 Such an assumption is not in general necessary.

4 We have used the word contest in chapters 2, 4 and 11 to indicate a game in which the players choose their strategies without the opportunity of any pre-play communication. A bargaining contest is one in which the negotiation strategies chosen by the players can be classified either as compatible (in which case a contract is signed) or else as incompatible (in which case a pre-determined outcome called the status quo results).

Nash's simple demand game (see section 3 of chapter 4). They then observe that Nash's axiom scheme (see section 4 of chapter 2) may be regarded as a set of criteria for the selection of one of these equilibria as the solution of the game.[5] Next they assert their intention of adopting a similiar modus operandi in respect of 'bargaining games of incomplete information'. Having noted that such games will in general admit of a multiplicity of equilibria,[6] they then propose an axiom scheme based on that of Nash which they regard as providing a set of criteria for determining which of the available equilibria is to be regarded as the solution of the game.

Harsanyi and Selten next show that the use of their axiom scheme reduces to finding the vector $(\mathbf{x}, \mathbf{y})$ at which the product

$$\prod_{i \in I} (x_i - \xi_i)^{\lambda_i} \prod_{j \in J} (y_j - \eta_j)^{\mu_j} \tag{8.1}$$

is maximized subject to the constraint that $(\mathbf{x}, \mathbf{y}) \in X$. Some explanation of the notation is necessary. The set X consists of the convex hull of the collection of all equilibrium payoff vectors $(\mathbf{x}, \mathbf{y})$. An equilibrium payoff vector $(\mathbf{x}, \mathbf{y})$ summarizes for a given equilibrium the payoff x_i that a player $i \in I$ will receive if chosen as negotiator I and the payoff y_j that a player $j \in J$ will receive if chosen as negotiator II. The numbers ξ_i and η_j represent the status quo payoffs for players i and j respectively (i.e. what they will receive if they are not chosen as negotiators or, if chosen, fail to reach agreement). If $(\mathbf{x}, \mathbf{y})$ is the vector which maximizes the product (8.1), the players are assumed to select an equilibrium with the help of a jointly observable random event which assigns probabilities to the equilibria in such a way that $(\mathbf{x}, \mathbf{y})$ is the expected payoff vector. (An agreement to choose the solution of the game in this way will be 'self-policing' in that no player will wish to deviate from the agreed solution strategies because only equilibria are considered.)

One may summarize Harsanyi and Selten's conclusions by saying that they identify the solution outcome of a two-person bargaining game of incomplete information with an asymmetric Nash bargaining solution (see chapter 2 and Roth 1980) of a multi-person bargaining game of complete information. The word 'asymmetric' here means that the usual Nash product is replaced by a product of the form (8.1) in which the numbers λ_i and μ_j may be interpreted as the 'bargaining powers' for players i and j respectively.[7]

5 Such an interpretation of Nash's work is consistent with the view expressed in chapter 2 that Nash's axiomatic scheme is best viewed as implicitly embodying a convention to be used consistently in selecting a solution from the available equilibria. But it seems rather doubtful that Nash regarded his axiomatic system as relevant *only* to his simple demand game.

6 Here and throughout, our discussion of the Harsanyi–Selten paper fudges the question of what the authors do and do not regard as admissible equilibria.

7 Harsanyi uses the term 'bargaining power' in a different sense (1977).

Our aim in this chapter is to establish a result of the same type as that summarized in the last paragraph. We shall not however seek to justify our conclusions by reference to a system of axioms. Instead, we shall look at some very primitive bargaining models based on Nash's simple demand game and deal with the problem of multiple equilibria (in so far as this is possible) by using Nash's 'smoothing technique'. Although this smoothing technique allows one to eliminate large numbers of equilibria, it does not, unfortunately, leave a unique candidate for the solution equilibrium in the case of games of incomplete information (except under special circumstances – e.g. if only one type I player exists). As a result, we are only able to identify the solution outcome of a game of incomplete information of the type we consider with the asymmetric Nash bargaining solution obtained by maximizing

$$\prod_{i \in I} (x_i - \xi_i)^{\lambda_i} \prod_{j \in J} (y_j - \eta_j)^{\mu_j}$$

over an appropriate set in those cases for which it is possible to assert that the solution equilibrium of the game of incomplete information is non-revealing.

For appropriate games whose solution equilibrium is non-revealing, the analysis offered in this chapter is supportive of the Harsanyi–Selten theory in that it leads to the same result as they obtain but via a quite different argument. On the other hand, if there exist games of the type we consider which have a revealing (or partially revealing) equilibrium solution, then our analysis would seem to weigh against the argument in part III of the Harsanyi–Selten paper.[8] It should be noted however that the smoothing technique restricts the revealing equilibria as well as the non-revealing equilibria and there may well be reasons why the remaining revealing equilibria can be eliminated as candidates for the solution of the game given the presence of the remaining non-revealing equilibrium (for example, if the latter Pareto-dominates each of the former). Possibly such reasons can be given in general but it is enough for the purposes of this chapter to remark that such reasons can sometimes be given.

A more precise statement of the results obtained in this chapter is left to section 5 et seq. At this stage we shall only continue the remarks of the preceding paragraph with the observation that the class of bargaining problems to which our analysis is relevant is not without interest. I am told, for example, that the economics faculty at the Massachusetts Institute of Technology (MIT) makes decisions over new appointments at meetings which proceed continuously and indefinitely until a consensus is reached. Assuming

8 It would seem that generalized Nash bargaining solutions remain significant but that 'bargaining powers' other than the simple λ_i and μ_j are necessary. (See note 6 of section 6.)

that resignation threats and the like are disbarred, such a situation is not unsimilar to that described in section 9 and would therefore lead to a generalized Nash bargaining solution for wide ranges of the parameters assuming the faculty members negotiate rationally.

2 THE SMOOTHED NASH DEMAND GAME

In this and the next sections, we shall be concerned with some simple two-person bargaining games with complete information. The techniques described will be applied to games of incomplete information in section 5 and later sections.

In the simple Nash demand game, the two players simultaneously announce demands x and y. If $(x, y) \in \mathcal{X}$, where $\mathcal{X}$ is the payoff region (i.e. the set of feasible payoff vectors), then each player receives his demand. If $(x, y) \notin \mathcal{X}$, then each player receives zero payoff (i.e. the status quo is normalized at 0). We shall assume that $\mathcal{X}$ has the properties indicated in figure 8.1.[9] Then each point (x, y) on the arc AB corresponds to a Nash equilibrium of the simple demand game.

The fact that the simple demand game has many equilibria led Nash to introduce a 'smoothed' approximation to the game in which the problem of multiple equilibria vanishes. In the smoothed version of the game, the two

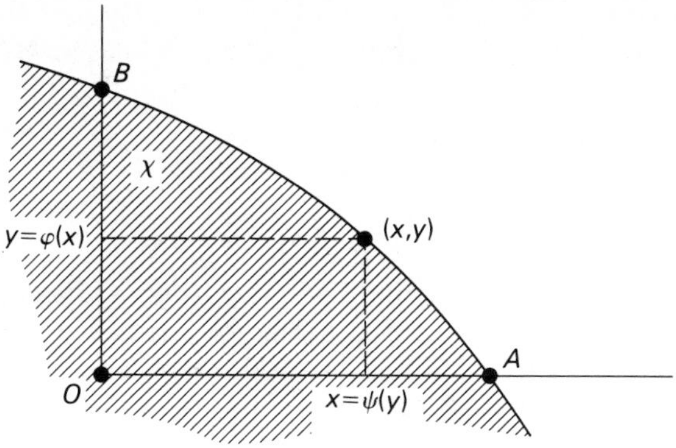

Figure 8.1

9 We assume that $\mathcal{X}$ is closed and convex, that $\mathbf{0} \in \mathcal{X}$ and that the intersection of $\mathcal{X}$ with the non-negative orthant is bounded. We also assume that $\mathbf{x} \in \mathcal{X}$ and $\mathbf{y} \leqslant \mathbf{x}$ implies that $\mathbf{y} = \mathbf{x}$ or else $\mathbf{y} \in \mathcal{X}$. Occasionally we restrict attention to the case when the functions ϕ and ψ which determine the boundary of $\mathcal{X}$ are differentiable.

players again simultaneously announce demands x and y but now a chance move intervenes as a result of which the players receive their demands with probability $p(x, y)$ or else their status quo payoffs (i.e. 0) with probability $1-p(x, y)$. The smoothed version may be regarded as an approximation to the original simple demand game if p approximates the function

$$\chi(x, y)=\begin{cases}1, (x, y) \in \mathscr{X} \\ 0, (x, y) \notin \mathscr{X}.\end{cases}$$

The smoothed game involves a certain amount of shared uncertainty on the part of the players. At the end of this section we mention some of the circumstances which might generate such uncertainties. For the moment, however, we shall simply stress that this is a game in which both players have the *same* information and concentrate on establishing that the problem of multiple equilibria does indeed disappear provided appropriate assumptions are made about the function $p: \mathbb{R}^2 \rightarrow [0, 1]$.

We shall consider in detail the easy case when p is differentiable, quasi-concave and strictly decreasing.[10] In note 2 we comment on the extent to which these conditions may be relaxed. Observe in particular that some smoothness condition on p is essential.

The first player seeks to choose x so as to maximize $xp(x, y)$ while the second player seeks to choose y so as to maximize $yp(x, y)$. If (x_0, y_0) is a Nash equilibrium, it therefore follows that

$$\left.\begin{aligned} x_0 p_x(x_0, y_0) + p(x_0, y_0) = 0 \\ y_0 p_y(x_0, y_0) + p(x_0, y_0) = 0 \end{aligned}\right\}$$

and so the tangent line to the contour $p(x, y) = p(x_0, y_0)$ at the point (x_0, y_0) is

$$\frac{x}{2x_0} + \frac{y}{2y_0} = 1$$

from which we deduce that (x_0, y_0) is the Nash bargaining solution for the bargaining game with payoff region $\mathscr{Y} = \{(x, y): p(x, y) \geqslant p(x_0, y_0)\}$ and status quo 0. Note that $\mathscr{Y}$ is convex because p is assumed quasi-concave (see figure 8.2).

We next introduce the requirement that the smooth game approximate the original simple demand game. The natural way to proceed is to suppose that the set where $p(x, y) \geqslant 1-\epsilon$ approximates $\mathscr{X}$ in an appropriate topology and that the set $\{(x, y): p(x, y) \leqslant \epsilon\}$ approximates the complement of $\mathscr{X}$. We may then conclude that $\mathscr{Y}$ approximates $\mathscr{X}$ provided that ϵ is suffi-

10 i.e. $x_1 \leqslant x_2$ and $y_1 \leqslant y_2 \Rightarrow p(x_1, y_1) > p(x_2, y_2)$ or $(x_1, y_1) = (x_2, y_2)$.

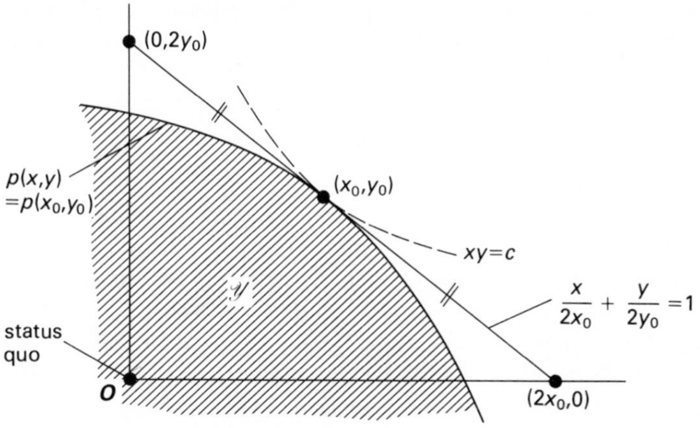

Figure 8.2

ciently small. But the Nash bargaining solution is continuous with respect to any natural topology. Thus (x_0, y_0) is necessarily an approximation to the Nash bargaining solution of the original simple demand game.

We therefore obtain the rather striking result that, given our hypotheses *every* Nash equilibrium of an approximating smoothed game is an approximation to the Nash bargaining solution of the original simple demand game. Or, to put the same thing another way, the solution of a bargaining game with a little shared uncertainty looks exactly like the Nash bargaining solution of a closely similar game without the shared uncertainty.

We return now to the question of the circumstances under which shared uncertainty of the type modelled in the smoothed game may occur. A detailed example in which the shared uncertainty arose over the time at which the players' demands were implemented was considered in section 4 of chapter 4. But many alternative examples suggest themselves. One might perhaps imagine the bereaved relatives of an intestate millionaire bargaining over the amount each is to receive before a precise valuation of the estate is available. The general idea is that the players do not know for sure what the payoff region is when they make their demands. Instead they regard this as a 'random variable' on whose probability distribution they are agreed. We expand upon this possibility in note 2. The essential point about such an interpretation is that one should think about the shared uncertainties concerning the payoff region as relating to small possible errors in the parameters which describe the bargaining situation. It may be, for example, that the extreme points of the payoff region $\mathscr{X}$ correspond to physical objects or events whose characteristics are known only to a limited accuracy. The collection of extreme points of $\mathcal{X}$ would then be a random vector and, to compute $p(x, y)$, it would first be necessary to identify those configurations of extreme points for which (x, y) lies in their convex hull and then to calculate the probability that one or other of these configurations is realized. Where the shared uncertainty is of a more fundamental nature, one cannot expect the results of this section to hold (see note 3).

Note 1. All of the above reasoning applies equally well with n players instead of two players provided that pre-play interaction is disbarred so that no question of coalition formation arises. The position over the formation of coalitions is rather peculiar in that the situation generates a version of Harsanyi's bargaining paradox (p. 203 of Harsanyi 1977). Suppose that the first and second of three players agree to coordinate their demands x and y and to redistribute their payoffs so that each receives $\frac{1}{2}(x+y)$ on those occasions when the status quo is avoided. (The assumption that such a redistribution is possible is, of course, quite a strong one.) The coalition of the first and second player then seeks to maximize $\frac{1}{2}(x+y)p(x, y, z)$ while the third player seeks to maximize $zp(x, y, z)$. If (x_0, y_0, z_0) is a Nash equilibrium of this essentially two-person game, we obtain that

$$\frac{x}{2(x_0+y_0)}+\frac{y}{2(x_0+y_0)}+\frac{z}{2z_0}=1$$

is the tangent plane to the contour $p(x, y, z)=p(x_0, y_0, z_0)$ at the point (x_0, y_0, z_0). In the case when the basic issue is to 'divide the dollar', it then follows that the first and second players will each receive approximately $\frac{1}{4}$ while the third player receives approximately $\frac{1}{2}$. If the coalition of the first and second player had not formed each would have received approximately $\frac{1}{3}$.

Note 2. Figure 8.3 makes it clear that with *no* conditions on the function $p: \mathbb{R}^2 \to [0, 1]$, *every* point on the Pareto boundary of $\mathcal{X}$ may appear as an approximation to a Nash equilibrium of an approximating smoothed game. (See also note 3.)

Nash's comments on this question are typically laconic but he remarks that the problem is not a serious one since the Nash bargaining solution is the only point of $\mathcal{X}$ which is *necessarily* approximated by Nash equilibria of approximating smoothed games. It is not clear, however, why Nash felt this fact to be significant. In any case, it seems to have little relevance to the point of view adopted in this chapter. For our interpretation it is important that *all* Nash equilibria of an approximating smoothed game approximate the Nash bargaining solution of the original game. We give here a set of sufficient conditions for this to be true. These are more general than those of section 2 and perhaps more natural.

We suppose that a random move, unobserved by both players, selects the actual payoff region from the collection of admissible regions (see foot-

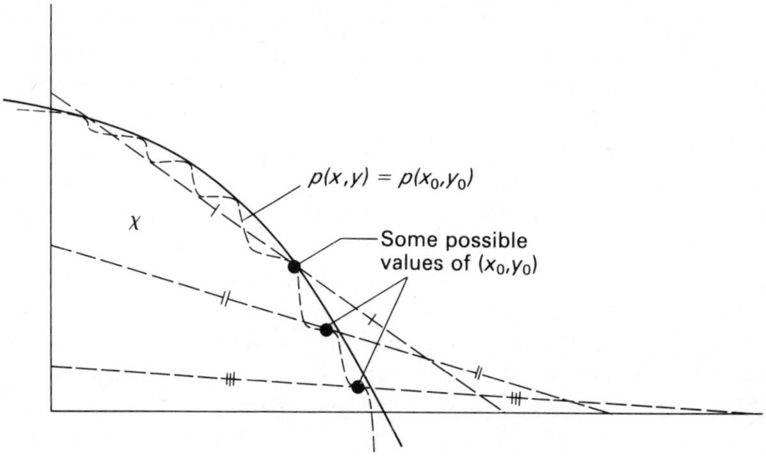

Figure 8.3

note 9). We are interested in payoff regions $\mathscr{Z}$ which approximate the original payoff region $\mathscr{X}$ and so we assume that the probability that the Hausdorff distance between $\mathscr{Z}$ and $\mathscr{X}$ exceeds $\epsilon > 0$ is at most ϵ. This guarantees that p is 'nearly' quasi-concave. Next we require a condition that makes p 'nearly' differentiable. For this purpose we consider the measure μ which has the property that

$$\iint_S d\mu(x, y)$$

is the probability that the Pareto boundary of $\mathscr{Z}$ passes through the set S. The natural assumption is that μ has a density function f which is continuous at every point. However, we prefer the weaker assumption that, for each $\boldsymbol{\xi}$, $f(\mathbf{x})$ tends to a limit as $\mathbf{x} \to \boldsymbol{\xi}$ through the closed orthant northwest of $\boldsymbol{\xi}$ and $f(\mathbf{x})$ also tends to a (possibly different) limit as $\mathbf{x} \to \boldsymbol{\xi}$ through the closed orthant southeast of $\boldsymbol{\xi}$. This weaker assumption allows us to encompass situations of the type considered in section 4 of chapter 4 in which p is not differentiable but is 'nearly' so. As explained in note 1, section 4 of chapter 4, some such requirement on p is essential.

Note 3. It is natural to ask what happens when the players share some uncertainty over what game they are playing in the case when the uncertainty does not only reflect the possible existence of small parametric errors but also contains a significant component reflecting structural uncertainties. Suppose, for example, that $p_1 : \mathbb{R}^2 \to [0, 1]$ and $p_2 : \mathbb{R}^2 \to [0, 1]$ are quasi-concave, differentiable, decreasing functions which determine smoothed demand games approximating the simple demand games with payoff regions $\mathscr{X}_1$ and $\mathscr{X}_2$ respectively. Then the smoothed demand game with $p: \mathbb{R}^2 \to [0, 1]$ defined by $p = \pi p_1 + (1 - \pi)p_2$ can be thought of as embodying the structural uncertainty as to whether the basic game has payoff region $\mathscr{X}_1$ or $\mathscr{X}_2$ (see figure 8.4).

Nash equilibria for the smoothed game determined by p will be approximations to the points $\mathbf{s}_0$, $\mathbf{s}_1$ and $\mathbf{s}_2$ which are respectively the Nash bargaining solutions of the simple demand games based on $\mathscr{X}_1 \cap \mathscr{X}_2$, $\mathscr{X}_1$ and $\mathscr{X}_2$ respectively. The smoothed game determined by p may be regarded as an approximation to the simple demand game in which the players make their demands without knowing the payoff region $\mathscr{X}$ but jointly attribute probability π to the event that $\mathscr{X}$ is $\mathscr{X}_1$ and $1-\pi$ to the event that $\mathscr{X}$ is $\mathscr{X}_2$. This unsmoothed game has an infinite set of Nash equilibria but the analysis above invites us to focus attention on $\mathbf{s}_0$, $\mathbf{s}_1$ and $\mathbf{s}_2$. Note that the payoffs at these equilibria are actually $\mathbf{s}_0$, $\pi\mathbf{s}_1$ and $(1-\pi)\mathbf{s}_2$. Thus, for many configurations of $\mathscr{X}_1$ and $\mathscr{X}_2$, the Nash equilibrium $\mathbf{s}_0$ will Pareto-dominate both equilibria $\mathbf{s}_1$ and $\mathbf{s}_2$. But, for other configurations this will not be the case and so the problem of deciding which of the Nash equilibria is to be identified as the solution remains. Possibly Harsanyi and Selten would advocate the implementation of the Nash bargaining solution of the game whose payoff region is the convex hull X of the set $\{\mathbf{0}, \mathbf{s}_0, \pi\mathbf{s}_1, (1-\pi)\mathbf{s}_2\}$. The selection of the solution equilibrium would then depend on a jointly observed random event.

Note 4. In chapter 11, a two-person bartering contest is described whose outcome space is the classical Edgeworth box. The players exchange unilateral demands but these are more complicated than in the Nash demand game in

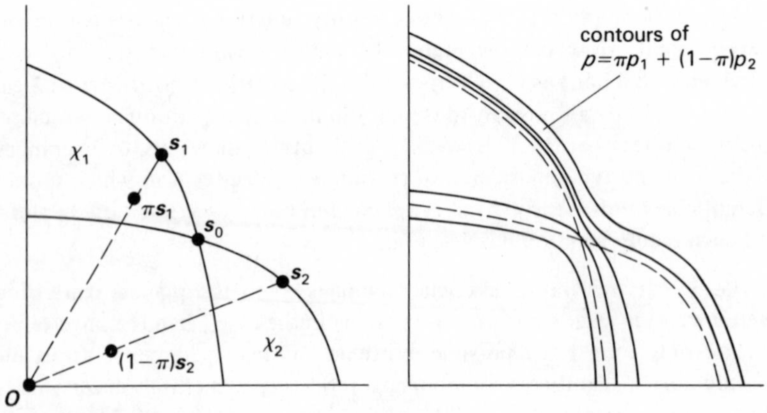

Figure 8.4

that a demand consists of a minimal price and a maximal quantity (of his own good) at which a player is willing to trade his good for that of his opponent. It is easily shown that the solution outcome is the competitive equilibrium (assuming this is unique and interior to the box). It should be noted that the solution payoffs are therefore *not* necessarily the Nash bargaining solution payoffs even when the payoff region $\mathscr{X}$ is entirely symmetric with respect to the two players. In chapter 11 this is attributed to the fact that the informational base for the bartering game differs from that envisaged in Nash's axioms in that communication takes place directly in terms of commodities to be exchanged and hence there is no good reason to suppose that the solution payoffs will depend *only* on $\mathscr{X}$ and the status quo $\boldsymbol{\xi}$. A direct comparison of the bartering game with Nash's simple demand game may help to clarify this point.

The set A in figure 8.5 represents the set of payoff pairs available for agreement in Nash's simple demand game when the first player demands utility a. The set B represents the set of payoff pairs available for agreement in the bartering game when the first player demands a minimal price of p. It is evidently the markedly different shapes of these two sets which leads to the different solution outcomes in the two different games. Note in particular that the relevant features of the set B are not easy to describe in terms of the payoff space exclusively.

The point of these remarks is to indicate the type of informational set-up for which the Nash demand game is an appropriate model. As an example, consider a game in which a set C of commodity bundles is available. The two players simultaneously announce demands for commodity bundles **c** and **d**. If $\mathbf{c}+\mathbf{d}\in C$, the players receive their demands. Otherwise the status

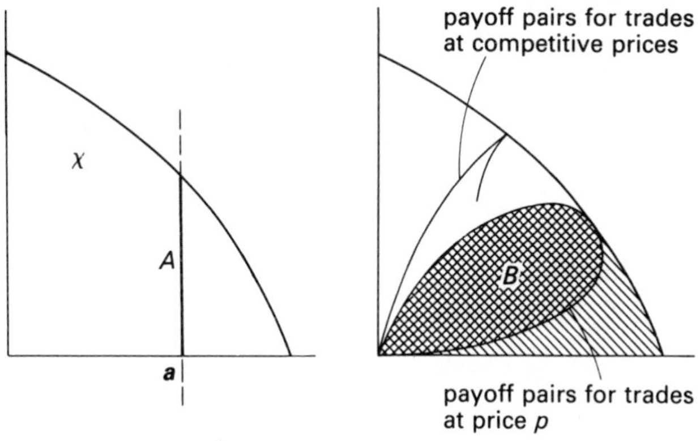

Figure 8.5

quo results. With natural assumptions on the set C and the players' preferences, a demand of **c** by the first player will restrict the payoff pairs available for agreement to a set of the form A. Hence such a game has the same strategic structure as Nash's demand game and the Nash bargaining solution is therefore an appropriate concept. Our bartering example on the other hand demonstrates that the Nash bargaining solution is *not* universally appropriate.

3 TREMBLING HAND DEMAND GAME

An alternative means by which some shared uncertainty may be introduced into Nash's simple demand game is by supposing that the players use a 'trembling hand' in choosing their demands from those available. More realistically, we may suppose that each player chooses a demand deterministically but that this is modified by a random error during transmission to the referee. Such an error may be due, for example, to rounding.

Formally, we assume that both players simultaneously announce demands x and y. An error term s with a continuous probability density function $f: \mathbb{R} \to [0, \infty)$ is then added to x and an independent error term t with a continuous probability density function $g: \mathbb{R}^2 \to [0, \infty)$ is added to y. Note that f and g are assumed not to depend on x or y. If $(x+s, y+t) \in \mathcal{X}$, then the first player receives $x+s$ and the second player receives $y+t$. If $(x+s, y+t) \notin \mathcal{X}$, then each player receives his status quo payoff (i.e. zero).

This game has a slightly different structure from that considered in section 2. It is not too surprising, however, that it should lead to similar results. We shall briefly sketch an appropriate argument under our usual hypotheses (see footnote 9) on $\mathcal{X}$ with the additional assumption that the boundary curve of $\mathcal{X}$ (given by $y = \phi(x)$ or $x = \psi(y)$) is differentiable.

We begin by writing

$$p(x,y) = \iint_{(x+s,\, y+t)\in\mathcal{X}} f(s)g(t)\,ds\,dt$$

$$= \int_{-\infty}^{\infty} f(s)ds \int_{-\infty}^{\phi(x+s)-y} g(t)dt.$$

Then $p(x,y)$ is the probability of agreement given that the first player chooses x and the second chooses y. The first player seeks to choose x so as to maximize

$$\iint_{(x+s,\, y+t)\in\mathcal{X}} (x+s)f(s)g(t)\,ds\,dt$$

$$= \int_{-\infty}^{\infty} (x+s)f(s)ds \int_{-\infty}^{\phi(x+s)-y} g(t)dt$$

Differentiating, we obtain that

$$0 = \int_{-\infty}^{\infty} f(s)ds \int_{-\infty}^{\phi(x+s)-y} g(t)dt$$

$$+ \int_{-\infty}^{\infty} (x+s)f(s)g(\phi(x+s)-y)\phi'(x+s)ds$$

$$= p + xp_x + \int_{-\infty}^{\infty} sf(s)g(\phi(x+s)-y)\phi'(x+s)ds$$

$$= p + xp_x + \int_{-\infty}^{\infty} (\psi(u)-x)f(\psi(u)-x)g(u-y)du.$$

If we now concentrate enough of the mass of the distributions represented by f and g at the origin, we can make the final term as small as we please. Thus, if (x_0, y_0) is a Nash equilibrium, then

$$\left.\begin{aligned} x_0 p_x(x_0, y_0) + p(x_0, y_0) \approx 0 \\ y_0 p_y(x_0, y_0) + p(x_0, y_0) \approx 0 \end{aligned}\right\}$$

One may then proceed as in section 2. In particular, if the function p is not pathological, then the Nash equilibria of an approximating trembling hand game approximate the Nash bargaining solution of the original simple demand game.

4 BARGAINING GAMES WITH ITERATED DEMANDS

It is argued in section 1 of chapter 4 that much of what takes place in real-life bargaining is irrelevant to a strictly game-theoretic analysis on the grounds that it consists of attempts by the bargainers to exploit irrationalities in their opponents. One should therefore expect a model which incorporates only the features of a situation which are relevant to a game-theoretic analysis to be very much simpler than would be necessary for a behavioural (i.e. positive) analysis. On the other hand, Nash's demand game has too simple a structure for it to be useful as a bargaining paradigm[11] without further evidence being

11 See chapter 11 for an example of a bargaining situation for which the Nash demand game is *not* an appropriate paradigm. Note however that this example posits a rather different informational framework from that contemplated in this chapter.

offered about the circumstances under which it captures the essential strategic essence of a bargaining situation. Note 4 treats one aspect of this question. Another aspect which is neglected by the demand games considered so far is the possibility that an agreement may evolve over time as a result of a succession of compromises. It is only, of course, where binding forward commitments are disbarred or restricted that such considerations are likely to be relevant. We shall argue however that, even in these cases, the Nash bargaining solution remains an appropriate concept provided the players discount their utilities over time at the same rate. Where their discount rates differ, an asymmetric Nash bargaining solution emerges. (It should be noted that matters cannot be expected to be so simple when incomplete information is involved since the introduction of a time element will allow the players the opportunity to exchange information.)

We shall examine the subgame perfect equilibria of games in which each of two players makes a demand at times $t_0, t_1, \ldots, t_n$ until agreement is obtained or else the final time period is reached in which case both players receive a 'status quo' payoff of zero. If the set of available payoff pairs at time t is denoted by $\mathscr{X}_t$, then we interpret agreement to mean that the pair (x_t, y_t) of demands made at time t lies in the set $\mathscr{X}_t$. In this case each player receives his demand. We assume that

$$s < t \Rightarrow \mathbf{0} \in \mathscr{X}_t \subseteq \mathscr{X}_s.$$

As in the simple demand game there exists an embarrassment of equilibria if both players are assumed fully informed about $\mathscr{X}_{t_0}, \mathscr{X}_{t_1}, \ldots, \mathscr{X}_{t_n}$. We therefore introduce some shared uncertainty at each stage along the lines discussed in section 2 for the one period case. A perfect equilibrium outcome will then be an approximation to the outcome $\boldsymbol{\sigma}$ obtained by first calculating the Nash bargaining solution $\boldsymbol{\sigma}_{t_n}$ for $\mathscr{X}_{t_n}$ with status quo $\mathbf{0}$, then calculating the Nash bargaining solution $\boldsymbol{\sigma}_{t_{n-1}}$ for $\mathscr{X}_{t_{n-1}}$ with status quo $\boldsymbol{\sigma}_{t_n}$ and so on until $\boldsymbol{\sigma}$ is finally obtained as the Nash bargaining solution $\boldsymbol{\sigma}_{t_0}$ of $\mathscr{X}_{t_0}$ with status quo $\boldsymbol{\sigma}_{t_1}$ (see figure 8.6).

In the general case, $\boldsymbol{\sigma}$ will obviously depend on $\mathscr{X}_{t_1}, \mathscr{X}_{t_2}, \ldots, \mathscr{X}_{t_n}$ as well as on $\mathscr{X} = \mathscr{X}_{t_0}$. If $\boldsymbol{\sigma}$ is to be characterizable as a (possibly asymmetric) Nash bargaining solution for $\mathscr{X}$ with status quo $\mathbf{0}$, it is therefore necessary that we restrict the choice of $\mathscr{X}_{t_1}, \mathscr{X}_{t_2}, \ldots, \mathscr{X}_{t_n}$ so that these appear as pre-determined functions of $\mathscr{X}$. One possibility would be to take $\mathscr{X}_{t_1}, \mathscr{X}_{t_2}, \ldots, \mathscr{X}_{t_n}$ to be constant. We shall however consider the case when

$$\mathscr{X}_t = \{(\delta_1^t x, \delta_2^t y) : (x, y) \in \mathscr{X}\}$$

where $t_0 = 0$ and the discount rates satisfy $0 < \delta_1 < 1$ and $0 < \delta_2 < 1$.

When $\delta_1 = \delta_2$, it is immediate that $\boldsymbol{\sigma}$ is simply the Nash bargaining solution for $\mathscr{X}$ with status quo $\mathbf{0}$ and hence the introduction of the time element has no strategic impact on the situation. When $\delta_1 \neq \delta_2$, $\boldsymbol{\sigma}$ is not the Nash bargain-

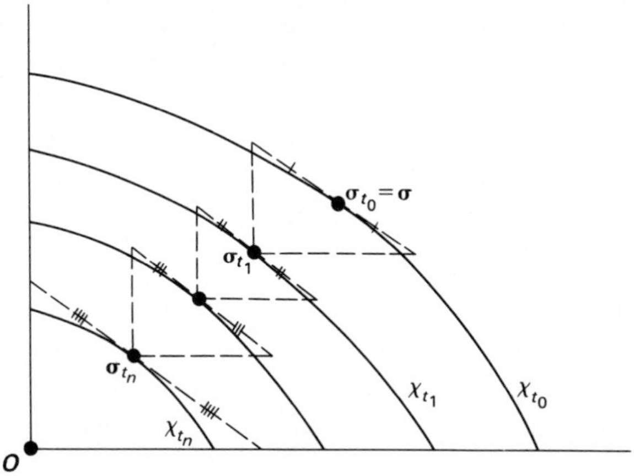

Figure 8.6

ing solution for $\mathscr{X}$ with status quo $\mathbf{0}$. This is not too surprising since there is then an asymmetry in the description of the players. We shall examine the case in which $t_k = kT$ where $T > 0$ and n is large. Writing $\Delta_1 = \delta_1^T$ and $\Delta_2 = \delta_2^T$, we shall show that $\boldsymbol{\sigma}$ is approximately the asymmetric Nash bargaining solution for $\mathscr{X}$ with status quo $\mathbf{0}$ in which the first player has 'bargaining power' $\tau = (1 - \Delta_2)/(2 - \Delta_1 - \Delta_2)$ and the second player has 'bargaining power' $1 - \tau = (1 - \Delta_1)/(2 - \Delta_1 - \Delta_2)$. (This means that $\boldsymbol{\sigma}$ is the value of $\mathbf{x}$ which maximizes $x_1^\tau x_2^{1-\tau}$ subject to $\mathbf{x} \geqslant \underline{\mathbf{0}}$ and $\mathbf{x} \in \mathscr{X}$.) See note 5 for some comments on the interesting case when $T \to 0+$.

Let $\mathscr{Y}$ denote any set which satisfies our usual conditions (see footnote 9) for a payoff region and let Q denote the intersection of the boundary of $\mathscr{Y}$ with the non-negative orthant. Write $P = \{(\Delta_1 x, \Delta_2 y) : (x, y) \in Q\}$ and define $M : Q \to P$ by $M(x, y) = (\Delta_1 x, \Delta_2 y)$. Next define $N : P \to Q$ by taking $N(\mathbf{x})$ to be the Nash bargaining solution for $\mathscr{Y}$ with status quo $\mathbf{x}$. We shall consider the function $f : P \to P$ defined by $f = N \circ M$.

It is easily shown that f has a unique fixed point $X = (X, Y)$ and that this satisfies the equations

$$\Delta_1^{-1} X = \tfrac{1}{2} \left\{ \Delta_1^{-1} a \left(1 - \frac{Y}{\Delta_2^{-1} b}\right) + X \right\}$$

$$\Delta_2^{-1} Y = \tfrac{1}{2} \left\{ \Delta_2^{-1} b \left(1 - \frac{X}{\Delta_1^{-1} a}\right) + Y \right\}$$

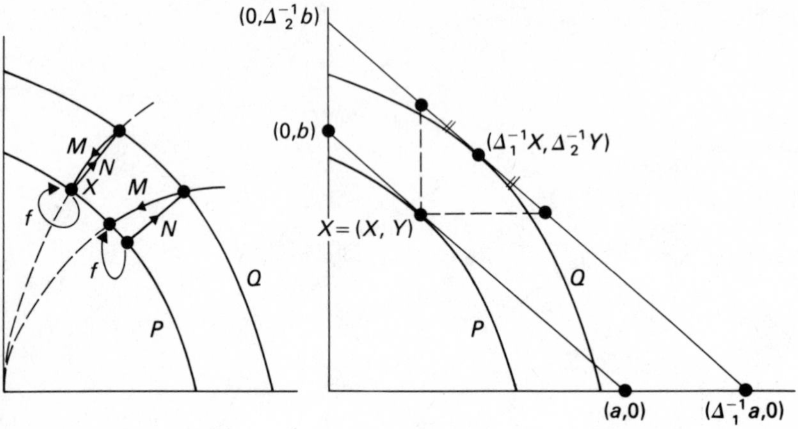

Figure 8.7

where a and b are as indicated in figure 8.7. We deduce that

$$\left.\begin{aligned} \frac{X}{a} &= \frac{1-\Delta_2}{2-\Delta_1-\Delta_2} \\ \frac{Y}{b} &= \frac{1-\Delta_1}{2-\Delta_1-\Delta_2} \end{aligned}\right\}$$

Next suppose that $\mathbf{x}_0$ is any given point on the set P and consider the sequence $\langle \mathbf{x}_k \rangle$ of points of P defined inductively by

$$\mathbf{x}_{k+1} = f(\mathbf{x}_k) \quad (k = 0, 1, \ldots).$$

This sequence has the property that $\mathbf{x}_{k+1}$ necessarily lies between $\mathbf{x}_k$ and $\mathbf{X}$. We deduce that $\langle \mathbf{x}_k \rangle$ converges. Since its limit must be a fixed point of f, it follows that $\mathbf{x}_k \to \mathbf{X}$ as $n \to \infty$. To check that $\mathbf{x}_{k+1}$ lies between $\mathbf{x}_k$ and $\mathbf{X}$, one need only observe that, in figure 8.8, the slope of the chord AB decreases as the x coordinate of A increases while the point G lies between D and F where EF is parallel to CD.

We now apply the above result with $\mathscr{Y} = \mathscr{X}_n$ and $\mathbf{x}_0 = \boldsymbol{\sigma}_{t_n}$. If $\boldsymbol{\sigma}$ is the asymmetric Nash bargaining solution for $\mathscr{X}$ with status quo $\mathbf{0}$ for which the first player has bargaining power τ_n, then the argument shows that τ_n is approximately $(1-\Delta_2)/(2-\Delta_1-\Delta_2)$ provided that n is sufficiently large.

Note 5. If we allow the time interval T in the above discussion to approximate zero, then the parameter $\tau = (1-\Delta_2)/(2-\Delta_1-\Delta_2)$ approximates

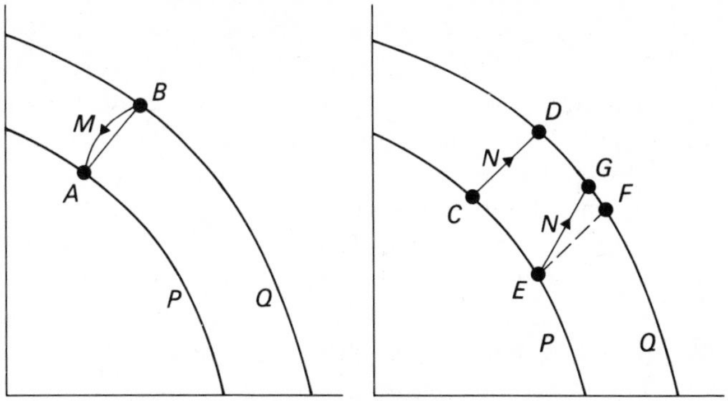

Figure 8.8

$$\frac{\log \delta_2}{\log \delta_1 + \log \delta_2}.$$

This should be compared with the analogous parameter obtained in section 5 of chapter 4 for the case with an infinite number of time periods in which the players take turns in making demands. It is encouraging for the stability of the analysis that the same result is obtained in both cases. Note incidentally that it is quite easy to adapt the analysis given above to the case of a finite number of time periods in which the players make alternating demands and results approximating those of Rubinstein (see Roth 1980 and section 5 of chapter 4) for the infinite case are obtained. It is not so easy, however, to analyse the case with an *infinite* number of time periods in each of which the players make simultaneous demands.

It is also of interest to note that $\tau \to \frac{1}{2}$ as $T \to +\infty$. Suppose, for example, that our players prefix the above game with a discussion of the timetable according to which successive demands are to be made. If this discussion consists of the players making demands T_1 and T_2 whereupon $T = \max\{T_1, T_2\}$, then the more impatient player will demand $T = +\infty$ and so gain parity.

5 SIMPLE NASH DEMANDS WITH INCOMPLETE INFORMATION

We now consider a version of the simple Nash demand game with incomplete information. As explained in section 1, it is appropriate to model such a two-person 'game of incomplete information' as a game (of complete information) with $m + n$ players. We suppose that there are m players of type I indexed with the elements i of the set I and that there are n players of type II indexed with the elements j of the set J. A random move selects player

$i \in I$ with probability λ_i to fill the role of negotiator I. An independent random move selects player $j \in J$ with probability μ_j to fill the role of negotiator II. Players of type I do not observe the random move which selects negotiator II: nor do players of type II observe the random move which selects negotiator I.

We next suppose that each player of type I chooses a real number x_i which represents the utility level which he will demand if chosen as negotiator I. Each player of type II similarly chooses a real number y_j. But we shall *not* assume as we did in section 2 that the negotiators simply announce these raw utility levels. This was a harmless assumption in section 2 (see note 4). But such an assumption in the current context would obscure the important difference between revealing and non-revealing equilibria. Instead we shall assume that the negotiators translate their chosen utility levels into physical terms and couch their demands in terms of these. To describe this process we require some further notation.

Each player $i \in I$ will be assumed to be equipped with a von Neumann and Morgenstern utility function $\phi_i : \$ \cup \{q\} \to \mathbb{R}$ and each player $j \in J$ with a similar utility function $\psi_j : \$ \cup \{q\} \to \mathbb{R}$. The object q will be identified with the status quo and, as always, we assume that $\phi_i(q) = \psi_j(q) = 0$. For simplicity, we suppose that $\$ = [0, 1]$ and assume that ϕ_i and ψ_j are strictly increasing, concave and differentiable on $\$$ with $\phi_i(0) \geqslant 0$ and $\psi_j(0) \geqslant 0$. The negotiators may now be assumed simultaneously to announce demands $a \in \$$ and $b \in \$$. If $a + b \leqslant 1$, each negotiator receives his demand: if $a + b > 1$, the status quo results. (This game is therefore 'divide the dollar' with incomplete information.) Assuming that negotiator I is player i, the demand $a \in \$$ will be chosen so that $x_i = \phi_i(a)$. Similarly, $b \in \$$ will be chosen so that $y_j = \psi_j(b)$.

It is immediately evident that if, for some given value of $a \in [0, 1]$, all players of type I choose their demands x_i so that $\phi_i(a) = x_i$ while all players of type II choose their demands y_j so that $\psi_j(1 - a) = y_j$, then the situation is in equilibrium. We shall call such an equilibrium *non-revealing*[12] because the announcements made by the negotiators reveal no information whatsoever about their identities. It is important to observe that *every* $a \in [0, 1]$ generates such a non-revealing equilibrium. As in the simple Nash demand game, we have found a large number of equilibria. Our primary purpose in section 6 will be to adapt the smoothing technique described in section 2 to the problem of selecting one of these equilibria.

So far, we have discussed non-revealing equilibria. But *revealing* equilibria will also exist in general if $m > 1$ or $n > 1$. As an example, consider the case $m = n = 2$. Suppose the first player of type I demands $x_1 = \phi_1(a)$ while the

12 There may also be non-revealing equilibria in which each player chooses a *mixed* strategy. But we shall disregard these as it seems sensible to consider them as pure equilibria in a game with more players.

second player of type I demands $X_2 = \phi_2(A)$. At the same time, the first player of type II demands $y_1 = \psi_1(b)$ while the second player of type II demands $Y_2 = \psi_2(B)$. If $0 \leqslant a \leqslant A \leqslant 1$, $b = 1 - a$ and $B = 1 - A$, then the players obtain the payoffs listed below:

$$\text{Type 1:} \begin{cases} 1: x_1 \\ 2: X_2\mu_2 \end{cases} \qquad \text{Type 2:} \begin{cases} 1: y_1\lambda_1 \\ 2: Y_2 \end{cases}$$

Is this situation in equilibrium? If it is worth a player's while to switch his strategy choice, then it will be worth his while to adopt the strategy choice of the other player of the same type. In the case of the first player of type I, he will improve on his payoff by switching in this way only if $X_1\mu_2 > x_1$. Similar considerations lead to the following necessary conditions for the situation described above to be in equilibrium:

$$X_1\mu_2 \leqslant x_1$$

$$x_2 \leqslant X_2\mu_2$$

$$Y_1 \leqslant y_1\lambda_1$$

$$y_2\lambda_1 \leqslant Y_2$$

- i.e. $\phi_2(a)/\phi_2(A) \leqslant \mu_2 \leqslant \phi_1(a)/\phi_1(A)$ and $\psi_1(B)/\psi_1(b) \leqslant \lambda_1 \leqslant \psi_2(B)/\psi_2(b)$.

Note that under certain circumstances these conditions will be satisfied for *no* values of $a < A$. On other occasions, such an equilibrium will lead to an outcome which is Pareto-dominated by the outcome which results from a non-revealing equilibrium. In any case, in this chapter we shall restrict our attention to those cases in which revealing equilibria can be eliminated as possible candidates for the solution of the game.

6 SMOOTHED DEMAND GAME WITH INCOMPLETE INFORMATION

In the previous section we described a version of the simple Nash demand game for the case of incomplete information. As in the case of complete information, we found an embarrassingly large set of equilibria. In this section we shall describe a version of the smoothed Nash demand game (see section 2) for the case of incomplete information. The aim is to eliminate as many equilibria as possible.

The model is the same as that of section 5 except that, after the negotiators have announced their demands $a \in \$$ and $b \in \$$, there follows a chance move. As a result of this chance move, the two negotiators receive their demands with probability $\pi(a, b)$ or else the status quo results with probability $1 - \pi(a, b)$. It is important that all players are assumed to be familiar with the probability function π - i.e. the additional chance move introduced here represents a measure of *shared* uncertainty on the part of the players. The function $\pi : \$ \times \$ \to [0, 1]$ will be assumed to be differentiable, decreasing (see footnote 10) and quasi-concave.

We shall be interested in studying the case in which the smoothed demand game described above is an approximation to the simple demand game of section 5. In this case π will approximate the function $\chi : \$ \times \$ \to [0, 1]$ defined by

$$\chi(a, b) = \begin{cases} 1, & a + b \leqslant 1 \\ 0, & a + b > 1. \end{cases}$$

A player $i \in I$ seeks to choose $x_i = \phi_i(a_i)$ so as to maximize

$$\phi_i(a_i) \sum_{j \in J} \mu_j \pi(a_i, b_j)$$

while a player $j \in J$ seeks to choose $y_j = \psi_j(b_j)$ so as to maximize

$$\psi_j(b_j) \sum_{i \in I} \lambda_i \pi(a_i, b_j).$$

Conditions for a Nash equilibrium are therefore

$$\phi_i'(a_i) \sum_{j \in J} \mu_j \pi(a_i, b_j) + \phi_i(a_i) \sum_{j \in J} \mu_j \pi_a(a_i, b_j) = 0$$

$$\psi_j'(b_j) \sum_{i \in I} \lambda_i \pi(a_i, b_j) + \psi_j(b_j) \sum_{i \in I} \lambda_i \pi_b(a_i, b_j) = 0.$$

The probability that agreement is reached given that the vector of demanded payoffs is $(\mathbf{x}, \mathbf{y})$ is equal to

$$p(\mathbf{x}, \mathbf{y}) = \sum_{i \in I} \sum_{j \in J} \lambda_i \mu_j \pi(a_i, b_j).$$

In terms of this quantity we may write the necessary equations for a Nash equilibrium as

$$\left.\begin{aligned} -\frac{x_i}{\lambda_i} \frac{\partial p}{\partial x_i} &= \sum_{j \in J} \mu_j \pi(a_i, b_j) \\ -\frac{y_j}{\mu_j} \frac{\partial p}{\partial y_j} &= \sum_{i \in I} \lambda_i \pi(a_i, b_j). \end{aligned}\right\}$$

As explained in section 5, our interest is in the non-revealing equilibria of the simple demand game. We therefore examine equilibria of the smoothed game which are approximately non-revealing[13] - i.e. equilibria for which $a_i \doteq a(i \in I)$ and $b_j \doteq b(j \in J)$ for some a and b. The imposition of this constraint yields that

$$\left.\begin{aligned} -\frac{x_i}{\lambda_i}\frac{\partial p}{\partial x_i} &\doteq \pi(a, b) \\ -\frac{y_j}{\mu_j}\frac{\partial p}{\partial y_j} &\doteq \pi(a, b). \end{aligned}\right\}$$

Finally note that, if the smoothed version approximates the simple version sufficiently closely, then we shall have that $\pi(a_i, b_j) \doteq 1$ for each $i \in I$ and $j \in J$. It then follows, as in section 2, that the payoff vector $(\mathbf{x}_0, \mathbf{y}_0)$ resulting from an approximately non-revealing equilibrium of an approximating smoothed demand game is an approximation to the point at which

$$\prod_{i \in I} x_i^{\lambda_i} \prod_{j \in J} y_j^{\mu_j}$$

is maximized subject to the constraint that $(\mathbf{x}, \mathbf{y}) \in \mathscr{X}$ where

$$\mathscr{X} = \{(\mathbf{x}, \mathbf{y}) : \exists a \in \$, \forall i \in I, \forall j \in J, \phi_i(a) = x_i \text{ and } \psi_j(1 - a) = y_j\}$$

- i.e. the feasible set consists of the vectors of payoff demands at non-revealing equilibria of the simple game.

Note 6. When studying revealing equilibria, the above analysis leads us to equations of the form

$$\frac{x_i}{\lambda_i}\frac{\partial p}{\partial x_i} \doteq \sum_{j \in J_i} \mu_j$$

$$\frac{y_i}{\mu_i}\frac{\partial p}{\partial y_i} \doteq \sum_{i \in I_j} \lambda_i.$$

For example, in the case of an approximation to the revealing equilibrium studied in section 5:

$$\frac{x_1}{\lambda_1}\frac{\partial p}{\partial x_1} = \mu_1 \pi(a, b) + \mu_2 \pi(a, B) \doteq 1$$

$$\frac{x_2}{\lambda_2}\frac{\partial p}{\partial x_2} = \mu_1 \pi(A, b) + \mu_2 \pi(A, B) \doteq \mu_2$$

13 Note that, with a trembling hand interpretation, such equilibria in the smoothed version will be difficult to distinguish from actual non-revealing equilibria in the smoothed version.

$$\frac{y_1}{\mu_1}\frac{\partial p}{\partial y_1} = \lambda_1\pi(a,b) + \lambda_2\pi(A,b) \doteqdot \lambda_1$$

$$\frac{y_2}{\mu_2}\frac{p}{y_2} = \lambda_1\pi(A,B) + \lambda_2\pi(A,B) \doteqdot 1.$$

Hence the smoothing technique selects that equilibrium (x_1, x_2, y_1, y_2) from the set of revealing equilibria of the type discussed in section 5 which maximizes

$$x_1^{\lambda_1}x_2^{\mu_2\lambda_2}y_1^{\lambda_1\mu_1}y_2^{\mu_2}$$

Generalized Nash bargaining solutions therefore remain relevant when revealing equilibria are studied but one must adjust the 'bargaining powers' assigned to the players.

Of course, the conditions given above are just local conditions. Also to be taken into account are the 'global conditions' given in section 5. If the latter are never satisfied there will be *no* revealing equilibria of the type considered here. Where revealing equilibria do exist it may well happen that the equilibrium selected by the smoothing technique may be Pareto-dominated by the non-revealing equilibrium selected by the smoothing technique (see section 7). In such cases, the analysis offered in this chapter supports the use of the weighted Nash product introduced by Harsanyi and Selten (1972) and obtained by alternative means in section 6. But cases exist (see section 7 again) for which there seem no immediate reasons for eliminating revealing equilibria and, if it happens that the players do operate a convention which selects a revealing equilibrium in some of these cases, then it would appear that this selection would not be characterizable in the manner described by Harsanyi and Selten. At the very least, these considerations suggest that there is scope for further research in this area.

Note 7. In section 5 and section 6, we discussed 'non-revealing equilibria' in a rather special context. In a more general framework, one may suppose that the payoff region is $\mathscr{X}_{ij}$ when negotiator I is player $i \in I$ and negotiator II is player $j \in J$. The analogue of a non-revealing equilibrium in this setting is a demand vector $(\mathbf{x}, \mathbf{y})$ which has the property that no player will profit by deviating from his demand even if he learns the identity of the opposing negotiator. Figure 8.9 illustrates such a configuration for the case $I = \{1, 2\}$ and $J = \{1, 2\}$.

Note that the existence of equilibria of this type constrains the sets $\mathscr{X}_{ij}$. If, for example, every x_1 in some range is the first coordinate of a 'non-revealing equilibrium', then the Pareto-frontier of $\mathscr{X}_{22}$ will be determined for a corresponding range by the Pareto-frontiers of the sets $\mathscr{X}_{11}$, $\mathscr{X}_{12}$ and $\mathscr{X}_{21}$. If the sets $\mathscr{X}_{ij}$ are arbitrary, one would expect to find only a small finite number (possibly zero) of 'non-revealing equilibria'.

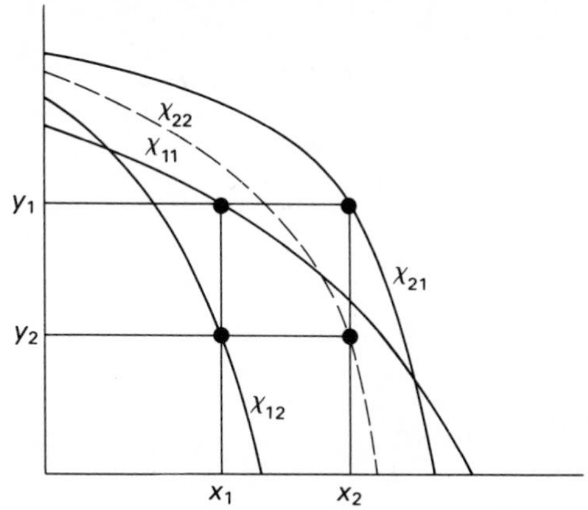

Figure 8.9

7 SPLIT THE DIFFERENCE

A commonly observed practical resolution of 'divide the dollar' problems consists of 'splitting the difference'. This means that each player receives 50¢ from the \$1 available. When is this explicable as the consequence of optimal play by rational bargainers? Certainly, the Nash bargaining solution will yield this result when the players' utilities for money are identical and so the bargaining problem is symmetric. However, in practice this is seldom likely to be true and, even when true, neither player is likely to be aware of the fact. A more satisfactory framework is to suppose that the players are drawn at random from identical populations and to analyse the problem as a bargaining game of incomplete information.

To illustrate the points made in section 5 and section 6 we analyse a simple special case. With the model and notation of section 5 we take $I = \{1, 2\}$, $J = \{1, 2\}$, $\lambda_1 = \mu_2 = \nu$ and $\lambda_2 = \mu_1 = 1 - \nu$. We also assume the special forms

$$\phi_1(a) = \psi_2(a) = a^\gamma \quad (0 \leqslant a \leqslant 1)$$

$$\phi_2(a) = \psi_1(a) = a \quad (0 \leqslant a \leqslant 1)$$

for the utility functions where $0 < \gamma < 1$. The non-revealing equilibrium selected by the smoothing technique is found by maximizing

$$a^{\nu\gamma} a^{1-\nu} (1-a)^{1-\nu} (1-a)^{\nu\gamma}$$

for $0 \leqslant a \leqslant 1$. Differentiating, we obtain that

$$\frac{\nu\gamma}{a}+\frac{1-\nu}{a}-\frac{1-\nu}{1-a}-\frac{\nu\gamma}{1-a}=0$$

and so

$$a=\tfrac{1}{2}$$

which is not surprising since both negotiators are chosen from identical populations.

Next consider a revealing equilibrium of the type described in section 5. (Note that other revealing equilibria never exist in this case.) Necessary conditions are:

$$\left.\begin{aligned}\frac{a}{A}\leqslant \nu\leqslant\left(\frac{a}{A}\right)^{\gamma}\\ \frac{1-A}{1-a}\leqslant \nu\leqslant\left(\frac{1-A}{1-a}\right)^{\gamma}.\end{aligned}\right\}\tag{8.2}$$

The smoothing technique selects the revealing equilibrium which maximizes

$$a^{\nu\gamma}A^{(1-\nu)\nu}(1-a)^{(1-\nu)\nu}(1-A)^{\nu\gamma}$$

for $0\leqslant a<A\leqslant 1$. Differentiating, we obtain

$$\frac{(1-\nu)\nu}{A}=\frac{\nu\gamma}{1-A};\quad \frac{\nu\gamma}{a}=\frac{(1-\nu)\nu}{1-a}$$

and so

$$a=\frac{\gamma}{1-\nu+\gamma};\quad A=\frac{1-\nu}{1-\nu+\gamma}.\tag{8.3}$$

Observe that these values are sometimes inconsistent with (8.2), for example, if $\gamma>1/4$. On other occasions, all players may prefer their 'split the difference' outcomes. This always happens, for example, if $\gamma>1/8$ or $\nu<1/2$. But circumstances also exist in which the smoothing technique selects a viable revealing equilibrium – i.e. (8.2) and (8.3) hold and also

$$\frac{\nu(1-\nu)}{\gamma+1-\nu}>\frac{1}{2}$$

which means that the second player of type I and the first player of type II prefer their revealing equilibrium payoffs to the 'split the difference' payoff. Suitable values for γ and ν are $\gamma=1/10$ and $\nu=3/4$.

We conclude that 'split the difference' is the solution of the game for a wide range of the parameters involved but that the analysis is inadequate to maintain this conclusion for all values of the parameters. It is possible that one should look for some insight in the difficult cases to the situation in which a large number of similar games of the type considered here are to be played. It need not be assumed that negotiators are always drawn from the same populations but it will be assumed that players have a record of previous decisions made by the player with whom they find themselves negotiating. In choosing a negotiation strategy in a given game, a player will then have to consider not only his prospective payoff in the current game but also the effect on his future payoffs of any information his strategy choice may reveal. The 'split the difference' equilibrium is clearly stable in the face of such considerations: a revealing equilibrium is equally clearly not viable even in the early stages since a weak player (e.g. the first player of type I) would find it profitable to mimic a strong player (e.g. the second player of type II) and hence be wrongly identified thereafter as a strong player.

8 SIGNALLING

In their fundamental paper, Harsanyi and Selten (1972) emphasize the importance of signalling in bargaining games of incomplete information. In the one-period models considered above, however, the players have no opportunity to signal (i.e. to communicate information about themselves to the other negotiator). It is therefore natural to ask what the impact on our analysis would be if opportunities for signalling are introduced into the model. In this section and the next we shall discuss this question very briefly for the simplest possible case – i.e. when negotiator I may be one of two players (labelled 1 and 2) but negotiator II is always player 3. Our particular interest will be in the circumstances under which the players choose *not* to signal.

We use $\mathscr{X}_i$ to denote the payoff region when negotiator I is player i ($i = 1, 2$). As always, status quo payoffs are fixed at 0. To begin with, we examine the simple demand game based on this structure. Each player chooses a utility demand (x, y and z) and the negotiators receive their demands if these turn out to be jointly feasible. The manner in which the negotiators communicate their demands is important in this context. We assume that negotiator II simply announces his demand z while negotiator I announces the maximal payoff to II consistent with I's demand. This means that player 1 will announce $\phi_1(x)$, where $z = \phi_1(x)$ determines the Pareto-boundary of $\mathscr{X}_1$. Similarly player 2 announces $\phi_2(y)$. With this understanding, *all* equilibria for the simple demand game are non-revealing because it will be optimal for both 1 and 2 to announce the same number as player 3. Moreover, given that 1 and 2 are announcing the same number it will be optimal for 3 to announce this number also.

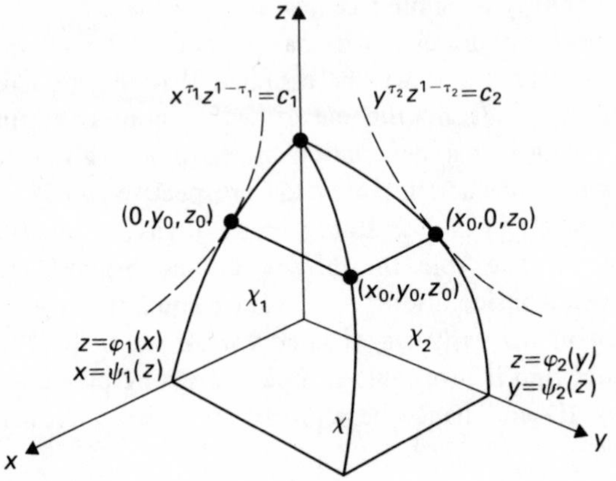

Figure 8.10

Figure 8.10 illustrates a typical equilibrium (x_0, y_0, z_0). Denote the set of all such equilibria by $\mathscr{X}$. If λ represents the probability that player 1 will be chosen to fill the role of negotiator I, then the smoothing technique selects the equilibrium (x_0, y_0, z_0) at which

$$x^{\lambda}y^{1-\lambda}z$$

is maximized subject to the constraint $(x, y, z) \in \mathscr{X}$.

It is of some interest to calculate what 'bargaining power' τ_1 we should need to attribute to player 1 when bargaining under complete information with player 3 in order that he receive the same payoff as that obtained at the non-revealing equilibrium described above. The quantities λ, τ_1 and τ_2 satisfy the equations

$$\frac{\tau_1\psi_1'(z_0)}{\psi_1(z_0)} + \frac{(1-\tau_1)}{z_0} = 0$$

$$\tau_2\frac{\psi_2'(z_0)}{\psi_2(z_0)} + \frac{(1-\tau_2)}{z_0} = 0$$

$$\lambda\frac{\psi_1'(z_0)}{\psi_1(z_0)} + (1-\lambda)\frac{\psi_2'(z_0)}{\psi_2(z_0)} + \frac{1}{z_0} = 0$$

and so

$$\begin{vmatrix} \tau_1 & 0 & 1-\tau_1 \\ 0 & \tau_2 & 1-\tau_2 \\ \lambda & 1-\lambda & 1 \end{vmatrix} = 0.$$

It follows that either $\tau_1 = \tau_2 = \frac{1}{2}$ or else

$$\lambda = \frac{\tau_1(1-2\tau_2)}{\tau_1 - \tau_2} \quad \text{and} \quad 1-\lambda = \frac{\tau_2(1-2\tau_1)}{\tau_2 - \tau_1}. \tag{8.4}$$

The bargaining powers τ_1 and τ_2 in this equation are, of course, only notional quantities. If players 1 and 3 actually bargain under complete information, the analysis of this chapter would assign each bargaining power $\frac{1}{2}$ - i.e. the classical Nash bargaining solution would apply. Given that player 1 has been chosen as negotiator I, he will therefore prefer this information to be conveyed to negotiator II if $\tau_1 < \frac{1}{2}$. On the other hand, if $\tau_1 > \frac{1}{2}$, he would prefer II to remain uncertain about the identity of I. Note that, if $0 < \lambda < 1$, then equations (8.4) imply that $\tau_2 > \tau_1 \Rightarrow \tau_1 < \frac{1}{2}$ and $\tau_2 > \frac{1}{2}$. We conclude from these calculations that either both player 1 and player 2 will be indifferent about whether or not their identity is revealed or else one of them will prefer his identity to be revealed while the other will prefer his identity to remain concealed. In what follows we shall assume that player 1 would like his identity revealed ($\tau_1 < \frac{1}{2}$) while player 2 would like his identity concealed ($\tau_2 > \frac{1}{2}$). Thus 1 is a 'strong' player while 2 is a 'weak' player.

Suppose that, preliminary to the demand game considered above, negotiator I has the opportunity to signal his identity to negotiator II. If such signals are costless, then the opportunity to signal will make little difference to the situation since, if player 3 were to adopt a strategy which led him to identify negotiator I as player 1 after certain signals, it would clearly be profitable for player 2 to send one of these signals and hence be falsely identified as a strong player. Given that signals must involve costs of some kind if they are to be significant, the simplest useful model would seem to be that in which negotiator I begins by destroying some object of value as his preliminary signal and then both negotiators make their demands as before. For example, negotiator I may begin by ostentatiously burning a banknote of large denomination. Alternatively, if one observes that the following analysis is not materially affected if the object destroyed by negotiator I also has value to player 3, one may interpret the signal as taking the form of a strike preliminary to bargaining over wages.[14]

In figure 8.11 N_1 and N_2 are the classical Nash bargaining solutions for $\mathcal{X}_1$ and $\mathcal{X}_2$ respectively with status quos at the origin. The point (x_0, y_0, z_0) is the non-revealing equilibrium obtained previously. The points M_1 or M_2 arise when player 3 wrongly identifies negotiator I and this misidentification is known to the player filling the role of negotiator I.

14 Although it must be borne in mind that this model allows no opportunity for the players to threaten each other and hence much of the interplay associated with real-life strike action is absent.

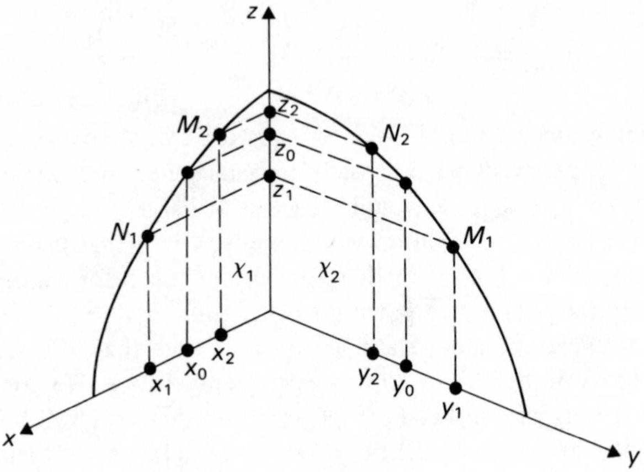

Figure 8.11

For simplicity, we shall assume that the possible signals which negotiator I may emit are indexed with real numbers $s \geqslant 0$ and that, if player 1 sends signal s and receives a payoff x as a result of the exchange of demands which follows, then his total utility is $x - s$. Under similar circumstances we assume that player 2 receives total utility $y - s$. (In general, of course, a player would like to emit a signal which is inexpensive for himself but which would be costly for his alter ego but our assumptions preclude this – see note 9.)

With these hypotheses, it is significant whether or not

$$y_1 - y_2 \leqslant x_1 - x_2. \tag{8.5}$$

If the inequality is satisfied, we may choose a signal S such that

$$\left.\begin{aligned} x_1 - S &\geqslant x_2 \\ y_1 - S &\leqslant y_2. \end{aligned}\right\}$$

The first condition means that it is worthwhile for player 1 to emit the signal S in order to be correctly identified rather than to be wrongly identified as player 2. The second condition means that it is not worthwhile for player 2 to emit the signal S in order to be wrongly identified as player 1 rather than to be correctly identified.

The following is a set of subgame-perfect equilibrium strategies in the case when inequality (8.5) is satisfied:

Player 1: Emit signal $s_1 = S$.
If $s_1 < S$, demand x_2.
If $s_1 \geqslant S$, demand x_1.

Player 2: Emit signal $s_2 = 0$.
If $s_2 < S$, demand y_2.
If $s_2 \geqslant S$, demand y_1.

Player 3: If observe signal $s < S$, demand z_2.
If observe signal $s \geqslant S$, demand z_1.

This equilibrium is a revealing equilibrium. It is also easy to identify a non-revealing equilibrium. Appropriate strategies are:

Player 1: Emit signal $s_1 = 0$.
Always demand x_0.

Player 2: Emit signal $s_2 = 0$.
Always demand y_0.

Player 3: Always demand z_0.

Other subgame-perfect equilibria can also be found. We are thus confronted again with the problem of multiple equilibria. As in previous sections of this chapter, we shall attempt to resolve this problem by means of the smoothing technique.

Into the model described above we now introduce a small amount of shared uncertainty. Suppose as before that negotiator I pays an amount s to transmit a signal but that the signal actually received by negotiator II is t. We shall suppose that t has probability density function $f(s, t)$. All players are assumed to observe t and all players know the function f but only negotiator I knows the value of s. Our interest is in the case when the level of uncertainty is small which we interpret to mean that the graph of f as a function of t resembles that of figure 8.12. (As in the 'trembling hand' version of the straightforward demand game of section 3, one could require negotiator I to select s but have to pay t without much change.)

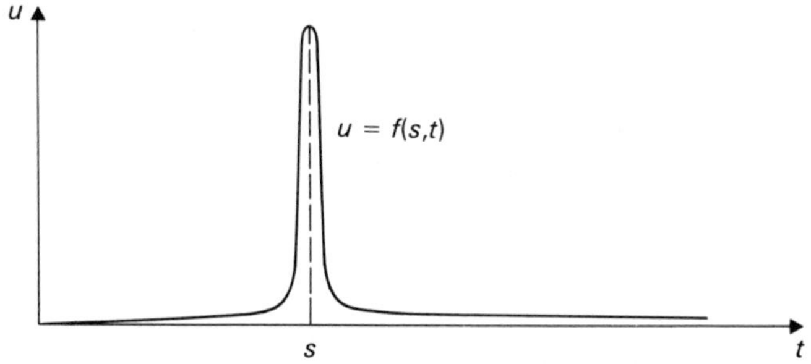

Figure 8.12

Given the existence of the uncertainty described above, what conclusion will player 3 draw from the observation that $t = T$ if he entertains the hypothesis that player 1 will make the signal $s = s_1$ while player 2 will make the signal $s = s_2$? Let the probability player 3 assigns to the event that negotiator I is player 1 under these conditions be $\mu = \mu(s_1, s_2, T)$. If $s_1 = s_2$, it is clear that $\mu = \lambda$. If $s_1 \neq s_2$, then the appropriate value of μ is obtained by an application of Bayes' theorem. This yields that

$$\mu = \frac{\lambda f(s_1, T)}{\lambda f(s_1, T) + (1 - \lambda) f(s_2, T)}.$$

At an equilibrium, player 3 will anticipate s_1 and s_2 correctly and all players will calculate the same value of μ after T has been observed. In the (smoothed) demand game which follows, the players will then demand (approximately) $\bar{x}_0 = \bar{x}_0(s_1, s_2, t)$, $\bar{y}_0 = \bar{y}_0(s_1, s_2, t)$ and $\bar{z}_0 = \bar{z}_0(s_1, s_2, t)$ respectively, where $(\bar{x}_0, \bar{y}_0, \bar{z}_0)$ is obtained by maximizing $x^{\mu} y^{1-\mu} z$ over $\mathscr{X}$. Figure 8.13 illustrates the situation.

In what follows we consider the special case in which, for a given s, the random variable t is normally distributed with mean s and variance σ^2. This particularization is not essential to the argument[15] but simplifies the question of what happens to the graph of $u = \mu(s_1, s_2, t)$ as the degree of uncertainty diminishes. If $s_1 > s_2$, $\mu(s_1, s_2, t)$ approaches

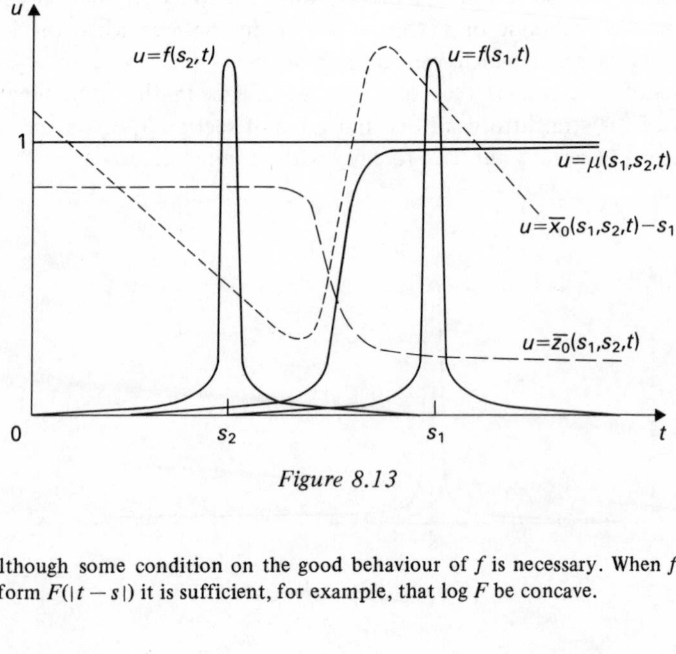

Figure 8.13

15 Although some condition on the good behaviour of f is necessary. When $f(s, t)$ is of the form $F(|t - s|)$ it is sufficient, for example, that $\log F$ be concave.

$$\chi(s_1, s_2, t) = \begin{cases} 0, & t < \frac{1}{2}(s_1 + s_2) \\ \lambda, & t = \frac{1}{2}(s_1 + s_2) \\ 1, & t > \frac{1}{2}(s_1 + s_2) \end{cases}$$

as $\sigma \to 0+$. (The case $s_1 < s_2$ is similar but uninteresting since then player 2 could obviously improve his payoff by copying player 1. Thus no equilibrium is possible with $s_1 < s_2$.)

The fact that player 3 reacts to the signal which he receives in such a simple manner means that only rather simple equilibria can occur in this model. In particular, for a fixed $\epsilon > 0$, there cannot exist an equilibrium with $s_1 - s_2 > \epsilon$ when σ is sufficiently small because player 1 could then pay $\epsilon/2$ less for his signal without significantly altering the probability of his being correctly identified. Thus, at an equilibrium, s_1 and s_2 must be approximately equal when σ is small. Also, player 3's reaction probability μ must be approximately λ for a signal in the vicinity of s_1 and s_2. Otherwise player 2 could obtain a large increase in the probability of his being identified as player 1 by emitting an only slightly more expensive signal.

It follows that when σ is small, an equilibrium will be approximately non-revealing.[16] Moreover, if we eliminate equilibria which are Pareto-dominated by another equilibrium, then we are left only with non-revealing equilibria in which the signals emitted cost very little.

The smoothing technique therefore selects only non-revealing equilibria and our generalized Nash bargaining solution remains valid even with a signalling facility of the type described. More precisely, the solution of an approximating smoothed signalling game is approximately non-revealing (see footnote 16) with almost costless signals.

Note 8. It would be of interest to know the conditions under which the solution of an approximating smoothed signalling game can be unambiguously identified with the exactly non-revealing equilibrium with zero signals (i.e. $s_1 = 0, s_2 = 0$ and $\mu(s_1, s_2, t) = \lambda$).

Note 9. Our assumptions in section 8 on the manner in which the signals emitted by negotiator I affect the players' utilities are rather simplistic. A more satisfactory analysis would associate payoff regions $\mathscr{X}_1(s)$ and $\mathscr{X}_2(s)$ with each possible signal.

Note 10. A familiar problem when discussing perfect equilibria in games of imperfect information is to decide what 'conjecture' it is appropriate to

16 In the following sense. An observer of the game not privy to the identity of negotiator I will assign the a priori probability of λ to the event that negotiator I is player 1 before the game takes place. After he has observed the strategy choices he will replace λ by an a posteriori probability μ. We regard an equilibrium as 'approximately non-revealing' if the probability that μ differs significantly from λ is small.

assign to a player at an information set which the player knows will be reached with zero probability if the other players choose their equilibrium strategies. Harsanyi and Selten advocate a 'trembling hand principle' which attributes arrival at such an information set to 'errors' of small probability on the part of the players at preceding decision points. It is possible to re-interpret the discussion of section 8 in terms of this principle rather than to explain the signalling uncertainty as 'noise' in the transmission system. However, we would be rather unhappy with such a re-interpretation, attractive though it is, since there seems no compelling reason why we should assume that player 3 will conjecture that the 'errors' made by players 1 and 2 should have probability distributions of the type used in section 8. A popular alternative, for example, would be to suppose that player 3 regards any signal other than s_1 and s_2 as conveying no information about the identity of players 1 and 2 on the grounds that each is equally likely to make a 'mistake'. But whatever defence may be offered for the latter alternative, it clearly must collapse if there is the slightest uncertainty about the preferences held by players 1 and 2. We would also like to draw attention to yet another alternative. This is best explained by first examining a simpler example.

Figure 8.14 shows two copies of the extensive form of a certain three-person game. The left-hand part of the figure represents one equilibrium while the right-hand side indicates another distinct equilibrium. The equilibrium choices for each player at a decision node are indicated by a double line.

In the right-hand case, player 3 is supposed to deduce nothing from the fact that he finds himself with an opportunity to play and hence maximizes expected utility on the assumption that player 1 was chosen with probability 0.6 and player 2 with probability 0.4. But, is it reasonable to argue in this way given that player 1 has a strictly dominating strategy? Would it not be more reasonable to argue as follows:

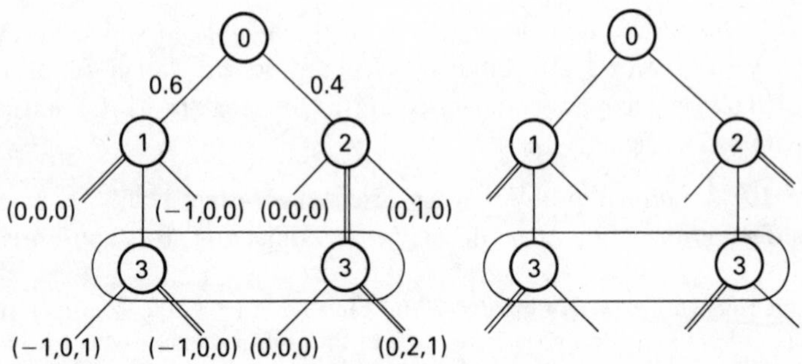

Figure 8.14

1. Player 1 has a strictly dominating strategy: therefore he will play this.
2. If player 3 finds that he has an opportunity to play it 'therefore' must be that his opponent is player 2.
3. Thus player 3 should choose the strategy indicated in the left-hand case.
4. Thus player 2 should choose the strategy indicated for him in the left-hand case.

We now return to our original model of section 8 and assume that inequality (8.5) is satisfied. Our purpose is to argue in favour of the *revealing* equilibrium of section 8 on lines analogous to those of the preceding paragraph. We first assume that, once he has received a signal, player 3 will assign some probability μ to the event that negotiator I is player 1, that players 1 and 2 guess μ correctly and that all make their optimal demands in the demand game which follows. Taking these assumptions as given, player 2 now observes that he can *guarantee* at least y_2 by emitting the signal $s_2 = 0$ and then demanding y_2. Hence this strategy is necessarily as good as any which begins by emitting the signal $s_2 = S$ because the most that player 2 could get is y_1 and $y_1 - S \leqslant y_2$. This is true *however* player 3 forms his conjectures. When player 3 is deciding what conjectures to form, he may therefore reasonably argue that player 2 will definitely *not* emit a signal $s_2 \geqslant S$ if $y_1 - S < y_2$. If such a signal is observed, it must therefore have come from player 1.

Perhaps the lesson to be learned is that one cannot meaningfully discuss the conjectures that players will form in a vacuum: hence our preference for regarding the uncertainty in the signalling model of section 8 as transmission 'noise' since this assumption determines the conjectures unambiguously. A more satisfactory treatment would also model 'noise' in the information players 1, 2 and 3 have about each others' preferences etc. In such a framework there would again always be a 'correct' conjecture for players at each decision point, but of course, there would be vastly more players to consider. It certainly does seem likely, however, that the essence of the argument of the previous paragraph might survive in such a setting and therefore that the result of section 8 is not stable in the presence of such further uncertainties.

Note 11. It is natural to ask in respect of the bargaining model of section 8 whether or not it is true that $z_0 \leqslant \lambda z_1 + (1 - \lambda) z_2$. If the inequality holds, then player 3 will wish to be informed of the identity of negotiator I before demands are made provided that the fact that he has this information is known and believed by all. If the inequality does *not* hold, then player 3 will be prepared to pay in order to *prevent* the identity of negotiator I becoming general knowledge. It would be of interest to know whether the latter circumstances can ever actually occur – i.e. is it *always* true that $z_0 \leqslant \lambda z_1 + (1-\lambda)z_2$?

9 BARGAINING OVER TIME WITH INCOMPLETE INFORMATION

In this section we shall assume as in section 8 that negotiator I is either player 1 or player 2 and that negotiator II is always player 3. However, instead of allowing the players to signal directly before demands are made, we shall suppose that the negotiators make a sequence of simultaneous demands as described in section 4. In principle, this set-up provides negotiator I with the opportunity to signal his identity by the choice of demands he makes. For example, player 1 may choose to make an absurdly high demand initially with the aim of establishing that he is not player 2. The cost of so doing will be that he has to wait until the second round before reaching agreement. His reward will be that the final agreement will be an improvement on what he would obtain if unidentified. As in section 8, we shall chiefly be concerned with the case when players choose *not* to signal. Mixed strategies are not considered.

We begin with the three-person demand game described in section 8 and extend this to be a two-period model by supposing that this game is only reached if there is disagreement in a preliminary demand game which has the same structure as the original game except that $\mathscr{X}_1$ and $\mathscr{X}_2$ are replaced by $\mathscr{Y}_1 = \{(\Delta_1^{-1}x, \Delta_3^{-1}z) : (x, z) \in \mathscr{X}_1\}$ and $\mathscr{Y}_2 = \{(\Delta_2^{-1}y, \Delta_3^{-1}z) : (y, z) \in \mathscr{X}_2\}$, where the factors Δ_1, Δ_2 and Δ_3 satisfy $0 < \Delta_1 < 1$, $0 < \Delta_2 < 1$ and $0 < \Delta_3 < 1$.

Figure 8.15 is an elaboration of figure 8.11. The point P_1 is the Nash bargaining solution for the game between players 1 and 3 with payoff region $\mathscr{Y}_1$ and status quo N_1. Similarly, P_2 is the Nash bargaining solution for the game between players 2 and 3 with payoff region $\mathscr{Y}_2$ and status quo N_2.

Can there be a revealing equilibrium? Suppose that players 1, 2 and 3 choose initial announcements ζ_1, ζ_2 and ζ_3 respectively. (As in section 8, these announcements refer to the utility player 3 is to receive at the proposed agreement point.) With a trembling hand story of the type given in section 3 and section 8, we can determine the 'conjecture' that player 3 will make given the announcement he observes on the assumption that he anticipates ζ_1 and ζ_2 correctly. In particular, if $\zeta_1 < \zeta_2$ and the degree of uncertainty is sufficiently small, player 3 will identify negotiator I accurately with probability nearly 1. We therefore know what will happen in the second demand game should agreement not be reached in the first demand game. The final outcome will be approximately N_1 or N_2 depending on the identity of negotiator I. We can therefore apply the analysis of section 2 to derive some conclusions about the equilibrium values of ζ_1, ζ_2 and ζ_3. Assuming

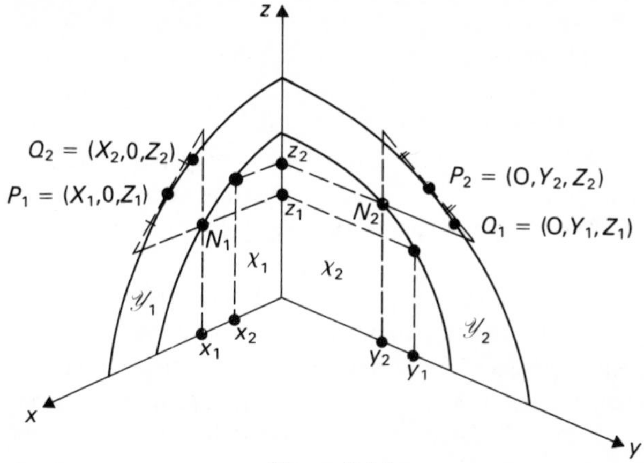

Figure 8.15

as before that $\zeta_1 < \zeta_2$ and the degree of uncertainty is sufficiently small, player 3 will act as though he is either playing against player 1 for certain or else against player 2 for certain. The first possibility can be eliminated since then player 2 would switch his announcement to ζ_1. In the case of the second possibility, it must be that ζ_2 and ζ_3 are approximately equal to Z_2 (which is the third coordinate of P_2 in figure 8.15). It follows that, if there exists an equilibrium of the type we are concerned with, it requires the players to adopt strategies which approximate the following:

Player 1. Choose an opening announcement of $\zeta_1 > Z_2 + \epsilon$. If the actual opening announcement observed by player 3 is ζ and disagreement occurs at stage 1, always demand x_1 at stage 2 if $\zeta > Z_2 + \frac{1}{2}\epsilon$ and x_2 if $\zeta < Z_2 + \frac{1}{2}\epsilon$.

Player 2. Choose an opening announcement of $\zeta_2 = Z_2$. If the actual opening announcement observed by player 3 is ζ and disagreement occurs at stage 1, always demand y_1 at stage 2 if $\zeta > Z_2 + \frac{1}{2}\epsilon$ and y_2 if $\zeta < Z_2 + \frac{1}{2}\epsilon$.

Player 3. Choose an opening announcement of $\zeta_3 = Z_2$. If the actual opening announcement observed by player 3 is ζ and disagreement occurs at stage 1, always demand z_1 at stage 2 if $\zeta > Z_2 + \frac{1}{2}\epsilon$ and z_2 if $\zeta < Z_2 + \frac{1}{2}\epsilon$.

The consequent payoffs to the players are:

Player 1: x_1.

Player 2: Y_2.

Player 3: $\lambda z_1 + (1 - \lambda) Z_2$.

Note that this situation is in equilibrium only if $Y_2 \geqslant y_1$ since player 2 can achieve a payoff of y_1 by copying player 1. But if Δ_2 is sufficiently close to 1, we shall have that $Y_2 < y_1$ and hence no equilibrium will exist with $\zeta_1 > \zeta_2 + \epsilon$. Equilibria with $\zeta_2 > \zeta_1 + \epsilon$ can never exist. If the time period between successive demands is sufficiently small it follows that we shall be left only with approximately non-revealing equilibria and an inductive argument yields the same result in the n period case.

The smoothing technique therefore selects only non-revealing equilibria in the iterated demand case provided that the time period between successive demands is sufficiently small and so the use of a generalized Nash bargaining solution remains valid.

It should be noted however that where the players discount at different rates, these will affect the 'bargaining powers' to be assigned to the players. As an example, consider the situation described in section 4 where the negotiators make demands at times $t_0, t_1, t_2, \ldots, t_n$ where n is very large. A fixed point can be calculated as in section 4 for the case $t_k = kT$ where $T > 0$. If the respective discount rates are δ_1, δ_2 and δ_3, the 'bargaining powers' so obtained are

$$\left.\begin{aligned} \tau_1 &= \frac{k\lambda}{1-\Delta_1} \\ \tau_2 &= \frac{k(1-\lambda)}{1-\Delta_2} \\ \tau_3 &= \frac{k}{1-\Delta_2} \end{aligned}\right\}$$

where $\Delta_i = \delta_i^T$ $(i = 1, 2, 3)$. The constant k is determined by the requirement that $\tau_1 + \tau_2 + \tau_3 = 1$. In view of our observations on signalling in this section, this result is only of real interest in the case when T is small. In the limiting case when $T \to 0+$ we obtain that

$$\left.\begin{aligned} \tau_1 &= k\lambda\rho_1 \\ \tau_2 &= k(1-\lambda)\rho_2 \\ \tau_3 &= k\rho_3 \end{aligned}\right\}$$

where $\rho_i = -(\log \delta_i)^{-1}$ and k is chosen so that $\tau_1 + \tau_2 + \tau_3 = 1$.

Thus, if demands are made continuously over time, the smoothing technique selects the payoff vector (x, y, z) which maximizes

$$x^{\lambda\rho_1} y^{(1-\lambda)\rho_2} z^{\rho_3}$$

over the set

$$\mathcal{X} = \{(x, y, z) : (x, z) \in \mathcal{X}_1 \quad \text{and} \quad (y, z) \in \mathcal{X}_2\}$$

where $\mathscr{X}_1$ is the payoff region at time 0 if player 1 is negotiator I and $\mathscr{X}_2$ is the payoff region at time 0 if player 2 is negotiator I.

Note that the effect of a player's impatience on his 'bargaining power' in this situation is very simple. One has only to multiply by the index

$$\rho_i = -(\log \delta_i)^{-1},$$

where δ_i is his discount factor.

Note 12. We have chosen to confine attention in section 8 and section 9 to three-player games. However, similar considerations are relevant in n-player games although then one has the additional problem that an equilibrium in the basic demand game is not necessarily non-revealing.

10 UNCERTAIN DISCOUNT RATES

The rate at which a player discounts utility is not a parameter which one is likely to know with any degree of accuracy. What influence will such uncertainty exercise on the analysis of section 4 in which we considered a two-person bargaining game with iterated demands?

If uncertainty exists about the appropriate discount rates, then it will be necessary to model the situation as a game with many players. All players who may be selected as negotiator I will have the same utility function over possibilities available at time 0 but will have different discount rates. Similarly for type II players. One can then embark on an analysis of the type considered in sections 5 and following. For the case in which demands are made continuously over time, an argument of the type described in section 9 will yield that each player of type I will receive (approximately) x and each player of type II will receive (approximately) y where (x, y) maximizes

$$x^{r_1} y^{r_2}$$

over the appropriate set $\mathscr{X}$. The numbers r_1 and r_2 are the 'expected bargaining powers' of the two negotiators – i.e.

$$\left.\begin{aligned} r_1 &= \mathscr{E}\rho_1 \\ r_2 &= \mathscr{E}\rho_2 \end{aligned}\right\}$$

where the random variable ρ_1 takes the value $-(\log \delta_i)^{-1}$ if negotiator I is player i (with discount rate δ_i) and similarly for the random variable ρ_2.

The problem therefore has a simple solution. One simply calculates the expected bargaining powers r_1 and r_2 for each negotiator and then pretends that each negotiator has this bargaining power for certain.

11 AXIOMS

Authors who write on Nash bargaining usually emphasize the fact that the Nash bargaining solution is characterized by a certain system of axioms. As far as the asymmetric Nash bargaining solutions considered in this chapter are concerned, it is fairly easy to give an appropriate axiomatic characterization. Indeed, all that is required to obtain a point (x, y) characterized by the fact that it maximizes

$$\prod_{i \in I} (x_i - \xi_i)^{b_i} \prod_{j \in J} (y_j - \eta_j)^{c_j}$$

for *some* constants b_i and c_j is the system of axioms given by Roth (1980) (see also chapter 2). These are much the same as those of Nash except that 'symmetry' is abandoned and 'Pareto-efficiency' is replaced by a condition even more innocent in appearance.

To obtain a product of the form of (8.1) in section 1, appropriate symmetry conditions are required. In particular, we require that the solution outcome does not depend on how the negotiators and players are labelled. (Thus two players who are identical apart from their labels will receive the same 'bargaining powers'.) Finally one requires an axiom which says essentially that a situation in which a particular player is chosen with probability m/n is the same as one in which m identical players are each chosen with probability $1/n$.

These remarks should be read as a postscript to the discussion of Harsanyi and Selten (1972 part III) who are more precise about the axioms outlined in the previous paragraph. I hope, however, that at least some readers will be convinced that an exclusively axiomatic approach to problems of the type studied in this chapter is inadequate and that there is no substitute for the examination of specific bargaining models.

REFERENCES

Harsanyi, J. C. 1977: *Rational Behaviour and Bargaining Equilibria in Games and Social Situations.* CUP, Cambridge.

Harsanyi, J. C. and Selten, R. 1972: A generalised Nash solution for two person bargaining games with incomplete information. *Management Science*, **18**, 80–106.

Nash, J. F. 1950: The bargaining problem. *Econometrica*, **18**, 155–62.

Nash, J. F. 1951: Non-cooperative games. *Annals of Mathematics*, **54**, 286–95.

Nash, J. F. 1953: Two-person cooperative games. *Econometrica*, **21**, 128–40.

Roth, A. 1980: Axiomatic models of bargaining. Lecture Notes in Economics and Mathematical Systems 170, Springer-Verlag, New York.

Selten, R. 1974: Bargaining under incomplete information. Institute of Mathematics Working Paper No. 20, University of Bielefeld.

9

A Bargaining Model with Incomplete Information about Time Preferences

A. Rubinstein

The chapter studies a strategic sequential bargaining game with incomplete information. Two players have to reach an agreement on the partition of a pie. Each player, in turn, has to make a proposal on how the pie should be divided. After one player has made an offer, the other must decide either to accept it or to reject it and continue the bargaining. Player 2 is one of two types, and player 1 does not know what type player 2 actually is.

A class of sequential equilibria (called bargaining sequential equilibria) is characterized for this game. The main theorem proves the (typical) uniqueness of the bargaining sequential equilibrium. It specifies a clear connection between the equilibrium and player 1's initial belief about his opponent's type.[1]

1 INTRODUCTION

One of the most basic human situations is one in which two individuals have to reach an agreement, chosen from among several possibilities. Traditional bargaining theory seeks to indicate one of the agreements as the expected (or desired) outcome on the sole basis of the set of possible agreements and on the point of non-agreement. Usually, the solution is characterized by a set of axioms. (For a survey of the axiomatic models of bargaining, see Roth 1979.)

Much of the recent work on bargaining aims at explaining the outcome of a bargaining situation using additional information about the time preference of the parties and the bargaining procedure. Such models are associated with the strategic approach. The players' negotiating manoeuvres are moves in a

1 This research began when I was a research fellow at Nuffield College, Oxford. It was continued at Bell Laboratories and at the Institute for Advanced Studies of the Hebrew University. I am grateful to these three institutions for their hospitality.

I owe thanks to many friends with whom I have discussed this subject during the past 5 years. Specifically I would like to thank Asher Wolinsky, Leo Simon and Ed Green for many valuable remarks; Avner Shaked, whose idea helped me simplify the proof of the main theorem; and an associate editor and referee of *Econometrica* for their extraordinarily helpful remarks.

non-cooperative game that describes the procedure of the bargaining; non-cooperative solutions to the game are explored.

The strategic approach also seeks to combine axiomatic cooperative solutions and non-cooperative solutions. Roger Myerson recently named this task the 'Nash program'. Though Nash is usually associated with the axiomatic approach, he was the first to suggest that this approach must be complemented by a non-cooperative game (see Nash 1953).

In chapter 3, I analysed the following bargaining model, using the strategic approach. Two players have to reach an agreement on the partition of a pie of size 1. Each player in turn has to make a proposal on the division of the pie. After one player has made an offer, the other must decide either to accept it or to reject it and continue the bargaining. The players have preference relations which are defined on the set of ordered pairs (s, t), which is interpreted as agreement on partition s at time t. Several properties are assumed: 'pie' is desirable, 'time' is valuable, the preference is continuous and stationary and the larger the portion of the pie the more 'compensation' a player needs to consider a delay of one period immaterial.

The set of outcomes of Nash equilibria for this game includes almost every possible agreement, and the bargaining in a Nash equilibrium may last beyond the first round of negotiations. The Nash equilibria are protected by threats, such as responding to a deviation by insisting on receiving the whole pie. It is very useful to apply the concept of (subgame) perfect equilibrium (see Selten 1965) which requires that the players' strategies induce an equilibrium in any subgame. This usually leads to a single solution, which is characterized by a pair of partitions, P_1 and P_2, that satisfies (a) player 1 is indifferent between 'P_2 today' and 'P_1 tomorrow', and (b) player 2 is indifferent between 'P_1 today' and 'P_2 tomorrow'. When a unique pair of P_1 and P_2 satisfies the above statements, the only perfect equilibrium partition is P_1 when 1 starts the bargaining and P_2 when 2 starts the bargaining. The structure of the unique perfect equilibrium is as follows: player 1 (2) always suggests $P_1(P_2)$ and player 2 (1) accepts any offer which is better for him than $P_1(P_2)$. For example, when each player has a fixed discounting factor δ, the perfect equilibrium outcome gives player 1 $1/(1 + \delta)$ if he starts the negotiation, and $\delta/(1 + \delta)$ if 2 starts. Both partitions tend to the equal partition when δ tends to 1. It is clear that the method demonstrated by this model is useful for analysing other bargaining procedures and other properties of time preferences. Additional properties of the model are studied by Binmore in chapter 6.

A critical assumption of the model is that each player has complete information about the other's preference. This assumption makes it less surprising that, typically, the bargaining in perfect equilibrium ends in the first period, although it could continue endlessly.

When incomplete information exists, new elements appear; a player may try to conclude from the other player's moves who his opponent really is; the other player may try to cheat him by leading him to believe that he is tougher than he actually is. In general terms, the incomplete information model enables us to address the issues of 'reputation building', signalling and self-selection mechanisms. Here continuation of bargaining beyond the first period becomes more likely.

A number of works have appeared on bargaining with incomplete information. For example, Harsanyi and Selten (1972) present a generalized Nash solution for two-person bargaining games with incomplete information. Myerson (1984) presents another generalization. Both solutions are characterized by sets of axioms. Finite-horizon bargaining games with incomplete information are treated by Fudenberg and Tirole (1983), Sobel and Takahashi (1983) and Ordover and Rubinstein (1982). In addition, Cramton (1982), Perry (in press) and Fudenberg et al. (1985) analyse seller–buyer infinite-horizon bargaining games in which reservation prices are uncertain, but time preferences are known.

The present chapter attempts to explain bargaining with incomplete information by investigating the model introduced in chapter 3, with one additional element: player 2 may be of one of two types: 2_w (weak) or 2_s (strong). The types differ in their time-preferences (for example, higher or lower discounting factors). Player 1 adopts an initial belief regarding the identity of player 2.

The difficulties of extending the notion of perfect equilibrium to games with incomplete information have been discussed extensively in the last few years (see Kreps and Wilson 1982). Generally speaking, the problem is that in order to check the optimality of 1's strategy after a given sequence of moves, we must verify his beliefs. Therefore, the notion of sequential equilibrium includes player 1's beliefs both on and off the equilibrium path.

The set of sequential equilibria for this game is very large. The freedom to choose new conjectures off the equilibrium path enables the players to establish credibility for 'too many' threats and thus to support 'too many' equilibria. I suggest that many of the sequential equilibria are unreasonable. These can be eliminated by making additional requirements on beliefs and on equilibrium behaviour. Indeed, as Kreps and Wilson claim, '... the formation [of sequential equilibria] in terms of players' beliefs gives the analyst a tool for choosing among sequential equilibria' (1982, p. 884).

The quite reasonable, additional requirements are described in section 5. The key requirement (*B*-1) makes it possible for type 2_s to screen himself: if player 2 rejects an offer made by player 1 and makes a counteroffer which is worse than 1's offer for the weak type but better for the strong type, then 1 concludes that player 2 is strong (2_s).

The main theorem of this chapter specifies a clear connection between the unique bargaining sequential equilibrium and 1's initial belief that 2 is 2_w, denoted by ω_0. The theorem states that there exists a cut-off point ω^* such that if ω_0 is strictly below ω^*, player 1 gives up: he offers the partition he would have offered if he thought he was playing against 2_s, and player 2, whatever his type, accepts it. If ω_0 is strictly above ω^*, some continuation of the bargaining is possible. In equilibrium, player 1 offers P_1, 2_w would accept it, and 2_s would reject it and offer P_2, which is accepted by 1. The partitions P_1 and P_2 are the only ones satisfying: (a) player 1 is indifferent between 'P_2 today' and the lottery of 'P_1 tomorrow' with probability ω_0 and 'P_2 after tomorrow' with probability $1-\omega_0$; (b) player 2_w is indifferent between 'P_1 today' and 'P_2 tomorrow'.

2 THE BARGAINING MODEL

Two players, 1 and 2, are bargaining on the partition of a pie. The pie will be partitioned only after the players reach an agreement. Each player in turn makes an offer and his opponent may agree to the offer. 'Y', or reject it, 'N'. Acceptance of an offer ends the bargaining. After rejection, the rejecting player has to make a counteroffer and so on without any given limit. There are no rules which bind the players to previous offers they have made.

Formally, let $S = [0, 1]$. A partition of the pie is identified with a number s in S by interpreting s as the proportion of the pie that 1 receives (2 receives $1-s$).

A strategy specifies the offer that a player makes whenever it is his turn to make an offer, and his reaction to any offer made by his opponent. A strategy includes the player's plans even after a series of moves that are inconsistent with the strategy itself.

Let F denote the set of all strategies available to the player who starts the bargaining. Formally, F is the set of all sequences of functions $f = \{f^t\}_{t=0}^{\infty}$, when for t even, $f^t: S^{t-1} \to S$ and for t odd, $f^t: S^t \to \{Y, N\}$, where S^t is the set of all sequences of length t of elements of S. (For example, $f^2(s^0, s^1)$ is a player's offer at time 2 assuming that he offered s^0, his opponent rejected it and made the offer s^1, which was rejected by the other player.) Similarly, G is the set of all strategies for a player whose first move is a response to the other player's offer. Note that mixed strategies are not allowed.

A typical outcome of the game is a pair (s, t), which is interpreted as agreement on the partition s in period t. Perpetual disagreement is denoted by $(0, \infty)$.

The outcome function of the game $P(f, g)$, then, takes the value (s, t) if two players who adopt strategies f and g reach an agreement s at period t, and the value $(0, \infty)$ if they do not reach an agreement.

The players have preference relations $\succsim_1$ and $\succsim_2$ on the set of pairs $S \times T = [0, 1] \times \{0, 1, 2, \ldots\}$. It is assumed that $\succsim_1$ satisfies the following assumptions (analogous assumptions are assumed about $\succsim_2$; recall that according to our notation, player 2 receives the fraction $1-x$ in the partition x):

Assumption (A-0). The relation is complete, reflexive and transitive.

Assumption (A-1). 'Pie' is desirable: If $x > y$, *then* $(x, t) \succ_1 (y, t)$.

Assumption (A-2). 'Time' is valuable: If $x > 0$ *and* $t_2 > t_1$, *then* $(x, t_1) \succ_1 (x, t_2)$.

Assumption (A-3). Continuity: The graph of $\succsim_1$ *is closed in* $(S \times T) \times (S \times T)$.

Assumption (A-4). Stationarity: $(x, t) \succsim_1 (y, t+1)$ *iff* $(x, 0) \succsim_1 (y, 1)$.

Assumption (A-5). The increasing compensation property: If

$$(x, 0) \sim_1 (x + \epsilon(x), 1),$$

then ϵ *is strictly increasing.*

For the sake of simplicity I make the following assumption:

Assumption (A-6). For every x there is $d(x) \in [0, 1]$, *such that* $(x, 1) \sim_1 (d(x), 0)$.

This implies that $(0, t) \sim_1 (0, 0)$ for all t and I assume further that $(0, \infty) \sim_1 (0, 0)$.

The present value of (x, t) is the t-fold composition of d with itself. Notice that by (A-1)–(A-3) and (A-6) the function d is increasing, continuous and satisfies $d(0) = 0$ and $d(x) < x$ for $x > 0$. By assumption (A-5), $x - d(x)$ is an increasing function and by assumption (A-4) the function d gives us all the information about the relation $\succsim_1$.

The most important family of relations satisfying the above assumptions are those induced by a utility function $x\delta^t$. The number $0 < \delta < 1$ is interpreted as the fixed discounting factor.

It is shown in Fishburn and Rubinstein (1982) that time preferences satisfying (A-0)–(A-4) and (A-6) can be represented by a utility function of the form $u(x)\delta^t$. A sufficient (but not necessary) condition for the preference to satisfy assumption (A-5) is that it be representable by $u(x)\delta^t$ where u is concave. The fixed bargaining costs preference (induced from a utility function $x - ct$) does not satisfy assumptions (A-5) and (A-6).

In the current chapter player 1's preference must be extended to refer to the lotteries of elements in $S \times T$. (The symbol $\omega 0_1 \oplus (1 - \omega) 0_2$ stands for the lottery which provides the outcome 0_1 with probability ω, and outcome 0_2 with probability $(1 - \omega)$.)

Assumption (A-7). A player maximizes the expectation of $u(x)\delta^t$ *for some concave function* $u(x)$ *and some* δ.

The new element in this chapter is the extension of the analysis to situations of incomplete information. Player 1's preference is known by player 2, but player 2 may possess one of the preferences $\succcurlyeq_w$ (weak) or $\succcurlyeq_s$ (strong). If 2 holds $\succcurlyeq_w(\succcurlyeq_s)$ it is said that he is of type $2w(2_s)$.

It is assumed that 2_w is more impatient than 2_s, that is:

Assumption (C-1). If $x \neq 1$ *and* $(y, 1) \sim_w (x, 0)$, *then* $(y, 1) \succ_s (x, 0)$.

With fixed discounting factors δ_w and δ_s, this assumption means that $\delta_s > \delta_w$.

In chapter 3 a full characterization of the perfect equilibrium outcomes of this game with complete information is presented. We summarize these results here.

Under the current assumptions about the time preferences there is a unique $(x^*, y^*) \in S^2$ satisfying

$$(x^*, 1) \sim_1 (y^*, 0) \quad \text{and} \quad (y^*, 1) \sim_2 (x^*, 0).$$

Denote the pair (x^*, y^*) by $\Delta(\succcurlyeq_1, \succcurlyeq_2)$. It was proved in chapter 3 that the only perfect equilibrium partition is x^* when 1 starts the bargaining, and y^* when 2 starts the bargaining and in either case the negotiation ends in the first period. Denote

$$\Delta(\succcurlyeq_1, \succcurlyeq_w) = (V_w, \hat{V}_w),$$

$$\Delta(\succcurlyeq_1, \succcurlyeq_s) = (V_s, \hat{V}_s).$$

Since 2_w is more impatient than 2_s, $V_w > V_s$ and $\hat{V}_w > \hat{V}_s$.

Figure 9.1 is useful at several subsequent points. The curve which includes the origin represents the present value to player 1 of the partition x tomorrow, that is, it includes all the pairs (x, y) such that $(x, 1) \sim_1 (y, 0)$. The other two curves represent the present value to player 2 of y tomorrow, depending on whether he is weak or strong. That is, they include the pairs (x, y) such that $(y, 1) \sim_s (x, 0)$ and $(y, 1) \sim_w (x, 0)$. The intersections of those curves with player 1's present value curve are $(V_s, \hat{V}_s)$ and $(V_w, \hat{V}_w)$.

Example. When the players have fixed discounting factors δ_1 and δ_2,

$$\Delta = \left(\frac{1-\delta_2}{1-\delta_1\delta_2}, \delta_1 \frac{1-\delta_2}{1-\delta_1\delta_2}\right).$$

In chapter 4, Binmore shows that where the δ_i are derived from the continuous discounting formula $x(e^{-r_i\tau})^t$ (τ is the length of one period of bargaining), then

$$\lim_{\tau \to 0} \Delta = \left(\frac{r_2}{r_1 + r_2}, \frac{r_1}{r_1 + r_2}\right).$$

In other words, when the time interval tends to 0, player 1 gets $r_2/r_1 + r_2$, regardless of whether he is the first or the second player to make an offer.

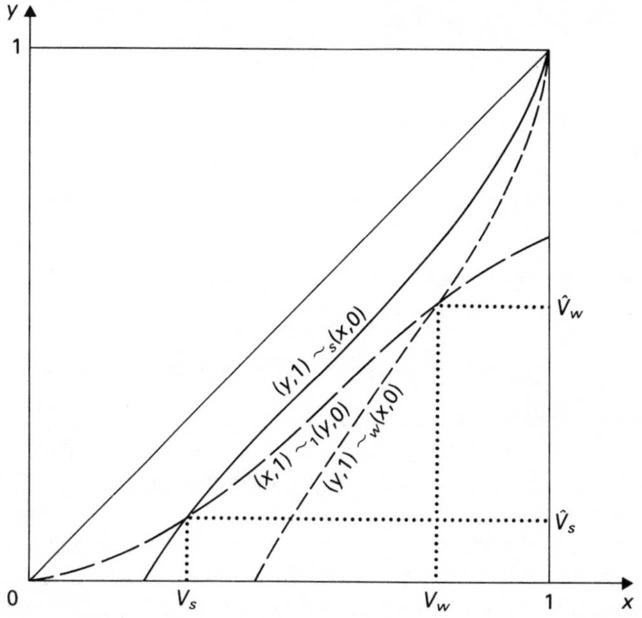

Figure 9.1

In particular, if the players have identical time preferences, the solution coincides with the equal partition.

The last assumption about the time preferences is the following:

Assumption (C-2). $(\hat{V}_w, 0) \prec_w (V_s, 1)$.

By (C-1), $(\hat{V}_w, 0) \prec_w (\hat{V}_s, 0)$. Here it is further assumed that type 2_w prefers the complete information partition between 1 and 2_s even if player 1 starts the bargaining, and there is a delay of one period in the agreement. Assumption (C-2) excludes the possibility that in equilibrium 2_w sorts himself by making an offer z satisfying $(z, 0) \succcurlyeq_1 (V_w, 1)$ (and thus $(z, 0) \preccurlyeq_w (\hat{V}_w, 0)$) and $(z, 0) \succcurlyeq_w (V_s, 1)$ (see proposition 4). Notice that if the players have discounting factors $\delta_i = e^{-r_i \tau}$ (see the previous example), then assumption (C-2) is satisfied whenever $r_w > r_s$ and τ is small enough.

3 NASH EQUILIBRIA

Section 2 defines a game with incomplete information. The usual solution concept for such games is Harsanyi's Bayesian Nash equilibrium (see Harsanyi

1967). Let (f, g, h) be a triple of strategies for 1, 2_w, and 2_s respectively. The outcome of the play of (f, g, h) is

$$P(f, g, h) = \langle P(f, g), P(f, h) \rangle.$$

This means that the outcome of the game is a pair of outcomes, one each for the cases that 2 is actually 2_w or 2_s.

Definition. A triple $(\hat{f}, \hat{g}, \hat{h}) \in F \times G \times G$ is a Nash equilibrium if there is no $f \in F$ such that

$$\omega_0 P(f, \hat{g}) \oplus (1 - \omega_0) P(f, \hat{h}) \succ_1 \omega_0 P(\hat{f}, \hat{g}) \oplus (1 - \omega_0) P(\hat{f}, \hat{h}),$$

and there is no $g \in G$ or $h \in G$ such that

$$P(\hat{f}, g) \succ_w P(\hat{f}, \hat{g}) \quad \text{or} \quad P(\hat{f}, h) \succ_s P(\hat{f}, \hat{h}).$$

(A similar definition is suitable for $(\hat{f}, \hat{g}, \hat{h}) \in G \times F \times F$.)

In other words, 1's strategy has to be a best response against 2_w's and 2_s's plans 'weighted' by ω_0 and $1 - \omega_0$, respectively, and $\hat{g}$ and $\hat{h}$ have to be best responses in the usual sense.

As in the complete information case, the set of Nash equilibria in this model is very large. Proposition 1 provides a complete characterization of the set of Nash equilibrium outcomes in this model. In particular, proposition 1 implies that, for every partition P, $\langle (P, 0), (P, 0) \rangle$ is a Nash equilibrium outcome.

Proposition 1. *$\langle (P_w, t_w), (P_s, t_s) \rangle = \langle 0_w, 0_s \rangle$ is a Nash equilibrium outcome if and only if*

$$0_w \succcurlyeq_w 0_s \quad \text{and} \quad 0_s \succcurlyeq_s 0_w.$$

Proof. If $\langle 0_w, 0_s \rangle$ satisfies the conditions, then the following is a description of a Nash equilibrium: both players demand the entire pie and reject every offer except at periods t_w and t_s when players 1 and 2_w, and 1 and 2_s, offer and accept P_w and P_s, respectively.

Obviously the conditions $0_w \succcurlyeq_w 0_s$ and $0_s \succcurlyeq_s 0_w$ are necessary. *QED*

4 SEQUENTIAL EQUILIBRIUM

The basic idea of sequential equilibrium is similar to the idea of perfect equilibrium: the players' strategies are best responses not only at the starting point, but at any decision node. For player 1, the test whether a strategy is the best response depends on 1's belief that 2 is 2_w. Therefore, a sequential equilibrium includes the method of updating l's belief that 2 is 2_w.

Define a belief system to be a sequence $\boldsymbol{\omega} = (\omega^t)_{t=-1,1,3,\ldots}$, such that $\omega^t = \omega_0$ and, for $t \geqslant 1$, $\omega^t : S^t \to [0, 1]$. $\omega^t(s^1, \ldots, s^t)$ is 2's belief that 2 is 2_w, after the sequence of rejected offers $s^1, \ldots, s^{t-1}$ and after 2 made offer s^t.

The formal definition of $(f, g, h, \boldsymbol{\omega})$ being a sequential equilibrium is messy but intuitively straightforward. The belief system must be consistent with the Bayesian formula. Moreover, after unexpected behaviour by player 2, player 1 makes a new conjecture regarding 2's type; the equilibrium behaviour of player 1 must be consistent with the new conjecture as long as no new deviation is observed. Player 1 does not change his belief about player 2 as a result of any deviation of his own.

For formal definition, let us use the notation $\mathbf{s}^t = (s^0 \ldots s^t)$ and the brief notation $\omega^t = \omega^t(\mathbf{s}^t)$, $f^t = f^t(\mathbf{s}^t)$, or $f^t(\mathbf{s}^{t-1})$ according to the evenness of t, and similarly define g^t and h^t.

The Bayesian requirements are:

a If $g^{T-1} = h^{T-1} = N$ and $g^T = h^T = s^T$, then $\omega^T = \omega^{T-2}$.

b If $0 < \omega^{T-2} < 1$, $g^{T-1} = N$ and $g^T = s^T$, and either $h^{T-1} = Y$ or $h^T \neq s^T$, then $\omega^T = 1$.

c If $0 < \omega^{T-2} < 1$, $h^{T-1} = N$, $h^T = s^T$ and either $g^{T-1} = Y$ or $g^T \neq s^T$, then $\omega^T = 0$,

Finally I add to the definition of sequential equilibrium another constraint which does not follow from Kreps-Wilson's definition. If 1 concludes with probability 1 that player 2 is of a certain type, he continues to hold this belief whatever occurs after he comes to this conclusion; in other words, if he concludes that $\omega^t = 0$ or $\omega^t = 1$, he continues to play the game as if it were a game with complete information against 2_s or 2_w respectively.

Formally, if $\omega^{T-2} = 1$, then $\omega^T = 1$, and if $\omega^{T-2} = 0$, then $\omega^T = 0$.

The next proposition can be proved using the arguments presented in chapter 3.

Proposition 2. *Let* $(f, g, h, \boldsymbol{\omega})$ *be a sequential equilibrium in the game when* 1 *starts; then* $P(f, h) \succcurlyeq_1 (V_{s_2}, 0)$ *and* $P(f, g) \preccurlyeq_1 (V_w, 0)$; *and if* 2 *starts,* $P(f, h) \succcurlyeq_1 (\hat{V}_s, 0)$ *and* $P(f, g) \preccurlyeq_1 (\hat{V}_w, 0)$.

Thus, a sequential equilibrium outcome cannot be better (worse) for player 1 than the perfect equilibrium outcome in the complete information bargaining game with $2_w(2_s)$.

The next proposition demonstrates that the set of sequential equilibrium outcomes is very large.

The sequential equilibrium, which is constructed in the proof, is supported by player 1's belief that a deviation must be of type 2_w. Whenever there is a deviation from the equilibrium path, player 1 concludes that he is playing against 2_w. Such conjectures serve as threats against player 2. The definition of sequential equilibrium that was suggested to exclude non-credible threats does not exclude threatening by beliefs.

Proposition 3. *For every* ω_0*, and for all* x^* *satisfying* $(\hat{V}_s, 1) \succcurlyeq_w (x^*, 0)$ *and* $(x^*, 0) \succcurlyeq_s (\hat{V}_w, 1)$*, either* $\langle (x^*, 0), (x^*, 0) \rangle$ *or* $\langle (x^*, 0), (y^*, 1) \rangle$ *(where* $(y^*, 1) \sim_w (x^*, 0)$*) is a sequential equilibrium outcome.*

Proof. Let x^* satisfy $(\hat{V}_s, 1) \succcurlyeq_w (x^*, 0)$ and $(x^*, 0) \succcurlyeq_s (\hat{V}_w, 1)$. Let y^* and z^* satisfy $(y^*, 1) \sim_w (x^*, 0)$ and $(z^*, 0) \sim_s (y^*, 1)$. Let $\tilde{y}$ and $\tilde{z}$ satisfy $(\tilde{y}, 0) \sim_1 (x^*, 1)$ and $(\tilde{z}, 0) \sim_w (\tilde{y}, 1)$. Let us distinguish between the following two comprehensive (and non-exclusive) cases:

Case I: $\omega_0 (x^*, 0) \oplus (1 - \omega_0)(y^*, 1) \succcurlyeq_1 (z^*, 0)$.
Case II: $\omega_0 (\tilde{z}, 0) \oplus (1 - \omega_0)(\tilde{y}, 1) \preccurlyeq_1 (x^*, 0)$.

By proposition 7, at least one of the following inequalities is true:

$$\omega_0 (x^*, 1) \oplus (1 - \omega_0)(y^*, 2) \succcurlyeq_1 (y^*, 0),$$

$$(\tilde{y}, 0) \succcurlyeq_1 \omega_0 (\tilde{z}, 1) \oplus (1 - \omega_0)(\tilde{y}, 2).$$

Combining $(y^*, 0) \succcurlyeq_1 (z^*, 1)$ and $(x^*, 1) \sim_1 (\tilde{y}, 0)$ with the stationarity of $\succcurlyeq_1$, we get that for all ω_0 either case I or case II must be true.

Figure 9.2 is a description of a sequential equilibrium for case I with the outcome $\langle (x^*, 0), (y^*, 1) \rangle$:

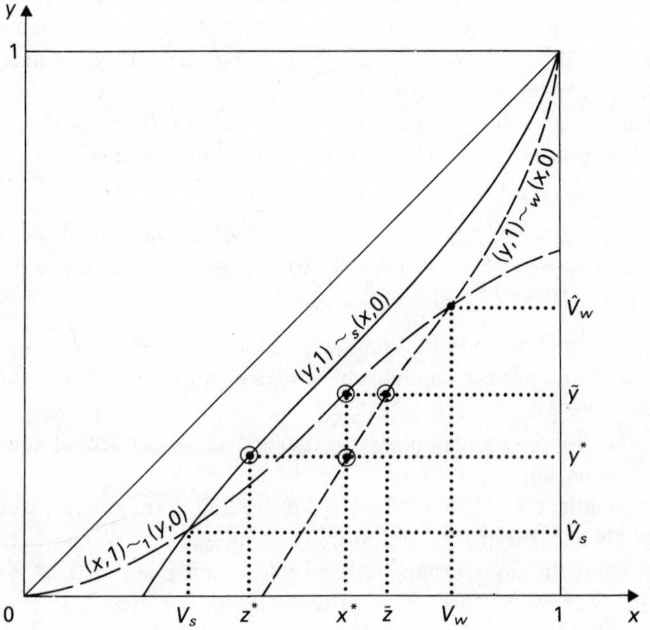

Figure 9.2

Unless 1 changes his belief, he always demands x^* and agrees at most to y^*; type 2_w accepts only any offer below x^* and demands y^*; type 2_s accepts offer x satisfying $(x, 0) \succsim_s (y^*, 1)$ and always offers y^*. On condition that 2 made a move unplanned by any of the types, 1 changes ω^t to 1 and the continuation is as in the complete information game between 1 and 2_w. Clearly the only change for a profitable deviation of 1 would be to make a lower demand like z^* which would be accepted by both types. But since $\omega_0(x^*, 0) \oplus (1 - \omega_0)(y^*, 1) \succsim_1 (z^*, 0)$, this is not profitable. Checking the other conditions of the sequential equilibrium condition is straightforward.

The following is a description of a sequential equilibrium for case II with the outcome $\langle(x^*, 0), (x^*, 0)\rangle$:

Unless 1 changes his belief, he always offers x^* and accepts y only if $(y, 0) \succsim_1 (x^*, 1)$, $(y \geqslant \tilde{y})$. Both 2_w and 2_s agree to settle for x only if $(x, 0)$ is preferable (according to $\succsim_w$ and $\succsim_s$ respectively) to $(\tilde{y}, 1)$. They always offer $\tilde{y}$. After 2 made a move unexpected from both types, 1 changes ω^t to 1 and the continuation is as in the complete information game between 1 and 2_w.

Notice that if 1 demands $x > x^*$, 2_s rejects it because $(\tilde{y}, 1) \succ_s (x^*, 0)$. Thus the most that 1 can achieve by deviating is that 2_w would agree to $\tilde{z}$ and 2_s would offer $\tilde{y}$. However since $\omega_0(\tilde{z}, 0) \oplus (1 - \omega_0)(\tilde{y}, 1) \precsim_1 (x^*, 0)$, this is not profitable. The requirement that $(x^*, 0) \succsim_s (\hat{V}_w, 1)$ is important to assure that 2_s will not gain by rejecting x^*. *QED*

5 BARGAINING SEQUENTIAL EQUILIBRIUM

Section 4 shows that it might be desirable to place additional requirements on the belief systems in a sequential equilibrium $(f, g, h, \boldsymbol{\omega})$.

In order to demonstrate the first assumption, (B-1), imagine that 1 makes an offer x and 2 rejects it and offers y which satisfies that $(y, 1) \succsim_s (x, 0)$ and $(x, 0) \succ_w (y, 1)$. That is, 2 presents a counteroffer which is better for 2_s and worse for 2_w than the original offer x. Then, we assume, 1 concludes that he is playing against 2_s.

This assumption is related to an element which is missing from most studies in game theory: we tend to conclude facts from other people's behaviour even when the unexpected occurs. (B-1) is this type of inference. The effect of (B-1) is to exclude sequential equilibria like those constructed in the proof of proposition 3.

Assumption (B-1). $\boldsymbol{\omega}$ *is such that if* $\omega^{t-2}(s^{t-2}) \neq 1$, $(s^t, 1) \succsim_s (s^{t-1}, 0)$, *and* $(s^{t-1}, 0) \succ_w (s^t, 1)$, *then* $\omega^t(s^1 \ldots s^{t-1}, s^t) = 0$.

The next assumption states that 2's insistence cannot be an indication that 2 is more likely to be 2_w. Assume that 2 rejects an offer x and suggests an offer y, such that

$$(y, 1) \succcurlyeq_w (x, 0) \quad \text{and} \quad (y, 1) \succcurlyeq_s (x, 0).$$

Then when 1 updates his belief, his subjective probability that he is playing against 2_w does not increase.

Assumption (B-2). If $(s^t, 1) \succcurlyeq_s (s^{t-1}, 0)$ *and* $(s^t, 1) \succcurlyeq_w (s^{t-1}, 0)$, *then* $\omega^t(s^t) \leqslant \omega^{t-2}(s^{t-2})$.

The next two assumptions place direct restrictions on the players' equilibrium behaviour, rather than describing their beliefs or preferences.

Assumption (B-3) is a 'tie-breaking' assumption. If player 1 has been offered a partition x and after rejecting it he expects to reach an agreement whereby he is indifferent to x, then 1 accepts x.

Assumption (B-3). If $\langle P(f_{|s^1 \ldots s^t}, g_{|s^1 \ldots s^t}, h_{|s^1 \ldots s^t}), 1\rangle \sim_1 (s^t, 0)$, *then* $f(s^t) = Y$. ($f_{|s^1 \ldots s^t}$ *is the residual strategy of* f *after the history* $s^1 \ldots s^t$.)

The last assumption is that player 2 never makes an offer lower than $\hat{V}_s$.

Assumption (B-4). Whenever it is 2's turn to make an offer,

$$g^t \geqslant \hat{V}_s \quad \textit{and} \quad h^t \geqslant \hat{V}_s.$$

Notice that by proposition 2, player 2 rejects any offer which is lower than $\hat{V}_s$. Still, the players could use such offers as a communication method. Assumption (B-4) excludes this possibility (see figure 9.3).

I would also like to mention the *Econometrica* associate editor's suggestion that assumption (B-4) can be replaced by a requirement on the beliefs that if 2_w is supposed to accept an offer and 2_s is supposed to reject it, then whatever is 2's offer, 1 concludes from 2's rejection that he is playing against 2_s. The reader can verify that the alternative assumption plays the same role as assumption (B-4) in the only place it is used (the proof of proposition 6).

Definition. (f, g, h, ω) is a *bargaining sequential equilibrium* if it is a sequential equilibrium and satisfies (B-1)-(B-4).

For the following examination of some of the properties of bargaining sequential equilibrium, let (f, g, h, ω) be a bargaining sequential equilibrium:

Proposition 4. *Whenever it is 2's turn to make an offer,* 2_w *and* 2_s *make the same offer.*

Proof. Assume that there is a history after which 2_w and 2_s make the offers y and z, respectively, and that $y \neq z$. If 1 accepts both y and z, then the type making the higher offer will deviate to the lower offer.

If 1 rejects both y and z, then in the next period he believes he knows which type 2 is, and he offers V_w or V_s accordingly. Then type 2_w will gain by offering z.

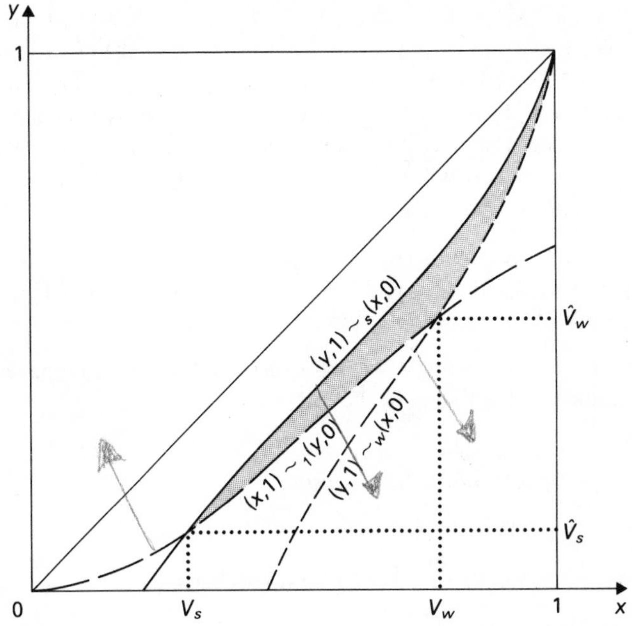

Figure 9.3

If 1 accepts only z, then the outcome against 2_w will be V_w in the next period. By definition of sequential equilibrium, $(z, 0) \succcurlyeq_s (V_w, 1)$. By assumption (C-1), $(z, 0) \succ_w (V_w, 1)$. Thus, 2_w can deviate and gain by offering z.

If 1 accepts y and rejects z, it must be that $(y, 0) \succcurlyeq_1 (V_w, 1)$ and $(y, 0) \succcurlyeq_w (V_s, 1)$, in contradiction to assumption (C-2). *QED*

Proposition 5. *If both 2_w and 2_s accept an offer x, then $x \leqslant V_s$.*

Proof. If $x > V_s$, then 2_s may deviate, say 'N' and suggest y at the next period, where $(y, 1) \succ_s (x, 0)$, $(y, 1) \prec_w (x, 0)$, and $(y, 0) \succ_1 (V_s, 1)$.

Any y that satisfies (x, y) is in the shaded area in figure 9.3 will do. By assumption (B-1), player 1 concludes that 2 is 2_s, and since $(y, 0) \succ_1 (V_s, 1)$, he accepts the offer y; since $(y, 1) \succ_s (x, 0)$, 2_s gains by the deviation. When $x = 1$ there is no (x, y) in the shaded area of the figure. However, $(\hat{V}_w, 1) \succ_s (1, 0)$ and 2_s could gain by offering and accepting $\hat{V}_w$. *QED*

Proposition 6. *If 2_w accepts offer x and 2_s rejects it, then 2_s makes an offer y such that $(y, 1) \sim_w (x, 0)$, and player 1 accepts it.*

Proof. If 1 rejects y then he offers V_s in the next period. Therefore, $(V_s, 1) \succcurlyeq_1 (y, 0)$, and by assumption (B-3) $(V_s, 1) \succ_1 (y, 0)$ which means $y < \hat{V}_s$, contradicting (B-4). Thus, 1 accepts y, which implies $(y, 0) \succsim_1 (V_s, 1)$. It must be that $(x, 0) \succsim_w (y, 1)$. If $(x, 0) \succ_w (y, 1)$ and $y > \hat{V}_s$, then 2_s gains by decreasing the offer to $y - \epsilon$ (for $\epsilon > 0$ small enough). This lower offer persuades 1 that 2 is 2_s and 1 accepts it because $(y - \epsilon, 0) \succ_1 (V_s, 1)$.

If $(x, 0) \succ_w (y, 1)$ and $y = \hat{V}_s$, then 1 may deviate and demand $x + \epsilon$. Type 2_w will accept the offer if ϵ is small enough to satisfy $(x + \epsilon) \succ_w (y, 1)$, and 2_s will reject it and will offer $\hat{V}_s$, the same offer he intended to make before the deviation. Thus 1 gains by the deviation. *QED*

Remark. Unless we assume (B-3) and (B-4) we can get additional equilibria, where 1 offers x, 2_w accepts x, 2_s rejects it and offers y ($y < \hat{V}_s$), which 1 rejects in favour of the agreement V_s in the next period.

From propositions 4, 5 and 6 we may conclude that a bargaining sequential equilibrium must end with one of the following:

(T-1): *1 offers x, 2_w accepts it and 2_s rejects it; 2_s offers y at the next period and 1 accepts it. The offer y satisfies*

$$(x, 0) \sim_w (y, 1),$$

$$(y, 0) \succsim_1 (V_s, 1) \quad (i.e., y \geqslant \hat{V}_s),$$

and

$$\omega_0(x, 0) \oplus (1 - \omega_0)(y, 1) \succsim_1 (V_s, 0).$$

(T-2): *1 offers V_s and 2 accepts it.*

(T-3): *2_w and 2_s offer y and 1 accepts it.*

6 THE POINT (x^ω, y^ω)

Let $z(x)$ be the z satisfying $(x, 0) \sim_w (z, 1)$ if such a z exists. Define for every $0 \leqslant \omega \leqslant 1$,

$$d^\omega(x) = y \quad \text{where} \quad (y, 0) \sim_1 \omega(x, 1) \oplus (1 - \omega)(z(x), 2).$$

Thus, $d^\omega(x)$ is the minimum that 1 would now agree to accept from 2 if he expects a bargaining sequential equilibrium in the subgame starting the next period, where his agreement with 2_w is x, and his agreement with 2_s is $z(x)$ one period later.

The function d^ω has several straightforward properties. It is continuous, it is strictly increasing, it satisfies $d^\omega(x) < d_1(x) < x$ for $x \geqslant V_s$, where $(d_1(x), 0) \sim_1 (x, 1)$. It satisfies $d^{\omega_1}(x) \geqslant d^{\omega_2}(x)$ if $\omega_1 > \omega_2$.

Let d_w be the graph of the present value function for 2_w (i.e. $(d_w(y), 0) \sim_w (y, 1)$). For x_0 satisfying $(x_0, 0) \sim_w (0, 1)$, $d^\omega(x_0) =$

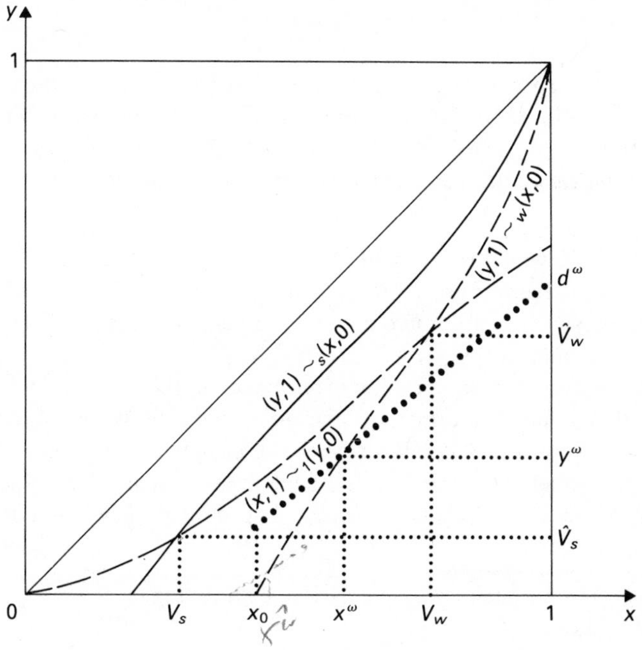

Figure 9.4

$\omega d_1(x_0) + (1-\omega_0)\cdot 0 \geqslant 0$ and therefore d^ω and d_w must have a non-empty intersection (see figure 9.4).

The next proposition states that the intersection of d^ω and d_w consists of a single point (x^ω, y^ω). This point plays an important role in the main theorem:

Proposition 7. *There is a unique point* (x, y) *such that* $(x, 0) \sim_w (y, 1)$ *and* $y = d^\omega(x)$.

Proof. Assume that both (x_1, y_1) and (x_2, y_2) satisfy the two equations and $x_1 < x_2$. By definition,

$$(y_i, 0) \sim_1 \omega(x_i, 1) \oplus (1-\omega)(y_i, 2) \quad \text{for} \quad i = 1, 2.$$

By (A-7), $(1-\omega)(u(y_i) - \delta^2 u(y_i)) = \omega(\delta u(x_i) - u(y_i))$. Therefore $y_2 > y_1$ implies that $\delta u(x_2) - u(y_2) > \delta u(x_1) - u(y_1)$ and $u(x_2) - u(x_1) > u(y_2) - u(y_1)$. The function u is concave and $x_i > y_i$. Therefore, $y_2 - y_1 > x_2 - x_1$, contradicting (A-5) as to $\succcurlyeq_w$. *QED*

7 THE MAIN RESULT

Theorem. *For a game starting with player* 1*'s offer*: (a): *If* ω_0 *is high enough such that* $y^{\omega_0} > \hat{V}_s$, *then the only bargaining sequential equilibrium outcome is* $\langle (x^{\omega_0}, 0), (y^{\omega_0}, 1) \rangle$. (b): *If* ω_0 *is low enough such that* $y^{\omega_0} < \hat{V}_s$, *then the only bargaining sequential equilibrium outcome is* $\langle (V_s, 0), (V_s, 0) \rangle$.

In the first case, player 1 offers x^{ω_0}, 2_w accepts this offer, and 2_s rejects it and offers y^{ω_0}, which is accepted. 2_w is indifferent between $(x^{\omega_0}, 0)$ and $(y^{\omega_0}, 1)$. In the second case, 1 offers V_s and both types 2_w and 2_s accept it. In the boundary between the two zones more than one bargaining sequential equilibrium outcome is possible.

Let us contrast the equilibrium determined in the theorem with that derived with complete information: the weak player is better off whereas the strong player suffers a disadvantage in the incomplete information game. It is not clear whether player 1 would benefit from having the information about player 2's type. Delays can occur but only if they result in some information transmission.

Example. Assume that 1, 2_w and 2_s have discounting factors δ, δ_w and δ_s, respectively. Then

$$V_s = \frac{1-\delta_s}{1-\delta\delta_s}, \quad V_w = \frac{1-\delta_w}{1-\delta\delta_w},$$

$$\hat{V}_s = \delta V_s, \quad \text{and} \quad \hat{V}_w = \delta V_w.$$

Here, if

$$\omega_0 > \frac{V_s - \delta^2 V_s}{1-\delta_w + \delta V_s(\delta_w - \delta)} = \omega^*$$

we are in case (a) of the theorem and

$$x^{\omega} = \frac{(1-\delta_w)(1-\delta^2(1-\omega))}{1-\delta^2(1-\omega)-\delta\delta_w\omega}.$$

Notice that $1 \geqslant \omega^*$ if and only if $V_w \geqslant V_s$.

Where the discounting factors are derived from the continuous discounting formula we get the following limit result when we tend the length of a period to zero: If $\omega_0 > 2r_s/(r_s + r_w)$, the outcome is $\omega_0 r_w/(\omega_0 r_w + (2-\omega_0)r)$, and if $\omega_0 < 2r_s/(r_s + r_w)$, the outcome is $r_s/(r + r_s)$.

The main difficulty here is to prove the uniqueness of the bargaining sequential equilibrium outcome. However, for a better insight of the result, I begin by describing a bargaining sequential equilibrium which induces the outcome $\langle(x^{\omega_0}, 0), (y^{\omega_0}, 1)\rangle$ if $y^{\omega_0} > \hat{V}_s$ and induces the outcome $\langle(V_s, 0), (V_s, 0)\rangle$ if $y^{\omega_0} < \hat{V}_s$.

1's *beliefs*: Player 1 does not change his initial belief unless player 2 has rejected an offer s^{t-1} and offers s^t such that $(s^{t-1}, 0) \succcurlyeq_w (s^t, 1)$; then he changes his belief to $\omega^t(\mathbf{s}^t) = 0$. (This belief system certainly satisfies (B-1)–(B-2).)

1's *strategy*: Let $\mathbf{s}^t$ be the 'empty' history or a history that ends with 2's offer s^t. Let $\omega = \omega^t(\mathbf{s}^t)$. If $y^\omega \leqslant \hat{V}_s$ (this case includes the possibility that $\omega = 0$), then 1 accepts s^t if and only if $s^t \geqslant \hat{V}_s$; otherwise he accepts s^t only if $s^t \geqslant y^\omega$. If 1 rejects s^t (or s^t is the empty history), then he makes the offer x^ω if $y^\omega > \hat{V}_s$ and the offer V_s if $y^\omega \leqslant \hat{V}_s$.

2_w's *strategy*: Let $\mathbf{s}^t$ be a history that ends with 1's offer s^t. If $y^\omega \leqslant \hat{V}_s$, then 2_w accepts s^t if and only if $(s^t, 0) \succcurlyeq_w (\hat{V}_s, 1)$. If $y^\omega > \hat{V}_s$ then 2_w accepts s^t only if $s^t \leqslant x^\omega$. In case 2_w rejects s^t he offers $\hat{V}_s$ if $y^\omega \leqslant \hat{V}_s$ and if $y^\omega > \hat{V}_s$ then he offers the lowest number from among y^ω and the y satisfying $(s^t, 0) \succcurlyeq_w (y, 1)$ and $y \geqslant \hat{V}_s$.

2_s's *strategy*: Player 2_s always makes the same offer that 2_w does. He accepts only V_s or less.

Whenever $\omega^t(\mathbf{s}^t) = 0$ or $\omega^t(\mathbf{s}^t) = 1$, player 1 follows the perfect equilibrium strategy described in chapter 3 for games with complete information between 1 and 2_s, or between 1 and 2_w. If $\omega^t = 0$, 2_s follows the perfect equilibrium strategy and 2_w chooses a perfect best-response strategy to 1's strategy (and similarly for the case $\omega^t = 1$).

8 PROOF OF THE UNIQUENESS OF THE BARGAINING SEQUENTIAL EQUILIBRIUM

Assume that $y^{\omega_0} > \hat{V}_s$; in what follows it is proved that the only bargaining sequential equilibrium outcome is $\langle(x^{\omega_0}, 0), (y^{\omega_0}, 1)\rangle$. The proof for the case that $y^{\omega_0} < \hat{V}_s$ is very similar.

Let BSE_1 be a short notation for a bargaining sequential equilibrium in the game where 1 starts the bargaining, and let BSE_2 be a short notation for a bargaining sequential equilibrium in a subgame in which 2 starts the bargaining, if an offer of 1 exists such that 2's reply 'N' and continuation according to the BSE_2 is a bargaining sequential equilibrium in the subgame starting after 1's offer. Given a number $0 \leqslant \omega \leqslant 1$, define:

$$A^\omega = \{u \mid (u, 0) \sim_w 0_w \text{ where } \langle 0_w, 0_s\rangle \text{ is a } BSE_1 \text{ outcome}\},$$

$$B^\omega = \{v \mid (v, 0) \sim_w 0_w \text{ where } \langle 0_w, 0_s\rangle \text{ is a } BSE_2 \text{ outcome}\}.$$

In section 7 it was proved that $x^{\omega_0} \in A^{\omega_0}$ and $y^{\omega_0} \in B^{\omega_0}$.

Lemma 1. *$\langle (V_s, 0)(V_s, 0)\rangle$ is not a BSE_1 outcome and $\langle (\hat{V}_s, 0)(\hat{V}_s, 0)\rangle$ is not a BSE_2 outcome.*

Proof. In any of these cases, player 1 would be able to deviate and offer a partition x such that $(x, 0) \sim_w (\hat{V}_s, 1)$. Type 2_w must accept it. Since $y^\omega > \hat{V}_s$, $\omega_0(x, 1) \oplus (1-\omega_0)(\hat{V}_s, 2) \succ_1 (\hat{V}_s, 0) \sim_1 (V_s, 1)$; thus $\omega_0(x, 0) \oplus (1-\omega_0)(\hat{V}_s, 1) \succ_1 (V_s, 0)$, and 1 gains by the deviation.

Lemma 2. *If $v \in B^\omega$ there exists a $u \in A^\omega$ such that $(u, 0) \sim_w (v, 1)$.*

Proof. Define the following BSE_1: Player 1 offers u, 2_w accepts it, 2_s rejects it and offers v at the next period. Type 2_w rejects any offer higher than u and 2_s rejects any offer higher than V_s. In the case that 1's offer is rejected, the continuation is like the original BSE_2 without a change in 1's belief.

Notice that $\omega_0(u, 1) \oplus (1-\omega_0)(v, 2) \succ_1 (\hat{V}_s, 0) \sim_1 (V_s, 1)$; thus $\omega_0(u, 1) \oplus (1-\omega_0)(v, 1) \succ_1 (V_s, 0)$ and 1 cannot gain by offering V_s.

Lemma 3. *Define $m_1 = \inf A^{\omega_0}$, $m_2 = \inf B^{\omega_0}$; then $(m_1, 0) \sim_w (m_2, 1)$.*

Proof. By lemma 2, $(m_1, 0) \succcurlyeq_w (m_2, 1)$. Assume $(m_1, 0) \succ_w (m_2, 1)$. Notice that $t_w = 0$ in every BSE_1 outcome $\langle 0_w, 0_s\rangle$ such that $0_w = (s_w, t_w)$ is $\succcurlyeq_w$ close enough to $(m_1, 0)$; otherwise $(s_w, t_w - 1) \sim_w (v, 0)$ for $v < m_2$, and $(s_w, t_w - 1)$ is a BSE_2 outcome and thus $v \in B^{\omega_0}$. If 1 deviates from the BSE_1 which yields the outcome $\langle 0_w, 0_s\rangle$, and demands x such that $(x, 0) \sim_w (m_2, 1)$, 2_w agrees and 1 gains since $x > m_1$ and, by lemma 1, the BSE_1 must be of type $(T-1)$. Thus $(m_1, 0) \sim_w (m_2, 1)$.

Lemma 4. $\omega_0(m_1, 1) \oplus (1-\omega_0)(z(m_1), 2) \sim_1 (m_2, 0)$. (*Recall that* $(x, 0) \sim_w (z(x), 1)$.)

Proof. The existence of bargaining sequential equilibrium described in section 5 implies that $m_1 \leqslant x^{\omega_0}$ and $m_2 \leqslant y^{\omega_0}$. Therefore by lemma 3

$$\omega_0(m_1, 1) \oplus (1-\omega_0)(z(m_1), 2) \succcurlyeq_1 (m_2, 0).$$

Assume that we have a strict inequality. A BSE_2 with an outcome $\langle 0_w, 0_s\rangle$ where 2_w is indifferent between $0_w = (s_w, t_w)$ and an outcome in which he receives a partition whose present value is quite close to m_2, satisfies that $t_w = 0$ (otherwise it contradicts lemma 3). Thus the BSE_2 has to be of type (T-3). Player 1 can deviate profitably by demanding m_1, accepting it from 2_w, and accepting $z(m_1)$ from 2_s.

Lemma 5. $m_1 = x^{\omega_0}$ *and* $m_2 = y^{\omega_0}$.

Proof. Use lemmas 3 and 4.

Lemma 6. *Define* $M_1 = \sup_{\omega \leqslant \omega_0} A^\omega$ *and* $M_2 = \sup_{\omega \leqslant \omega_0} B^\omega$; *then* $(M_1, 0) \preccurlyeq_w (M_2, 1)$.

Proof. Similar to the proof of lemma 2.

Lemma 7. $(M_1, 0) \sim_w (M_w, 1)$.

Proof. This is clear if $M_1 = x^{\omega_0}$. Assume $M_1 > x^{\omega_0}$ and $(M_1, 0) <_w (M_2, 1)$. Let $x > x^{\omega_0}$ be close enough to M_1 and $\epsilon > 0$ small enough such that $x \in A^{\omega}$ for some $\omega \leqslant \omega^0$, $z(x) - \epsilon > M_2$, and $(z(x) - \epsilon, 0) >_1 \omega_0(M_1, 1) + (1 - \omega_0)$ $(z(M_1), 2)$. There exists a BSE_1 such that 1 offers x and 2_w accepts it (otherwise it contradicts the definition of M_2). Assume 2 deviates and offers $z(x) - \epsilon$. Player 1 must reject it.

Let $\bar{\omega} \leqslant \omega_0$ be 1's belief that 2 is 2_w after the offer $z(x) - \epsilon$. Then

$$(z(x) - \epsilon, 0) <_1 \bar{\omega}(M_1, 1) \oplus (1 - \bar{\omega})(z(M_1), 2).$$

When rejecting $z(x) - \epsilon$, 1 must expect an outcome better than $z(x) - \epsilon$ of a bargaining sequential equilibrium in the subgame after rejection. The last inequality with lemma 1 implies that this bargaining sequential equilibrium must be of type (T-3). Therefore there is $y \in B^{\bar{\omega}}$ such that $(y, 2) <_1$ $(z(x) - \epsilon, 0)$. Since $z(x) - \epsilon > M_2$, this contradicts M_2's definition.

Lemma 8. $\omega_0(M_1, 1) \oplus (1 - \omega_0)(M_2, 2) \sim_1 (M_2, 0)$.

Proof. By lemma 7 and $M_1 \geqslant x^{\omega_0}$,

$$\omega_0(M_1, 1) \oplus (1 - \omega_0)(M_2, 2) \leqslant_1 (M_2, 0).$$

Assume $\omega_0(M_1, 1) + (1 - \omega_0)(M_2, 2) <_1 (M_2, 0)$. Let $y < v < M_2$ satisfy that v is the 2_w's present value of a BSE_2, $\omega_0(M_1, 1) \oplus (1 - \omega_0)(M_2, 2) <_1$ $(y, 0)$, and $(M_2, 2) <_1 (y, 0)$.

By the definition of BSE_2 there exists an offer of player 1, x, such that 2's reply 'N' and continuation according to the BSE_2 is a bargaining sequential equilibrium in the subgame starting with 1's offer. Then if player 2 deviates by refusing x and offering y then player 1 must reject y; otherwise it would be a profitable deviation. This contradicts the fact that by the choice of y, and since 1's belief that 2 is of type 2_w after 2's deviation is lower than ω_0, it is optimal for 1 to accept y.

Lemma 9. $M_1 = x^{\omega_0}$ *and* $M_2 = y^{\omega_0}$.

Proof. A conclusion of lemmas 7 and 8.

9 FINAL REMARKS

9.1 The length of negotiation

There is a significant difference between the bargaining sequential equilibria in the two cases where $y^{\omega_0} > \hat{V}_s$ and $y^{\omega_0} < \hat{V}_s$. When $y^{\omega_0} < \hat{V}_s$ the bargaining ends at the first period and there is no screening of player 2's type, if $y^{\omega_0} >$

$\hat{V}_s$, then the bargaining in equilibrium might continue into the second period. Further research is needed for clarifying the direction for the generalization of the current result.

9.2 The choice of conjectures

In Rubinstein 1985 I study the set of sequential equilibrium outcomes under several other assumptions about the choice of conjectures in the bargaining game with fixed bargaining costs. Specifically, I study the 'optimistic conjectures' according to which a deviation of player 2 always convinces player 1 that 2 is of type 2_w. The optimistic conjectures are very often used in the incomplete information bargaining literature since they serve as the best deterring conjectures.

9.3 The game starts with player 2's proposal

In the main theorem the bargaining sequential equilibria are characterized for the case that player 1 opens the game. If player 2 starts the bargaining and if $y^{\omega_0} > \hat{V}_s$, then $\langle (y^{\omega_0}, 0), (y^{\omega_0}, 0) \rangle$ is a BSE outcome and if $y^{\omega_0} < \hat{V}_s$, then $\langle (\hat{V}_s, 0), (\hat{V}_s, 0) \rangle$ is a BSE outcome. This can be verified from the construction of a BSE in the Theorem while noticing that the strategies, after 1 has demanded the whole pie and has been refused, are BSE in a subgame starting with player 2's proposal and initial belief ω_0. However, this is not the only BSE since (B-1)-(B-2) are not effective as restrictions on 1's belief after 2's first offer. By proposition 4, 2_w and 2_s make the same offer at the first period. We may extend (B-2) so that we require that if 2 was supposed to make the offer y and his offer s^0 was less than y, then $\omega^1(s^0) \leqslant \omega_0$. With the extension of (B-2) we get that if $y^{\omega_0} < \hat{V}_s$, then $\langle (\hat{V}_s, 0), (\hat{V}_s, 0) \rangle$ is the only BSE outcome. If $y^{\omega_0} > \hat{V}_s$, we get one additional BSE outcome $\langle (y^{\omega_0}, 1), (x^{\omega_0}, 2) \rangle$: player 2 starts with the offer $\hat{V}_s$, which is rejected by player 1 who continues as if he had started the game; if 2 makes an offer $s_0 > \hat{V}_s$, 1 concludes that 2 is 2_w and demands V_w.

9.4 Fixed bargaining costs

Assume that player 1's utility is $s - ct$ while the utilities of types 2_w and 2_s are $(1-s) - c_w t$ and $(1-s) - c_s t$. The preferences which are represented by these utilities do not satisfy assumptions (A-5), (A-6) but the same techniques which are used to prove the main theorem are useful to calculate the bargaining sequential equilibrium here:

a If

$$\omega_0 > \frac{2c}{c + c_w},$$

the only bargaining sequential equilibrium outcome is $\langle(1,0),(1-c_w,1)\rangle$.

b If

$$\frac{2c}{c+c_w} > \omega_0 > \frac{c+c_s}{c+c_w},$$

the only bargaining sequential equilibrium outcome is $\langle(c_w,0),(0,1)\rangle$.

c If

$$\frac{c+c_s}{c+c_w} > \omega_0,$$

the only bargaining sequential equilibrium outcome is $\langle(c_s,0),(c_s,0)\rangle$. (The appearance of the intermediate zone (2) is due to the fact that here $(V_s,1) <_1 (\hat{V}_s,0)$ and not $(V_s,1) \sim_1 (\hat{V}_s,0)$, as in the previous case.)

REFERENCES

Cramton, P. C. 1984: Bargaining with incomplete information: An infinite horizon model with continuous uncertainty. *Review of Economic Studies*, **51**, 579–93.

Fishburn, P. C. and Rubinstein. A, 1982: Time preference. *International Economic Review*, **23**, 719–36.

Fudenberg, D. and Tirole, J. 1983: Sequential bargaining with incomplete information. *Review of Economic Studies*, **50**, 221–47.

Fudenberg, D., Levine, D. and Tirole, J. 1985: Infinite-horizon models of bargaining with one-sided incomplete information. In: Roth, A. (ed.), *Game Theoretic Models of Bargaining*. University of Cambridge Press, Cambridge.

Harsanyi, J. C. 1967: Games with incomplete information played by 'Bayesian players'. *Management Science*, **14**, 159–82, 320–34 and 486–502.

Harsanyi, J. C. and Selten, R. 1972: A generalized Nash solution for two-person bargaining games with incomplete information. *Management Science*, **18**, 80–106.

Kreps, D. M. and Wilson, R. 1982: Sequential equilibria. *Econometrica*, **50**, 863–94.

Myerson, R. B. 1984: Two-person bargaining problems with incomplete information. *Econometrica*, **52**, 461–88.

Nash, J. F. 1953: Two-person cooperative games. *Econometrica*, **21**, 128–40.

Ordover, J. and Rubinstein, A. 1982: On bargaining, settling and litigating: A problem in multistage games with incomplete information. Mimeo, New York University.

Perry, M. 1985: A theory of price formation in bilateral situations. *Econometrica*.

Roth, A. E. 1979: Axiomatic models of bargaining. Lecture Notes in Economics and Mathematical Systems No. 170. Springer-Verlag, Berlin.

Rubinstein, A. 1985: The choice of conjectures in a bargaining game with incomplete information. In: Roth, A. (ed.), *Game Theoretic Models of Bargaining*. CUP, Cambridge.

Selten, R. 1965: Speiltheoretische Behandlung eines Oligopolmodels mit Nachfragetragheit. *Zeitschrift fur die Gesamte Staatswissenschaft*, **12**, 301–24 and 667–89.

Sobel, J. and Takahashi, I. 1983: A multi state model of bargaining. *Review of Economic Studies*, **50**, 411–26.

10

Distortion of Utilities and the Bargaining Problem

J. Sobel

Given two agents with von Neumann-Morgenstern utilities who wish to divide n commodities, consider the two-person non-cooperative game with strategies consisting of concave, increasing von Neumann–Morgenstern utility functions as well as rules to break ties and whose outcomes are some solution to the bargaining game determined by the strategies used. It is shown that, for a class of bargaining solutions which includes those of Nash and Raiffa, Kalai and Smorodinsky, any constrained equal-income competitive equilibrium allocation for the true utilities is a Nash equilibrium outcome for the non-cooperative game.[1]

1 INTRODUCTION

It is often the case in economic or game-theoretical models that predictions are based on information that is not observable. For example, Nash's (1950) theory of bargaining determines an outcome that depends on the bargainers' von Neumann-Morgenstern utility functions. Kurz (1977, 1980) has recently introduced a technique for analysing such models that yields predictions about the outcome of a game without relying on unobserved information. The technique of Kurz has been adopted by Crawford and Varian (1979) to analyse the outcomes of the Nash bargaining solution over the division of a single commodity. The purpose of this chapter is to extend the results of Crawford and Varian to bargaining over several commodities.

The approach used by Kurz (1977, 1980) and Crawford-Varian (1979) is to embed the original game into a non-cooperative *distortion game* in which the players' strategies consist of utility functions that may be distorted from their true utilities for strategic purposes. The outcomes are given by the solution to the underlying game determined by the reported utilities. If the

1 I am very grateful to Vincent Crawford, who suggested the problem considered in this chapter and made several crucial suggestions that helped to make its solution possible. A conversation with Theodore Groves was also of value.

Nash equilibria of the distortion game share common properties, then a description of the original game situation has been made without relying on information about the unobserved utility functions.

Kurz's (1977, 1980) papers are related to the work of Aumann and Kurz (1977a and b) on the determination of taxes in an exchange economy. Aumann and Kurz postulate a particular solution concept and then characterize the income tax schedules and allocations that result from it. Kurz (1977, 1980) observes that the Aumann and Kurz solution depends on the agents' von Neumann-Morgenstern utility functions. Since these functions are not directly observable, agents cannot be prevented from misrepresenting them if it is to their advantage to do so. Kurz therefore studies the game that results if each agent can report any utility function in an admissible class $\mathscr{U}$. $\mathscr{U}$ is intended to include all functions that are credible utilities for an agent. Kurz takes $\mathscr{U}$ to be the set of all von Neumann-Morgenstern utility functions that are increasing, concave and continuously differentiable. He shows (1977) that for the one-commodity Aumann-Kurz model (1977a), reporting any linear function in $\mathscr{U}$ is a dominant strategy for each agent. The marginal tax rate implied by the use of linear strategies is 50 per cent. Kurz generalizes this result to the n-commodity case (1980). Once again, players have dominant strategies that lead to a marginal tax rate of 50 per cent. In this case, however, the dominant strategy reported utility functions need not be linear. The significance of Kurz's results is that, regardless of the true preferences of the agents, the distortion game has a dominant strategy equilibrium that yields a Pareto-efficient outcome.

Crawford and Varian (1979) use the methods of Kurz to analyse the effect that distortion of utilities has on the solutions to bargaining games. Assuming that agents may report any concave, increasing utility function, they find that in Nash (1950) or Raiffa (1953)-Kalai-Smorodinsky (1975) bargaining over the division of a single good reporting linear utility functions constitutes a dominant strategy equilibrium. The allocation implied by the equilibrium reports is equal division. The purpose of this chapter is to generalize this result to include bargaining over more than one commodity.

The main link between the one-commodity bargaining game and its multi-commodity generalization has to do with the effect a player's attitude towards risk has on the utility he receives at the solution. A utility function U is said to be *more risk averse* than V if there is an increasing concave function k with $U = k(V)$; an agent is more risk averse than another agent if his utility function is more risk averse than the other agent's. Kihlstrom et al. (1979)[2] show that for a class of bargaining solutions that include the Nash (1950) and the Raiffa (1953)-Kalai-Smorodinsky (1975) solution, a player's utility increases as his opponent becomes more risk averse. This result, which is related to a theorem of Kannai (1977), makes it possible to deduce that players will

2 This result is presented in Roth (1979, pp. 38–48, 104–5).

report linear utilities in the one-commodity distortion game. This follows because all monotonic preferences defined over one commodity are (ordinally) equivalent. The Kihlstrom-Roth-Schmeidler results thus imply that the players will select the least risk-averse representation of these preferences. In the one-commodity case this will be a linear function. As long as the solution for the bargaining problem satisfies the axioms of Pareto optimality, symmetry and invariance with respect to affine transformations of utility, the linear strategies give rise to equal division.

The situation is made more complicated in the n-commodity case because there are many possible ordinal rankings of the outcomes. While the Kihlstrom, Roth and Schmeidler result restricts the possible strategies that the players will find advantageous to report, a broad class of possible distortions (including any increasing linear function) cannot be excluded on the basis of their theorem. Consequently, it is not surprising that the characterization of equilibria for the distortion game is less satisfactory for multi-commodity bargaining than for one-commodity bargaining. Furthermore, although the solution to the bargaining problem specifies unique utility levels for both players with respect to their reported utilities, in general, there will be more than one outcome that gives rise to these utility levels. Since there is no reason to expect that these outcomes are utility equivalent for the true utilities, the strategy space must be augmented by tie-breaking rules that provide for a selection from the solution correspondence.

In spite of the greater complexity, the Nash equilibria for the n-commodity distortion game have some attractive properties. For a class of bargaining solutions that include those of Nash and Raiffa-Kalai-Smorodinsky, I show that any constrained equal-income competitive equilibrium (EICE) allocation (a constrained equilibrium allocation reached when agents have equal initial endowments) for the true utilities is a Nash equilibrium allocation for the distortion game. The equilibrium strategy for both players is to report linear utilities with indifference surfaces parallel to the hyperplane that supports the EICE. These strategies will result in a set of allocations that solve the bargaining problem but the players will be able to agree on a most preferred outcome. Since agents are assumed to have concave utility functions, this guarantees the existence of Pareto-efficient Nash equilibria. Moreover, the EICE allocations are the only Nash outcomes provided both players are required to report linear utilities. On the other hand, there is no reason to expect the competitive allocations to be unique. Thus, dominant strategy equilibria are impossible. Also, in general, the distortion game has inefficient Nash equilibria. However, all of the equilibria are 'good' in a certain sense. Specifically, at any Nash equilibrium outcome, each agent prefers his allocation to that of the other agent.[3]

3 That no agent prefers another agent's allocation to his own is the definition of equity first used by Foley (1967).

The class of bargaining game solutions for which these results are valid is described in section 3. In addition to the axioms of Pareto optimality, symmetry and independence of positive affine transformations of utility, axioms common to the Raiffa-Kalai-Smorodinsky and Nash theories, the solution is required to satisfy another property. In a bargaining game in which the players' utility functions are normalized so that each player receives utility 0 at the disagreement point and utility 1 at his most preferred outcome, a solution satisfies the axiom of *symmetric monotonicity* if each player receives utility of at least $\frac{1}{2}$ at the solution. In section 3, it is shown that both the Nash and the Raiffa-Kalai-Smorodinsky solutions have this property. Furthermore, it is shown that symmetric monotonicity is guaranteed if the solution is symmetric, Pareto-optimal and *risk sensitive* as defined by Kihlstrom et al. (1979). A solution is risk sensitive if a player prefers to bargain against the more risk averse of two players.

The problem considered in this chapter may be viewed as an arbitration problem under ignorance. An arbitrator is assigned the task of determining a 'good' outcome to the bargaining game. A possible technique for the arbitrator would be to ask the players to report their utility functions and then determine an outcome to the resulting bargaining game according to a fixed solution concept. If the arbitrator has no knowledge about the players' true utilities except that they are in the set $\mathscr{U}$, then the agents will be playing the distortion game described here. Notice that if the arbitrator can restrict the reported preferences to be linear, EICE allocations are assured. This restriction may be attractive to the arbitrator because it reduces the information which needs to be reported to an $n-1$ dimensional vector (the constant marginal rates of substitution of each player, rather than entire utility functions) and the strategies used to resolve ties. The restriction may be acceptable to the agents because it guarantees EICE allocations.

Another model of arbitration under ignorance is discussed by Kalai and Rosenthal (1978). A cooperative two-player bi-matrix game is transformed into a non-cooperative game by an arbitrator. Players are asked to report a mixed strategy (threat) and two payoff matrices. The arbitrator then determines an outcome using a procedure that generalizes Nash's (1953) extended bargaining solution if the players report the same payoff matrices. If the players report different payoff matrices then they receive the threat outcome. Assuming that the players know the underlying cooperative game and that the arbitrator knows only the dimensions of its payoff matrices, Kalai and Rosenthal show that reporting the true payoff matrices and appropriate mixed strategies forms a Nash equilibrium for the arbitration game. Moreover, the equilibrium outcome is Pareto efficient and individually rational.

Both Kalai and Rosenthal (1978) and I assume that agents have perfect information about the game situation they are facing. The results would be more compelling if they remained valid under uncertainty. Suppose that the distortion game is being played with only linear strategies admissible, and the true utility functions of the agents are such that the EICE is unique. If player 1 is only slightly uncertain about his opponent's utility function (meaning that he knows with certainty that his opponent's utility function is 'close' to a specific function), then player 1 knows – under certain regularity assumptions – that the true distortion game has a unique Nash equilibrium outcome, and he knows approximately where it is. Thus, there is a possibility that an adjustment process could be designed that converges to the EICE. In the Kalai–Rosenthal model, there appears to be no restriction that can be made that would make the Nash equilibrium unique or even locally unique. Therefore, while under certainty the players in the Kalai–Rosenthal game are likely to report truthfully and reach a 'good' outcome, the introduction of a slight amount of uncertainty makes the argument for truthful reports lose much of its force: if the players' beliefs about the true game situation differ, then it is quite possible that reporting what are believed to be the true payoff matrices will lead to an equilibrium outcome inferior to the 'good' outcome. Thus, in the Kalai–Rosenthal model, unless the players know the underlying game with certainty, the existence of multiple equilibria makes it unlikely that an adjustment process converging to the 'good' Nash equilibria could be designed.

The requirement that players know each other's characteristics with certainty is a strong one. However, it appears that Pareto-efficient outcomes cannot be guaranteed in models in which the players are uncertain about their opponent's characteristics. In a model of arbitration under uncertainty, Myerson (1979) observes that the set of allocations that arise from incentive-compatible mechanisms (allocations Myerson calls *incentive feasible*) is strictly contained in the set of all feasible allocations. Myerson argues that the arbitrator should be satisfied with selecting a 'good' outcome which is undominated by any other incentive feasible allocation (but which may be Pareto-inefficient), and proves that this can be done.

Closely related to my results are those of Thomson (1979a and b). Thomson studies the Nash equilibria for the distortion game derived from a class of performance correspondences that yield individually rational and Pareto-efficient outcomes. Thomson (1979a) finds that if the reported utility functions are restricted to be twice continuously differentiable, concave and have the transferable utility (t.u.) property, then the Nash equilibria for the distortion game derived from the Shapley value with fixed initial endowments are exactly the constrained competitive allocations with respect to those endowments. This result is generalized to a broader class of performance correspondences in Thomson (1979b). As the Nash bargaining solution and the Shapley value, with appropriate disagreement outcomes, coincide under transferable utility, these results are quite similar to mine. The main differences fall into two classes: the nature of allowable utilities and the range of generality.

When strategy spaces consisting of utility functions with transferable utility are used, the class of admissible utility functions is then broad enough to eliminate the need for explicit tie-breaking rules. If a tie occurs, a player typically has another admissible strategy that will allow him to break the tie so as to receive his most preferred outcome. In equilibrium, ties will occur unless the original endowments are Pareto-efficient for the true preferences. However, these ties can be broken because both players are able to agree on a most preferred outcome. On the other hand, when reports must be t.u. utility functions, Nash equilibrium outcomes other than the constrained competitive equilibria for the economy do not occur.[4] Inefficient Nash equilibria for the distortion game derived from the Nash bargaining solution may occur if any smooth report is allowed; an example is given in section 5.

It seems unreasonable to require transferable utility reports, a priori. However, for a certain class of games this restriction may be justified. When tie-breaking rules are used, I can show that a player always has a linear best response for the Nash and Raiffa–Kalai–Smorodinsky distortion games. Thus, non-linear strategies are dominated by linear strategies. It is reasonable to assume that only linear strategies will be used in this situation. A similar analysis may make it possible to delete all non-transferable utility preferences in Thomson's model.

Thomson's results apply to a different range of solutions than do mine. They are more general in one direction: his results are valid for any number of players. The fact that the Nash equilibrium outcomes include the competitive allocations remains valid in my model for any number of bargainers; however, other Nash equilibria will exist in general. This difference is probably the result of the different strategy spaces. Also, Thomson's results apply to any initial endowments. However, the constrained competitive equilibria with respect to any initial endowments can be obtained in my framework by varying the disagreement outcomes.

Besides requiring Pareto efficiency with respect to the reported preferences and individual rationality with respect to the given initial endowments, the class of solutions for which Thomson's results are valid have the property that equilibrium strategies have a 'flatness' quality. Specifically, it is necessary that initial endowments be Pareto-efficient with respect to equilibrium strategies. This will be true, for example, if all players report linear utilities with indifference surfaces parallel to the hyperplane that supports the competitive allocation. The underlying solutions that I deal with yield Pareto-efficient outcomes. The flatness property is satisfied when only linear reports are allowed, but not in general. Symmetric monotonicity can be viewed as an individual rationality requirement: provided that reports are linear, this

4 These results require that the reported preferences be twice continuously differentiable. If reports are not smooth, there may be other equilibria.

assumption guarantees that outcomes are at least as good as equal division with respect to the reported utilities. The solutions I consider need not guarantee outcomes that are at least as preferred as equal division when non-linear strategies are allowed.

It should be emphasized that my results depend on strategy spaces that are different from the set of admissible utility functions. The results of Hurwicz (1972) guarantee that individually rational and Pareto-efficient allocations cannot coincide with the Nash equilibrium outcomes of a mechanism that has only preferences as strategies. Thus, my results depend in an essential way on the fact that a player's strategy includes tie-breaking rules as well as a utility function.

The distortion game is defined formally in section 2. In section 3, the class of bargaining solutions to be used is described. The main results are presented in section 4. Finally, section 5 further characterizes the equilibria of the Nash and Raiffa-Kalai-Smorodinsky distortion game.

2 DEFINITIONS AND NOTATION

Consider two agents with von Neumann-Morgenstern utility functions who are to divide a bundle of n commodities. Units are chosen so that there is exactly one unit of each commodity. Letting $\mathbf{a} \equiv (a, \ldots, a)$, an outcome will be an element of the set

$$T = \{x \in \mathrm{IR}^n : \mathbf{0} \leqslant x \leqslant \mathbf{1}\},$$

where agent 1 receives x and agent 2 receives $\mathbf{1} - x$. The true utility function of player 1 is denoted by u; of player 2, v. These functions are assumed to be concave and strictly increasing in T. Thus if $x, x' \in T$ and $x > x'$, then $u(x) > u(x')$ and $v(1 - x') > v(1 - x)$. The players report utilities that are restricted to lie in the class U, where U consists of those functions: $U: T \to [0, 1]$ such that (a) U is continuous, strictly increasing and concave in T; (b) U is normalized so that $U(\mathbf{0}) = 0$ and $U(\mathbf{1}) = 1$. The class of admissible utilities should include those functions that are credible representations of their true preferences. Thus, condition (a) is a regularity assumption on the range of potential players. The concavity assumption means that the agents cannot pretend to be risk lovers. Since the solutions to the bargaining problem to be discussed are independent of affine transformations, condition (b) is inessential.

The distortion game for the bargaining problem is played by each agent revealing a utility function in $\mathscr{U}$ and an element in a set $\mathscr{M}$ that will be used to resolve ties. Typically, U will denote the function revealed by player 1; V that of player 2. Given these reports, a set of outcomes $B(U, V)$ is selected. $B(U, V)$ is the set of allocations that give rise to a bargaining solution determined by U and V. The properties of solution concepts used to define B will be discussed in section 3. However, in order to define the distortion game, it is only necessary that $B(U, V)$ be a non-empty subset of T for all U and V in $\mathscr{U}$.

In order to completely characterize the strategies for the distortion game, the way in which a single element of B is selected must be described. In addition to a utility function, each player will report an element from a set $\mathscr{M}$. An outcome will then be selected by a function $\bar{B}$. Thus

$$\bar{B}: \mathscr{U} \times \mathscr{M} \times \mathscr{U} \times \mathscr{M} \to T$$

with $\bar{B}(U, f; V, g) \in B(U, V)$ for all U and $V \in \mathscr{U}$ and f and $g \in \mathscr{M}$. $(\mathscr{M}, \bar{B})$ will be called the *tie-breaking pair* associated with B.

Definition. The strategies $(U^*, f^*; V^*, g^*)$ constitute a *Nash equilibrium for the distortion game* determined by B with tie-breaking pair $(\mathscr{M}, \bar{B})$ if and only if (a) $(U^*, f^*; V^*, g^*) \in \mathscr{U} \times \mathscr{M} \times \mathscr{U} \times \mathscr{M}$; (b) $\bar{B}(U^*, f^*; V^*, g^*)$ solves: max $u(x)$ subject to $x \in \{\bar{B}(U, f; V^*, g^*) : (U, f) \in \mathscr{U} \times \mathscr{M}\}$; (c) $1 - \bar{B}(U^*, f^*; V^*, g^*)$ solves: max $v(y)$ subject to

$$1 - y \in \{\bar{B}(U^*, f^*; V, g) : (V, g) \in \mathscr{U} \times \mathscr{M}\}.$$

An appropriate choice of tie-breaking pair can allow the Nash equilibria for the distortion game to be characterized in terms of the reported utility functions and the correspondence B.

Definition. A tie-breaking pair $(\mathscr{M}, \bar{B})$ for the solution B is *unrestricted* if and only if, for all U and $V \in \mathscr{U}$ and f and $g \in \mathscr{M}$,

$$B(U, V) = \{\bar{B}(U, f; V, h) : h \in \mathscr{M}\} = \{\bar{B}(U, h; V, g) : h \in \mathscr{M}\}.$$

Suppose player two reports a utility function $V \in \mathscr{U}$ and a tie-breaking strategy $g \in \mathscr{M}$. If $(\mathscr{M}, \bar{B})$ is unrestricted then, for any $U \in \mathscr{U}$, player 1 has a tie-breaking strategy that can cause any outcome in $B(U, V)$ to be selected. This makes the following description of Nash equilibria possible.

Proposition 1. *If $(\mathscr{M}, \bar{B})$ is an unrestricted pair for a solution B then, for any U^* and $V^* \in U$, there exist f^* and $g^* \in \mathscr{M}$ such that $(U^*, f^*; V^*, g^*)$ is a Nash equilibrium for the distortion game determined by B with tie-breaking pair $(\mathscr{M}, \bar{B})$ if and only if there exists $x^* \in B(U^*, V^*)$ such that x^* solves:*

$$\max u(x) \text{ subject to } x \in \{B(U, V^*) : U \in \mathscr{U}\}$$

and $1 - x^*$ solves:

$$\max v(y) \text{ subject to } 1 - y \in \{B(U^*, V) : V \in \mathscr{U}\}.$$

Proof. If $(U^*, f^*; V^*, g^*)$ is a Nash equilibrium then the conditions are satisfied when $x^* = \bar{B}(U^*, f^*; V^*, g^*)$. This follows since $(\mathscr{M}, \bar{B})$ is unrestricted and so

$$\{\bar{B}(U, f; V^*, g^*) : (U, f) \in \mathscr{U} \times \mathscr{M}\} = \{B(U, V^*) : U \in \mathscr{U}\}$$

and

$$\{\bar{B}(U^*,f^*;V,g):(V,g)\in \mathscr{U}\times\mathscr{M}\}=\{B(U^*,V):V\in\mathscr{U}\}.$$

Conversely, if an x^* exists as described in the proposition then $(U^*, f^*; V^*, g^*)$ is a Nash equilibrium provided that f^* and g^* are selected so that $x^* = \bar{B}(U^*, f^*; V^*, g^*)$. This is possible because $(\mathscr{M}, \bar{B})$ is unrestricted. *QED*

Thus, if an unrestricted pair can be found to break ties, Nash equilibria can be characterized in terms of reported utility functions and the solution correspondence.

The next result constructs unrestricted tie-breaking rules.

Proposition 2. *Given any solution correspondence B, there exists an unrestricted pair $(\mathscr{M}, \bar{B})$ associated with B.*

Proof. Let $\mathscr{M} = \mathrm{IR}^n$ and define $\bar{B}$ by

$$\bar{B}(U,x;V,y)=\begin{cases}\frac{1}{2}(x+y) & \text{if } \frac{1}{2}(x+y)\in B(U,V),\\ \text{any element of } B(U,V) & \text{if } \frac{1}{2}(x+y)\notin B(U,V).\end{cases}$$

Clearly player 1 can obtain any $z \in B(U, V)$ given that player 2 is using strategy (V, y) by using a tie-breaking strategy x with $x = 2z - y$ and reporting the utility function U. Similarly, if player 1 reports the function U, player 2 can obtain any element of $B(U, V)$ by responding with the appropriate tie-breaking rule. *QED*

In this chapter all ties will be resolved using unrestricted tie-breaking rules. Proposition 2 says that this can be done, while proposition 1 provides a characterization of Nash equilibria when unrestricted tie-breaking rules are used. Notice that it is not necessary for the players to actually report complicated tie-breaking strategies, provided that they know that some unrestricted procedure to break ties exists. Proposition 1 guarantees that if $B(U^*, V^*)$ consists of more than one point for equilibrium strategies U^* and V^*, then both players can agree on a most preferred outcome. Thus there will be no difficulty in making a selection from $B(U^*, V^*)$. The existence of unrestricted pairs rules out the possibility that, at an equilibrium, players cannot agree on a selection from the solution correspondence.

The characterization of Nash equilibria given by proposition 1 will be used throughout this chapter. Thus, a Nash equilibrium will be described by a triple (U^*, V^*, x^*) satisfying the conditions of proposition 1. The vector x^* will be called the Nash equilibrium outcome or allocation.

In what follows, Nash equilibrium allocations will be related to certain competitive outcomes.

Definition. A constrained *equal-income competitive equilibrium* (EICE) is a pair, $(p^*; x^*)$ where (a) $p^* \in \mathrm{IR}^n$, $p^* \geqslant \mathbf{0}$, $p^* \neq \mathbf{0}$; (b) $x^* \in T$; (c) x^* solves:

$$\max u(x) \text{ subject to } p^* \cdot x \leqslant \tfrac{1}{2} p^* \cdot \mathbf{1} \quad \text{and} \quad x \in T;$$

(d) $\mathbf{1} - x^*$ solves:

$$\max v(y) \text{ subject to } p^* \cdot y \leqslant \tfrac{1}{2} p^* \cdot \mathbf{1} \quad \text{and} \quad y \in T.$$

In an equal-income competitive equilibrium, both agents make demands subject to a budget constraint only. A constrained equal-income competitive equilibrium requires that these demands do not exceed the total resources available.[5] It is well known that any equal-income competitive equilibrium is a constrained equal-income competitive equilibrium and that, provided preferences are convex, any interior constrained equal-income competitive equilibrium is an equal-income competitive equilibrium.

The vector x^* will be referred to as the competitive allocation.

On occasion, a vector $p = (p_1, \ldots, p_n)$ will be used to refer to the linear function from T to $\mathbb{R}$, where

$$p(x) \equiv p \cdot x \equiv \Sigma\, p_i x_i.$$

No confusion should arise.

3 SOLUTIONS TO BARGAINING GAMES

The underlying bargaining problem can be formulated as follows. A *bargaining game* is characterized by a pair (S, d), where: (a) $d = (d_1, d_2) \in \mathbb{R}^2$; (b) $S \subset \mathbb{R}^2$ is compact, convex and contains d as well as some point $x > d$.

The set S is interpreted as the set of feasible utility payoffs to the players. A point $x = (x_1, x_2)$ can be achieved if both players agree to it. In that case, player 1 receives x_1 and player 2 receives x_2. If the players are unable to agree, then the outcome d, called the disagreement outcome, is the result.

In what follows, the set S will depend on the reported utilities U and V, and will be defined as the set of feasible utility payoffs. That is,

$$S = S(U, V) = \{(x_1, x_2) : 0 \leqslant x_1 \leqslant U(t), 0 \leqslant x_2 \leqslant V(\mathbf{1} - t) \quad \text{for some} \quad t \in T\}.$$[6]

The disagreement outcome will always be taken to be $(0, 0) = (U(\mathbf{0}), V(\mathbf{0}))$.

5 The concept of constrained competitive equilibria was introduced by Hurwicz et al. (1978). It is the smallest extension of the competitive correspondence that can be implemented in Nash strategies.

6 The definition of $S(U, V)$ includes a free disposal assumption. Another (equivalent) definition would allow outcomes that do not distribute all of the commodities. That is,

$$S(U, V) = \{(a, b) : a = U(x), b = V(y), x, y \in T \quad \text{and} \quad x + y \leqslant \mathbf{1}\}.$$

Notice that when the functions U and V are in $\mathscr{U}$, the set S is compact, convex and contains a point $x > (0, 0)$. In fact, $(1, 0) \in S$, $(0, 1) \in S$, and $S \subset \{x_1, x_2) : 0 \leqslant x_1, x_2 \leqslant 1\}$. Such a game will be called 0-1 *normalized.*[7]

Nash (1950) introduced the concept of a *solution* to a bargaining game. A solution is a function f, defined on the class of all bargaining games with $f(S, d) = (f_1(S, d), f_2(S, d)) \in S$ for all pairs (S, d). Nash characterized a particular solution in terms of the following axioms.

Axiom 1 (Pareto efficiency). *If* $f(S, d) = x$ *and* $y \geqslant x$, *then either* $y = x$ *or* $y \notin S$.

Axiom 2 (symmetry). *If* (S, d) *is a symmetric game (that is,* $(x_1, x_2) \in S$ *if and only if* $(x_2, x_1) \in S$, *and* $d_1 = d_2$*) then* $f_1(S, d) = f_2(S, d)$.

Axiom 3 (independence of equivalent utility representations). *If* (S, d) *and* (S', d') *are bargaining games such that*

$$S' = \{(a_1x_1 + b_1, a_2x_2 + b_2) : (x_1, x_2) \in S\} \quad \textit{and}$$

$$d' = (a_1d_1 + b_1, a_2d_2 + b_2) \quad \textit{where} \quad a_1 \textit{ and } a_2 > 0, \quad \textit{then}$$

$$f(S', d') = (a_1 f_1(S, d) + b_1, a_2 f_2(S, d) + b_2).$$

Axiom 4 (independence of irrelevant alternatives). *If* (S, d) *and* (S', d) *are bargaining games such that* $S \subset S'$ *and* $f(S', d) \in S$, *then* $f(S, d) = f(S', d)$.

Nash's result (1950) was that axioms 1-4 characterize a solution, η. In terms of utilities, $U, V \in \mathscr{U}$,

$$N(U, V) = \{x \in T : x \in \arg\max \{U(y)V(1-y) : y \in T\}\},$$

is the set of allocations that give rise to the Nash solution to the bargaining game with disagreement outcome (0, 0). That is, for all $x \in N(U, V)$,

$$\eta(S(U, V), (0, 0)) = (U(x), V(1-x)).^8$$

Axiom 4 has been criticized for a variety of reasons (see, for example, Luce and Raiffa 1957). Another solution to the bargaining game has been presented by Raiffa (1979) and axiomatized by Kalai and Smorodinsky (1975). Kalai and Smorodinsky replace Nash's axiom 4 with an axiom of monotonicity. To state the axiom formally, it is necessary to define, for all bargaining games (S, d).

$$b_1(S) = \sup\{x_1 \in \mathbb{R} : \text{for some } x_2 \in \mathbb{R}\ (x_1, x_2) \in S\}$$

7 The normalization of $S(U, V)$ anticipates the axiom of independence of equivalent utility representations (axiom 3). With that axiom, any game can be taken to be 0-1 normalized without loss of generality.

8 The Nash solution gives rise to a single level of utility for each player. However, unless the utility functions are strictly concave, there may be several allocations that give rise to these utility levels.

and

$$b_2(S) = \sup\{x_2 \in \mathbb{R} : \text{for some } x_1 \in \mathbb{R}\ (x_1, x_2) \in S\}.$$

Also, let g_S be a function defined for $x_1 \leqslant b_1(S)$ as follows:

$$g_S(x_1) = x_2 \quad \text{if } (x_1, x_2) \in S \quad \text{and} \quad (x_1, x) \in S \quad \text{implies } x_2 \geqslant x,$$
$$= b_2(S) \quad \text{if no such } x_2 \text{ exists.}$$

Then $g_S(x)$ is the maximum player 2 can get if player 1 gets at least x. Because S is compact, $b_1(S)$ and $b_2(S)$ are finite and attained by points in S. I can now state the following axiom:

Axiom 5 (monotonicity). *If* (S, d) *and* (S', d) *are bargaining pairs such that*

$$b_1(S) = b_1(S') \quad \text{and} \quad g_S \leqslant g_{S'} \quad \text{then} \quad f_2(S, d) \leqslant f_2(S', d).$$

This axiom says that if the maximum feasible utility level that player 2 can obtain is increased for every utility level that player 1 may demand, then the utility level assigned to player 2 according to the solution should also be increased.

Kalai and Smorodinsky (1975) show that axioms 1, 2, 3 and 5 characterize a solution, ξ, to the bargaining problem. In terms of utilities, $U, V \in \mathscr{U}$,

$$K(U, V) = \{x \in T : x \in \{\arg\max U(y) : U(y) = V(1-y), y \in T\}\}$$

is the set of allocations that give rise to the Raiffa-Kalai-Smorodinsky (R-K-S) solution to the bargaining game with disagreement outcome (0, 0). That is, for all $x \in K(U, V)$,

$$\xi(S(U, V), (0, 0)) = (U(x), V(1-x)).$$

Many of the results to be presented on the equilibria to distortion games are valid for a class of solutions that include both the Nash bargaining solution and the R-K-S solution. The crucial property, in addition to axioms 1-3, seems to be the following.

Axiom 6 (symmetric monotonicity). *If* $(S, (0, 0))$ *is a* 0-1 *normalized bargaining game, then* $f(S, (0, 0)) \geqslant (\frac{1}{2}, \frac{1}{2})$.

Axiom 6 can be stated in a more general fashion, but the above formulation is sufficient for my purposes.

Any solution that satisfies axioms 1, 2, 3 and 6 will be called *admissible.* It turns out that the equilibria for the distortion game can be characterized provided that the underlying bargaining solution is admissible.

Axiom 6 is a weaker assumption than axiom 5. One consequence of axiom 5 is this context is that if $d \in S \subset S'$, then $f(S', d) \geqslant f(S, d)$. For any

0-1 normalized game S', the convex hull of (0, 1), (0, 0) and (1, 0), S, is contained in S'. Thus, whenever axioms 1, 2, 3 and 5 are satisfied by a solution f,

$$f(S', (0,0)) \geqslant f(S, (0,0)) = (\tfrac{1}{2}, \tfrac{1}{2}).$$

It follows that the R-K-S solution is admissible. The Nash bargaining solution is also admissible. To see this, it is convenient to present another axiom, which was introduced by Kihlstrom et al. (1979) in order to study the effect that a player's attitude towards risk has on solutions to the bargaining game.

Axiom 7 (risk sensitivity). *Suppose the bargaining game (S, d) is transformed into a game (S', d') by replacing player 2, say, with a more risk-averse player (that is, if $S = S(U, V)$ then $S' = S'(U, k(V))$ where k is increasing and concave); then $f_1(S', d') \geqslant f_1(S, d)$.*

Any solution f that satisfies axiom 7 describes a bargaining process in which it is advantageous to have a highly risk-averse opponent. The axioms that have been presented are related by the following result.

Lemma 1. *If f is a solution that satisfies axioms 1, 2 and 7, then f satisfies axioms 3 and 6.*

Proof. Kihlstrom et al. (1979) show that axioms 1 and 7 imply axiom 3. It therefore suffices to show that axioms 1, 2, 3 and 7 imply axiom 6. Let $S = \{(a, b) \in \mathrm{IR}^2 : a, b \geqslant 0, a + b \leqslant 1\}$; then $S = S(U, V)$ when $U(x) \equiv V(x) \equiv \frac{1}{n}\mathbf{1} \cdot x$. By axioms 1 and 2, $f(S, (0, 0)) = (\frac{1}{2}, \frac{1}{2})$. I will show that if S' is 0-1 normalized, then it can be obtained from S by replacing player 2 by a more risk-averse player. Then axiom 7 will imply $f_1(S', (0,0)) \geqslant \frac{1}{2}$. The lemma will follow by symmetry. Let ϕ be a parametrization of the Pareto-efficient set of S'. That is, suppose the northeast boundary of S' can be written as

$$P' = \{(a, \phi(a)) : 0 \leqslant a \leqslant 1\}.$$

Clearly ϕ is decreasing and concave. Further, because of the normalization, $\phi(0) = 1$ and $\phi(1) = 0$. Now let k be the function defined by $k(a) = \phi(1 - a)$. Then k is increasing, concave and satisfies $k(0) = 0$, $k(1) = 1$. Moreover,

$$P' = \{(a, k(1 - a)) : 0 \leqslant a \leqslant 1\}.$$

Thus k takes the Pareto set of S onto the Pareto set of S'. It follows that S' can be derived from player 1 using the strategy $L(x) \equiv \frac{1}{n}\mathbf{1} \cdot x$ and player 2 using the strategy $k(L)$. As noted earlier this is sufficient to prove the lemma.

QED

Kihlstrom et al. (1979) show that the R-K-S and the Nash solutions satisfy axiom 7. Therefore, these solutions are admissible.

4 MAIN RESULTS

In this section, it will be assumed that the distortion game is determined by an admissible solution to the bargaining game, f. Such a game will be referred to as an *admissible distortion game.* Associated with a solution f and functions U and $V \in \mathscr{U}$ there is a set $B(U, V)$ defined by

$$B(U, V) = \{x \in T : f(S(U, V), (0, 0)) = (U(x), V(\mathbf{1} - x))\}.$$

$B(U, V)$ is the set of outcomes giving rise to the utilities specified by the solution f.

The main theorem can now be stated.

Theorem 1. *If* $(p^*; x^*)$ *is an EICE for the true preferences, then* (p^*, p^*, x^*) *is a Nash equilibrium for any admissible distortion game.*

Proof. Since B is admissible, it follows that for all U and $V \in \mathscr{U}$,

$$x \in B(U, p^*) \quad \text{implies} \quad p^* \cdot (\mathbf{1} - x) \geqslant \tfrac{1}{2} \tag{10.1}$$

and

$$x \in B(p^*, V) \quad \text{implies} \quad p^* \cdot x \geqslant \tfrac{1}{2}. \tag{10.1$'$}$$

Moreover, by symmetry,

$$B(p^*, p^*) = \{x \in T : p^* \cdot x = \tfrac{1}{2}\}.$$

On the other hand, since $(p^*; x^*)$ is an EICE, x^* solves:

$$\max u(x) \quad \text{subject to} \quad p^* \cdot x \leqslant \tfrac{1}{2} \quad \text{and} \quad x \in T \tag{10.A}$$

and $\mathbf{1} - x^*$ solves:

$$\max v(y) \quad \text{subject to} \quad p^* \cdot y \leqslant \tfrac{1}{2} \quad \text{and} \quad y \in T. \tag{10.A$'$}$$

Since u and v are increasing, $p^* \cdot x^* = \tfrac{1}{2} = p^* \cdot (\mathbf{1} - x^*)$. It follows that $x^* \in B(p^*, p^*)$. Furthermore, since x^* solves (10.A), (10.1) implies that x^* solves:

$$\max u(x) \quad \text{subject to} \quad x \in B(U, p^*).$$

Similarly, combining (10.A′) and (10.1′) shows that $\mathbf{1} - x^*$ solves:

$$\max v(y) \quad \text{subject to} \quad \mathbf{1} - y \in B(p^*, V).$$

This establishes the theorem. *QED*

The EICE is attained as follows: each player reports linear preferences with indifference surfaces parallel to the supporting prices. The set of solutions to the bargaining problem then consists of an entire hyperplane. However, since the hyperplane supports the EICE, both agents can agree on a most preferred point (with respect to their true preferences). This point is a competitive allocation.

Theorem 1 has a partial converse.

Theorem 2. *If (p, q, x^*) is a Nash equilibrium for an admissible distortion game, and if p and q are linear, then x^* is an EICE allocation.*

Proof. Since $x^* \in B(p,q)$,

$$p \cdot x^* \geqslant \tfrac{1}{2} \quad \text{and} \quad q \cdot (\mathbf{1} - x^*) \geqslant \tfrac{1}{2}.$$

Also, since $B(q, q) = \{x \in T : q \cdot x = \frac{1}{2}\}$ and u is increasing, it follows that x^* solves:

$$\max u(x) \quad \text{subject to} \quad q \cdot x \leqslant \tfrac{1}{2} \quad \text{and} \quad x \in T.$$

Similarly, $\mathbf{1} - x^*$ solves:

$$\max v(y) \quad \text{subject to} \quad p \cdot y \leqslant \tfrac{1}{2} \quad \text{and} \quad y \in T$$

and

$$q \cdot x^* = \tfrac{1}{2} = p \cdot x^*. \tag{10.2}$$

To prove the theorem, it suffices to show that $p = q$. But this follows because x^* is a Pareto-efficient allocation with respect to the utilities p and q. Hence (2) implies that equal division must be Pareto-efficient with respect to the utilities p and q. Since p and q are normalized, $p = q$. *QED*

Informally theorem 2 can be explained as follows. The use of a linear strategy by player 2 restricts player 1 to outcomes x that satisfy $q \cdot x \leqslant \frac{1}{2}$. Thus, the way to guarantee the most preferred outcome in this set is to use the strategy q; in this way $B(q, q) = \{x : q \cdot x = \frac{1}{2}\}$. Similarly, the best response player 2 can make to player 1 is to use the same strategy. It follows that two linear strategies p and q can comprise an equilibrium only if $p = q$ and the players are able to agree on the most preferred outcome in $\{x : p \cdot x = \frac{1}{2}\}$.

Taken together, theorems 1 and 2 characterize the Nash equilibria for the distortion game if agents are restricted to linear strategies. In general, the equilibrium outcome of the distortion game cannot be guaranteed to be an EICE allocation. The next result shows that all equilibrium outcomes are envy-free. That is, each agent weakly prefers his allocation to the allocation of the other player.

Theorem 3. *If (U, V, x^*) is a Nash equilibrium for an admissible distortion game and the true utilities are concave, then*

$$u(x^*) \geqslant u(\tfrac{1}{2}) \geqslant u(\mathbf{1} - x^*) \quad \textit{and}$$
$$v(\mathbf{1} - x^*) \geqslant v(\tfrac{1}{2}) \geqslant v(x^*).$$

Proof. $u(x^*) \geqslant u(y)$ where y solves:

$$\max u(x) \quad \text{subject to} \quad x \in B(V, V).$$

Since $\frac{1}{2} \in B(V, V)$, $u(x^*) \geqslant u(\frac{1}{2})$. The concavity of u and the fact that $u(x^*) \geqslant u(\frac{1}{2})$ guarantee $u(\frac{1}{2}) \geqslant u(1 - x^*)$. Identical arguments establish the statements about v. *QED*

By using his opponent's strategy, a player can guarantee himself the outcome $\frac{1}{2}$. Thus, any Nash equilibrium must yield each an outcome at least as preferred as equal division.

Theorem 3 is true even if the underlying distortion game is not admissible. Since both players have the same strategy set, a player is able to use his opponent's strategy. Also, any solution B that is Pareto-efficient and symmetric satisfies $\frac{1}{2} \in B(V, V)$ for all V. It follows that a player is able to guarantee an outcome that is at least as preferred as equal division by using his opponent's strategy.

The following consequence of theorem 3 is immediate.

Corollary. *If equal division is efficient, then all Nash equilibria to the distortion game give player* 1 *utility* $u(\frac{1}{2})$, *and player* 2 *utility* $v(\frac{1}{2})$.

In particular, if the agents have identical preferences, equal division - or a utility equivalent allocation - will be the unique outcome of the distortion game.

5 NASH AND RAIFFA-KALAI-SMORODINSKY DISTORTION GAMES

The previous section proved existence of Nash equilibria for admissible distortion games. The characterization can be made more explicit if the nature of the solution to the bargaining problem is restricted. In this section, the Nash and the R-K-S solutions will be considered specifically.

For the results of this section, I shall assume that the true utilities are differentiable on the interior of T, and that reported utilities are twice continuously differentiable on the interior of T. I shall denote the partial derivative of a function U with respect to its ith argument by U_i. ∇U denotes the gradient vector $(U_1, \ldots, U_n)$, and second partial derivatives are denoted by U_{ij}.

The function, V, reported by agent 2 constrains the possible equilibria of the distortion game. In order for x to be a Nash equilibrium of the distortion game determined by the Nash bargaining solution, there must exist a $U \in \mathscr{U}$ such that x solves:

$$\max U(y)V(1 - y) \quad \text{subject to} \quad y \in T. \tag{10.B}$$

Since U and V are concave, the first order conditions associated with this maximization problem are necessary and sufficient. That is, x solves (10.B) if and only if, for all i,

$$U_i(x)V(\mathbf{1}-x)-U(x)V_i(\mathbf{1}-x)\leqslant 0 \quad \text{if} \quad x_i<1 \tag{10.3}$$

and

$$U_i(x)V(\mathbf{1}-x)-U(x)V_i(\mathbf{1}-x)\geqslant 0 \quad \text{if} \quad x_i>0. \tag{10.3$'$}$$

Hence,

$$x_i U_i(x)V(\mathbf{1}-x)\geqslant U(x)V_i(\mathbf{1}-x)x_i \quad \text{for all } i.$$

Summing and using the fact that, for all $y\in T$,

$$U(y)\geqslant \nabla U(y)\cdot y \quad \text{whenever} \quad U\in\mathscr{U},$$

it follows that

$$V(\mathbf{1}-x)\geqslant \nabla V(\mathbf{1}-x)\cdot x \tag{10.4}$$

for any potential Nash equilibrium allocation x.

Thus, given the report V, the best possible Nash outcome for player 1 is the solution, x^*, to:

$$\max u(x) \quad \text{subject to} \quad V(\mathbf{1}-x)\geqslant \nabla V(\mathbf{1}-x)\cdot x.$$

An identical argument can be used to deduce the restrictions a reported strategy U has on the possible outcomes for player 2 and these results can be used to characterize the equilibria of the distortion game.

Lemma 2. *Let U^*, $V^*\in\mathscr{U}$. Then (U^*, V^*, x^*) is a Nash equilibrium for the Nash distortion game if and only if*

$$x^*\in N(U^*, V^*),$$

x^* solves:

$$\max u(x) \quad \textit{subject to} \quad V^*(\mathbf{1}-x)\geqslant \nabla V^*(\mathbf{1}-x)\cdot x \quad \textit{and} \quad x\in T. \tag{10.C}$$

and $\mathbf{1}-x^*$ solves:

$$\max v(y) \quad \textit{subject to} \quad U^*(\mathbf{1}-y)\geqslant \nabla U^*(\mathbf{1}-y)\cdot y \quad \textit{and} \quad y\in T. \tag{10.C$'$}$$

Proof. If x^* solves problem (10.C), then x^* is the best outcome player 1 can obtain given that player 2 reports V^*; and if $\mathbf{1}-x^*$ solves (10.C$'$), then $\mathbf{1}-x^*$ is the best that player 2 can obtain given that player 1 reports U^*. Thus, (U^*, V^*, x^*) is a Nash equilibrium provided $x^*\in N(U^*, V^*)$.

To prove that the conditions are necessary, it suffices to show that, given V^*, player 1 can always report a utility function, U, so that the element in $P=\{z\in T: V^*(\mathbf{1}-z)\geqslant \nabla V^*(\mathbf{1}-z)\cdot z\}$ that he most prefers is contained in $N(U, V^*)$. Since $x^*\in P$ by (4), (U^*, V^*, x^*) will be a Nash equilibrium only if x^* solves (10.C). Suppose z^* solves:

$$\max u(z) \quad \text{subject to} \quad z\in P.$$

Then, since u is strictly increasing,

$$V^*(\mathbf{1}-z^*)=\nabla V^*(\mathbf{1}-z^*)\cdot z^*.$$

Let

$$U(x)=\nabla V^*(\mathbf{1}-z^*)\cdot x/\nabla V^*(\mathbf{1}-z^*)\cdot \mathbf{1}.$$

It is easy to check that $U\in \mathscr{U}$ and $z^*\in N(U, V^*)$. Therefore, the earlier comments guarantee that $u(x^*)=u(z^*)$ if (U^*, V^*, x^*) is a Nash equilibrium. A similar argument shows $\mathbf{1}-x^*$ solves (10.C′) and proves the lemma.

QED

Notice that a player can select a best response which is linear. In this sense, non-linear strategies are dominated. It is unlikely that dominated strategies will be used.

A similar result is true for the R-K-S distortion game.

Lemma 3. *A triple (U^*, V^*, x^*) is a Nash equilibrium for the R–K–S distortion game if and only if*

$$x^*\in K(U^*, V^*),$$

x^* solves

$$\max u(x) \quad \textit{subject to} \quad (\nabla V^*(\mathbf{1}-x)\cdot(\mathbf{1}-x))V^*(\mathbf{1}-x) \geqslant (1-V^*(\mathbf{1}-x))\nabla V^*(\mathbf{1}-x)\cdot x, \tag{10.D}$$

and $\mathbf{1}-x^*$ solves:

$$\max v(y) \quad \textit{subject to} \quad (\nabla U^*(\mathbf{1}-y)\cdot(\mathbf{1}-y))U^*(\mathbf{1}-y) \geqslant (1-U^*(\mathbf{1}-y))\nabla U^*(\mathbf{1}-y)\cdot y. \tag{10.D′}$$

The proof of lemma 3 is analogous to that of lemma 2, and is omitted.

In order to identify other possible Nash equilibria, implications of the necessary conditions given in the previous lemmas must be examined.

Lemma 4. *Suppose (U, V, x) is a Nash equilibrium for the distortion game determined by the Nash or R–K–S bargaining solution. Then*

$$x\cdot \nabla U(x)=(\mathbf{1}-x)\cdot \nabla U(x)=U(x), \tag{10.5}$$

$$(\mathbf{1}-x)\cdot \nabla V(\mathbf{1}-x)=x\cdot \nabla V(\mathbf{1}-x)=V(\mathbf{1}-x), \tag{10.5′}$$

$$U(\lambda x) = \lambda U(x) \quad for \quad 0 \leqslant \lambda \leqslant 1, \tag{10.6}$$

$$V(\lambda(1-x)) = \lambda V(1-x) \quad for \quad 0 \leqslant \lambda \leqslant 1, \tag{10.6'}$$

$$\nabla U(\lambda x) = \nabla U(x) \quad for \quad 0 \leqslant \lambda \leqslant 1, \tag{10.7}$$

$$\nabla V(\lambda(1-x)) = \nabla V(1-x) \quad for \quad 0 \leqslant \lambda \leqslant 1, \tag{10.7'}$$

$$\sum_i U_{ij}(x)x_i = 0 \quad for\ all\ j \quad and \tag{10.8}$$

$$\sum_i V_{ij}(1-x)(1-x_i) = 0 \quad for\ all\ j. \tag{10.8'}$$

The proof of lemma 4 is given in the Appendix. Properties (10.5) and (10.5′) are derived directly from lemma 2 or 3, the remaining properties follow for any elements of $\mathscr{U}$ satisfying (10.5) or (10.5′). The restrictions placed on the reported strategies at a Nash equilibrium can be interpreted in the context of the Nash bargaining solution. The Nash solution to the bargaining problem depends on the local properties of the reported preferences and of their derivatives. Properties (10.6), (10.6′), (10.7) and (10.7′) say that utility, as measured by the reported preferences, increases linearly along the segment connecting **0** to the outcome, the direction and magnitude of increase along the ray being constant.

The characterization of equilibrium strategies given in lemma 4 is suggested by the results of Kihlstrom et al. (1979) and Kannai (1977). Their results show that a player's utility at the Nash and R-K-S bargaining solutions increases as his opponent becomes more risk averse. Thus, one would expect equilibrium strategies to be 'least concave'[9] representations of some ordinal preferences. The fact that equilibrium strategies must be linearly homogeneous along a segment is consistent with this expectation.

Notice that, at least when there are only two commodities, (10.7) and (10.7′) imply Thomson's (1979) results. U is required to be of the form $U(x_1, x_2) = x_1 + W(x_2)$ with W concave, and (10.7) implies W is linear. Similarly (10.7′) requires that V must be linear. Therefore, the results of section 3 imply that the equilibria of the Nash and R-K-S distortion games coincide with the EICE's when reported strategies are smooth and there is transferable utility.

The next results are true for the distortion games determined by both the R-K-S and the Nash bargaining solutions. Proofs are given only in the Nash case.

9 A least concave utility function is a minimal element in the set of continuous, concave functions on T ordered by $\gtrsim$, where $U \gtrsim V$ if U is more risk averse than V.

Theorem 4. *If (U, V, x^*) is a Nash equilibrium for the distortion game, then $(\nabla U(x^*); x^*)$ is an EICE for the reported utilities.*

Proof. Lemma 4(10.6) guarantees that

$$\nabla U(x^*) \cdot x^* = \nabla U(x^*) \cdot \tfrac{1}{2} = \nabla U(x^*) \cdot (\mathbf{1} - x^*)$$

and clearly x^* solves:

$$\max U(x) \quad \text{subject to} \quad \nabla U(x^*) \cdot x \leqslant \nabla U(x^*) \cdot \tfrac{1}{2} \quad \text{and} \quad x \in T.$$

To show that $\mathbf{1} - x^*$ solves:

$$\max V(y) \quad \text{subject to} \quad \nabla U(x^*) \cdot y \leqslant \nabla U(x^*) \cdot \tfrac{1}{2} \quad \text{and} \quad y \in T,$$

it must be verified that $\mathbf{1} - x^*$ satisfies the first order conditions

$$V_i(\mathbf{1} - x^*) - \lambda U_i(x^*) \geqslant 0 \quad \text{if} \quad x_i^* < 1 \quad \text{and}$$

$$V_i(\mathbf{1} - x^*) - \lambda U_i(x^*) \leqslant 0 \quad \text{if} \quad x_i^* > 0.$$

Since $x^* \in N(U, V)$, these conditions follow from (10.3) and (10.3′) with $\lambda = V(\mathbf{1} - x^*)/U(x^*)$. *QED*

The next result is used to characterize the efficient Nash equilibria.

Lemma 5. *If (U, V, x^*) is a Nash equilibrium then*

$$\nabla u(x^*) \cdot x^* \geqslant \nabla u(x^*) \cdot \tfrac{1}{2} \quad \textit{and}$$

$$\nabla v(\mathbf{1} - x^*) \cdot (\mathbf{1} - x^*) \geqslant \nabla v(\mathbf{1} - x^*) \cdot \tfrac{1}{2}.$$

Proof. By Lemma 2, x^* must satisfy the first order conditions for the maximization problem (10.C). Thus, there is a $\mu > 0$, such that for all i,

$$u_i(x^*) - \mu\left[2V_i(\mathbf{1} - x^*) - \sum_j V_{ij}(\mathbf{1} - x^*)x_j^*\right] \geqslant 0 \quad \text{if} \quad x_i^* > 0,$$

and

$$u_i(x^*) - \mu\left[2V_i(\mathbf{1} - x^*) - \sum_j V_{ij}(\mathbf{1} - x^*)x_j^*\right] \leqslant 0 \quad \text{if} \quad x_i^* < 1.$$

By (10.8′) and the concavity of V it follows that

$$\nabla u(x^*) \cdot x^* \geqslant 2\mu \nabla V(\mathbf{1} - x^*) \cdot x^*$$

and

$$\nabla u(x^*) \cdot (\mathbf{1} - x^*) \leqslant 2\mu \nabla V(\mathbf{1} - x^*) \cdot (\mathbf{1} - x^*).$$

Thus,

$$\nabla u(x^*) \cdot x^* \geqslant \nabla u(x^*) \cdot \tfrac{1}{2},$$

since

$$\nabla V(1-x^*)\cdot x^* = \nabla V(1-x^*)\cdot(1-x^*)$$

by (10.5′). Similar arguments establish the inequality involving v. *QED*

If x^* were a competitive allocation, $\nabla u(x^*)$ could be taken to be the supporting prices. Lemma 5 then guarantees that the value of the allocation x^* at these competitive prices is at least as great as the value of the equal division allocation, $\frac{1}{2}$. The next consequence is therefore evident.

Theorem 5. *If x^* is an efficient Nash equilibrium, then x^* is an EICE.*

Proof. I assume, for convenience, that x^* is in the interior of T. In this case,

$$\nabla u(x^*) = \lambda \nabla v(1-x^*) \quad \text{for some} \quad \lambda > 0.$$

Thus, lemma 5 implies

$$\nabla u(x^*)\cdot x^* \geqslant \nabla u(x^*)\cdot \tfrac{1}{2} \tag{10.9}$$

and

$$\nabla u(x^*)\cdot(1-x^*) \geqslant \nabla u(x^*)\cdot \tfrac{1}{2}. \tag{10.9′}$$

It follows that (10.9) and (10.9′) must hold as equalities. Hence, ($\nabla u(x^*)$; x^*) is an EICE. When $x^* \in$ Boundary (T) there may be several equilibrium prices associated with it. The argument given above can be modified to show that one of these supporting prices makes the allocation x^* an EICE. *QED*

Inefficient Nash equilibria exist in non-pathological settings, if non-linear strategies are allowed. The following example shows that an EICE allocation may not be Pareto-superior to other Nash equilibria of the distortion game.

Example. Let $u(x, y) = x^{5/6}y^{1/6}$, $v(x, y) = x^{1/2}y^{1/2}$, $U(x, y) = (5x+3y)/8$ and $V(x, y) = x^{1/2}y^{1/2}$. Then u, v, U and $V \in \mathscr{U}$, and routine verification, using lemma 2, shows that (U, V, x^*) is a Nash equilibrium for the distortion game when $x^* = (3/5, 1/3)$. In this example, there is a unique EICE for the true preferences. It is $(p^*; y^*) = ((2/3, 1/3); (5/8, 1/4))$. A computation shows that $v(1-y^*) > v(1-x^*)$ and that $u(x^*) > u(y^*)$. Thus, the first player prefers the inefficient outcome x^* to the EICE outcome y^*. Also, the second player is worse off at the Nash equilibrium (u, v, x^*) even though he is reporting his true utility function.

The example also shows that the set of equilibria to the Nash distortion game is not equal to the set of equilibria of the R-K-S distortion game. It is easy to check $x^* \notin K(U, V)$ so (U, V, x^*) is not a Nash equilibrium for the R-K-S distortion game.

APPENDIX

This appendix will prove lemma 4 for the Nash distortion game. The proof for R-K-S bargaining is similar. A preliminary result must be established first.

Fact. Let $A = (a_{ij})$, $1 \leqslant i, j \leqslant n$ by a (symmetric) negative semi-definite matrix, and suppose $\Sigma_{i,j} a_{ij} x_i x_j = 0$ for some $x = (x_1, \ldots, x_n)$. Then, for all i, $\Sigma_j a_{ij} x_j = 0$.

Proof. Let P be an orthogonal matrix that diagonalizes A, and let $D = P'AP$ be a diagonal matrix with diagonal entries λ_i (P' denotes the transpose of P). By assumption, $\lambda_i \leqslant 0$ for all i. Finally, let $y = P'x$. It follows that $0 = x'Ax = y'Dy = \Sigma_i \lambda_i y_i^2$. Since $\lambda_i \leqslant 0$ for all i, $\lambda_i y_i = 0$ for all i. Therefore,

$$Ax = Dy = 0.$$

This proves the fact. *QED*

Lemma 4. *Suppose* (U, V, x) *is a Nash equilibrium for the distortion game. Then*

$$x \cdot \nabla U(x) = (1 - x) \cdot \nabla U(x) = U(x), \tag{10.5}$$

$$(1 - x) \cdot \nabla V(1 - x) = x \cdot \nabla V(1 - x) = V(1 - x), \tag{10.5$'$}$$

$$U(\lambda x) = \lambda U(x) \quad \textit{for} \quad 0 \leqslant \lambda \leqslant 1, \tag{10.6}$$

$$V(\lambda(1 - x)) = \lambda V(1 - x) \quad \textit{for} \quad 0 \leqslant \lambda \leqslant 1, \tag{10.6$'$}$$

$$\nabla U(\lambda x) = \nabla U(x) \quad \textit{for} \quad 0 \leqslant \lambda \leqslant 1, \tag{10.7}$$

$$\nabla V(\lambda(1 - x)) = \nabla V(1 - x) \quad \textit{for} \quad 0 \leqslant \lambda \leqslant 1, \tag{10.7$'$}$$

$$\sum_i U_{ij}(x) x_i = 0 \quad \textit{for all } j, \quad \textit{and} \tag{10.8}$$

$$\sum_i V_{ij}(1 - x)(1 - x_i) = 0 \quad \textit{for all } j. \tag{10.8$'$}$$

Proof. It follows from lemma 2 and the fact that the true utilities are strictly increasing, that

$$(1 - x) \cdot \nabla U(x) = U(x) \quad \text{and} \tag{10.A1}$$

$$x \cdot \nabla V(1 - x) = V(1 - x). \tag{10.A1$'$}$$

Also, since $x \in N(U, V)$, (10.3) and (10.3′) hold. Therefore, for all i,

$$(1-x_i)U_i(x)V(1-x) - U(x)V_i(1-x)(1-x_i) \leqslant 0$$

and

$$x_iU_i(x)V(1-x) - U(x)V_i(1-x)x_i \geqslant 0.$$

Summing, applying (10.A1), (10.A1′), and the concavity of U and V show that

$$U(x) = x \cdot \nabla U(x) \quad \text{and} \quad V(1-x) = (1-x) \cdot \nabla V(1-x).$$

This establishes (10.5) and (10.5′).

To show (10.6) observe that because of concavity and the fact that $U(x) = x \cdot \nabla U(x)$,

$$\begin{aligned} U(x) - U(\lambda x) &\geqslant \nabla U(x) \cdot (x - \lambda x) \\ &= U(x) - \lambda \nabla U(x) \cdot x \\ &\geqslant U(x) - \lambda \nabla U(\lambda x) \cdot x \\ &\geqslant U(x) - U(\lambda x) \end{aligned}$$

whenever $0 \leqslant \lambda \leqslant 1$. It follows that all of the inequalities above hold as equalities. Hence, for $0 \leqslant \lambda \leqslant 1$,

$$U(\lambda x) = \lambda \nabla U(x) \cdot x = \lambda \nabla U(\lambda x) \cdot x = \lambda U(x).$$

Thus, for $0 \leqslant \lambda \leqslant 1$,

$$\nabla U(\lambda x) \cdot x = U(x). \tag{10.A2}$$

Identity (10.A2) can be differentiated with respect to λ. This yields

$$\sum_{i,j} U_{ij}(\lambda x)x_ix_j = 0 \quad \text{for} \quad 0 \leqslant \lambda \leqslant 1.$$

The preliminary fact now implies that for all j and $0 \leqslant \lambda \leqslant 1$,

$$\sum_i U_{ij}(\lambda x)x_i = 0.$$

This establishes (10.8). Equation (10.7) follows since

$$\frac{d}{d\lambda}(U_j(\lambda x)) = \sum_i U_{ij}(\lambda x)x_i = 0.$$

Similar arguments establish (10.6′), (10.7′) and (10.8′). *QED*

REFERENCES

Aumann, R. J. 1977a: Power and taxes. *Econometrica*, **45**, 1137–61.

Aumann, R. J. and Kurz, M. 1977: Power and taxes in a multi-commodity economy. *Israel Journal of Mathematics*, **27**, 185-234.

Crawford, V. P. and Varian, H. R. 1979: Distortion of preferences and the Nash theory of bargaining. *Economics Letters*, **3**, 203–6.

Foley, D. K. 1967: Resource allocation and the public sector. *Yale Economic Essays*, **7**, 45-98.

Hurwicz, L. 1972: On informationally decentralized systems. In C. B. McGuire and R. Radner (eds), *Decision and Organization: A Volume in Honor of Jacob Marshak.* American Elsevier Publishing Company, New York.

Hurwicz, L., Maskin, E. and Postlewaite, A. 1978: Constrained Walrasian equilibria. Mimeo, University of Minnesota.

Kalai, E. and Rosenthal, R. W. 1978: Arbitration of two-party disputes under ignorance. *International Journal of Game Theory*, **7**, 65–72.

Kalai, E. and Smorodinsky, M. 1975: Other solutions to Nash's bargaining problem. *Econometrica*, **43**, 513–18.

Kannai, Y. 1977: Concavifiability and constructions of concave utility functions. *Journal of Mathematical Economics*, **4**, 1–56.

Kihlstrom, R. E., Roth, A. E. and Schmeidler, D. 1979: Risk aversion and solutions to Nash's bargaining problem. Mimeo, University of Illinois.

Kurz, M. 1977: Distortion of preferences, Income distribution, and the case for a linear income tax. *Journal of Economic Theory*, **14**, 291–8.

Kurz, M. 1980: Income distribution and distortion of preferences: The *l*-commodity case. *Journal of Economic Theory*, **22**, 99–106.

Luce, R. D. and Raiffa, H. 1957: *Games and Decisions: Introduction and Critical Survey*, John Wiley and Sons, New York.

Myerson, R. 1979: Incentive compatibility and the bargaining problem. *Econometrica*, **47**, 61–73.

Nash, J. 1950: The bargaining problem, *Econometrica*, **18**, 155–62.

Nash, J. 1953: Two-person cooperative games. *Econometrica*, **21**: 128–40.

Raiffa, H. 1953: Arbitration schemes for generalized two-person games. In H. W. Kuhn and A. W. Tucker (eds), *Contributions to the Theory of Games II*, Princeton University Press, Princeton, New Jersey.

Roth, A. 1979: *Axiomatic Models of Bargaining.* Springer-Verlag, Berlin and Heidelberg.

Thomson, W. 1979: The manipulability of the Shapley-value. University of Minnesota Discussion Paper No. 79-115.

Thomson, W. 1979: On the manipulability of resource allocation mechanisms designed to achieve individually-rational and Pareto-optimal outcomes. University of Minnesota Discussion Paper No. 79-116.

11

Nash Bargaining Theory III

K. Binmore

INTRODUCTION

In three remarkable papers written in the early 1950s, Nash outlined an approach to the theory of games in general and to the theory of bargaining in particular which remains of the greatest significance. This chapter, together with chapters 2 and 4 seeks to discuss and to extend the general approach to bargaining outlined in those papers.

Chapters 2 and 4 were devoted to the classical 'Nash bargaining solution'. Chapter 2 was concerned with Nash's axiomatic method and suggested a substitute for his 'independence of irrelevant alternatives' condition. Chapter 4 examined some specific bargaining models for which a non-cooperative analysis leads to the 'Nash bargaining solution'. In both chapters it was stressed that a strictly game-theoretic viewpoint was adopted (as opposed, for example, to a behavioural or ethical viewpoint).

In this chapter we propose to repeat the pattern of chapters 2 and 4 but in a different informational context. It should be emphasized that we do *not* regard the classical 'Nash bargaining solution' as being universally applicable to all two-person bargaining problems. Built into Nash's original assumptions is the proposition that the solution payoffs depend *only* on the von Neumann and Morgenstern utilities which the players attach to the various possible outcomes and depend not at all on the physical nature of these outcomes. This assumption was stated explicitly in chapter 2 as axiom B. The assumption seems harmless enough if it is understood that the players negotiate exclusively in terms of von Neumann and Morgenstern utility levels and, on the face of it, this seems a rational way for them to proceed since it is these utility levels which are immediately relevant to the players. Indeeed, Harsanyi (1977, p. 118) has formulated a 'rationality postulate' (B3) to this general effect and comments that his postulate rules out the possibility, for example, that the players' telephone numbers will be relevant variables in determining the players' solution payoffs. On p. 155, he gives a stronger version for the specific case of bargaining.

Our chief object in this chapter is to question assumptions of this type in the particular instance of the traditional problem of two agents trading in wheat and fish. Here it is natural to suppose that the players' statements to each other will be phrased in terms of quantities of wheat and fish: for example, 'I am willing to trade five loaves for two fishes or ten loaves for three fishes but under no circumstances will I trade two loaves for one fish'. One must also expect that contracts will be expressed directly in terms of quantities of wheat and fish to be delivered. Although, as always in game theory, the players are assumed to be fully aware of the other players' utilities there is no need for this to be true of any lawyers whose duty it may be to investigate possible charges of contract breaking. We shall refer to a problem of this type as a *bartering problem* to distinguish it from the set-up formulated by Nash which he calls a *bargaining problem.*

In section 3 we give a list of versions of Nash's axioms modified so as to take into account the natural informational structure for the simple bartering problem which is usually described in terms of the Edgeworth box. We shall then show that these axioms lead *not* to the classical 'Nash bargaining solution' but to the competitive equilibrium for the problem (provided, among other things, that this competitive equilibrium is unique). In general, of course, the competitive equilibrium yields different payoffs to the players from the 'Nash bargaining solution' as figure 11.1 indicates.

Our basic contention is that where extra informational structure exists beyond that envisaged by Nash, this may well be relevant and the solution outcome will reflect this fact. Even telephone numbers may be relevant under certain circumstances (especially if communication is telephonic). This is not to say that such extra informational structure is *necessarily* relevant. If agents bargaining over wheat and fish *do* communicate directly and exclusively in terms of utilities (as described, for example, in section 2 of chapter 2), then naturally the classical Nash theory *will* apply.

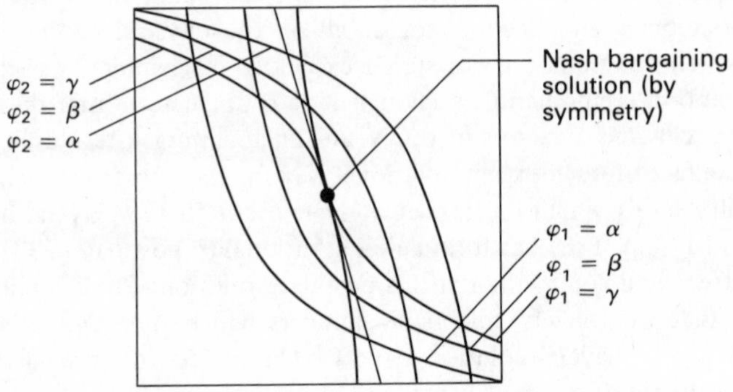

Figure 11.1

Having adapted Nash's axiomatic system to the natural informational structure for the bartering problem, we then turn to the question of finding a bargaining model to which the modified axiom system applies. As suggested in chapters 2 and 4, the fact that a particular 'solution concept' satisfies an axiom system is only significant if the axioms can be regarded as summarizing the salient features of a reasonably wide class of bargaining games. In the case of the original system of Nash, this class is typified by the simple 'demand game' described and elaborated in chapter 4. In the case of the modified system studied in this chapter, we offer as a typical bargaining game to which the system applies what we shall call a 'misrepresentation contest'. This is described and analysed in section 4. It is significant that the analysis yields a *unique* payoff pair which is both Pareto-efficient and implementable by a Nash equilibrium strategy pair (provided that there is a unique competitive equilibrium). Nash's demand game on the other hand usually has many such payoff pairs and some ingenuity is required in justifying the identification of the 'Nash bargaining solution' as being the payoff pair which will result from rational play.

This work on 'misrepresentation contests' in section 4 is to be compared with that of Hurwicz (1979) and Schmeidler (1980) who study non-cooperative implementation of competitive equilibria from a more abstract point of view.

2 THE BARTERING PROBLEM

Our intention in this chapter is to avoid technicalities in so far as this is possible. We therefore restrict our attention to what is essentially the traditional Edgeworth box with two agents and two commodities.

We use $\mathbb{R}_+$ to denote the set of non-negative real numbers. A point $(w,f) \in \mathbb{R}_+^2$ will be called a 'trade' and may be interpreted as representing an agreement that the first player dispatches $w \geqslant 0$ of wheat to the second player who simultaneously dispatches $f \geqslant 0$ of fish to the first player. The *no-trade point* is therefore fixed at $\mathbf{0}$.

Both players are assumed to prefer more of a commodity to less. It will also be assumed that their preferences are convex. We shall in fact assume that they are equipped with von Neumann and Morgenstern utility functions $\phi_1 : \mathbb{R}_+^2 \to \mathbb{R}$ and $\phi_2 : \mathbb{R}_+^2 \to \mathbb{R}$ which are quasi-concave, strictly increasing and upper semi-continuous so that $\{\mathbf{x} : \phi_1(\mathbf{x}) \geqslant \alpha\}$ and $\{\mathbf{x} : \phi_2(\mathbf{x}) \geqslant \alpha\}$ are closed and convex for all $\alpha \in \mathbb{R}$. As in chapter 2, we use the notation $\phi : \mathbb{R}_+^2 \to \mathbb{R}$ for the function defined by

$$\phi(\mathbf{x}) = (\phi_1(\mathbf{x}), \phi_2(\mathbf{x})).$$

It will be assumed that a *unique* competitive equilibrium $\boldsymbol{\gamma}$ exists for the above configuration and that this lies in the interior of $\mathbb{R}_+^2$.

We shall assume moreover that this competitive equilibrium is Pareto-efficient (with respect to lotteries over trades as well as trades).

Not all trades will be assumed to be feasible. The set of feasible trades will be denoted by $\mathscr{E}$ (for Edgeworth box). We shall allow $\mathscr{E}$ to be any subset of $\mathbb{R}^2_{\circ}$ which contains the point γ. However, the analysis goes through equally well if one restricts attention to the case $\mathscr{E} = [0, W] \times [0, F]$ (with $W \geqslant \gamma_1$ and $F \geqslant \gamma_2$). This corresponds to simple exchange where the first player begins with the commodity bundle $(W, 0)$ and the second player with $(0, F)$. The situation is illustrated in figure 11.2. Note that this Edgeworth box diagram is not drawn conventionally (as opposed, for example, to figure 11.1) in that it treats the players symmetrically. This has some technical advantages as well as facilitating comparisons with the traditional bargaining problem.

The preference structure $\phi|\mathscr{E}$ obtained by restricting ϕ to the set $\mathscr{E}$ of feasible trades is often referred to as a 'market game'. Implicit in this nomenclature is the understanding that the players have access to strategies which allow the joint implementation of feasible trades. We shall suppose that such a game exists and denote it by $G = G(\phi|\mathscr{E})$. In the case of simple exchange, suitable strategies consist of the unilateral dispatch of part of a player's initial endowment to the other. Since the dispatch of nothing dominates other strategies, the non-cooperative solution outcome is therefore the no-trade point in this case.

Our interest, however, is in possible cooperative solutions – i.e. in negotiated outcomes. We therefore suppose that G is embedded in a larger bargaining contest B. The general framework envisaged is described in section 2 of chapter 2 except that, in this chapter, we suppose that the players communicate directly in terms of wheat and fish rather than in terms of utilities. The status quo of the bargaining contest B will be taken to be the no-trade point $\mathbf{0}$.

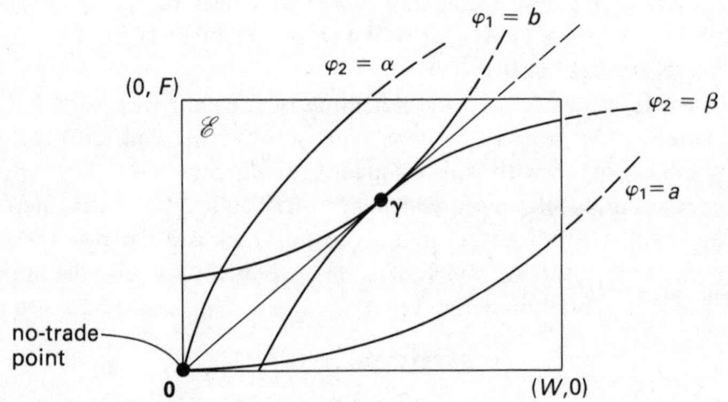

Figure 11.2

This completes the formal description of the bartering problem in so far as we intend to specify this in detail. In section 3 we offer an axiomatic analysis of the problem. In this analysis, each of Nash's formal axioms is replaced by a straightforward adaptation to the situation described above. (In this chapter we stick with Nash's 'independence of irrelevant alternatives' and leave 'convention consistency', introduced in chapter 2, for a later paper.) The only novelty of this treatment as compared with that of Nash is that we do not insist that the solution depend *only* on utility levels (axiom B of chapter 2) but allow it to depend also on the other information incorporated in the preference structure $\phi | \mathscr{E}$. In section 4 we consider a specific bargaining contest B whose solution outcome coincides with that predicted by the axiomatic analysis – i.e. it is the competitive equilibrium. In section 5 we examine the case of multiple competitive equilibria very briefly.

3 ADAPTATION OF THE NASH AXIOMS

In this section we give a list of axioms for the solution outcomes of the bargaining contests B described in section 2.

Axiom A. Each bargaining contest B has a solution and the implementation of the solution strategies results in a trade $\mathbf{s}$.

Note that this axiom excludes the possibility that the solution outcome is a lottery over trades. In fact our other assumptions are adequate to show that this cannot happen.

We shall use the notation $\mathscr{P}$ to denote the set of all preference structures $\phi | \mathscr{E}$ with the properties specified in section 2.

Axiom B. There exists a function $F: \mathscr{P} \to \mathbb{R}^2$ such that

$$F(\phi | \mathscr{E}) = \mathbf{s}.$$

Axiom B asserts that the solution outcome depends only on the informational structure built into the preference structure $\phi | \mathscr{E}$ and not on other extraneous information (such as telephone numbers). The corresponding implicit assumption of Nash (given explicitly as axiom B of chapter 2) is that there exists a function f for which $f(\phi(\mathscr{E}), \phi(\mathbf{0})) = \phi(\mathbf{s})$ – i.e. less information is required for the computation of the solution payoffs.

The remaining axioms deal with the mathematical properties of the function $F: \mathscr{P} \to \mathbb{R}^2$. Axioms A and B are concerned with the interpretation of this function. Each mathematical axiom below has been assigned the same number as its analogue in the version of Nash's axiom system given in chapter 2 to facilitate comparison.

Axiom 1 (feasibility)

$$\mathbf{s} \in \mathscr{E}$$

The first axiom needs no comment. The second axiom requires that the solution outcome be invariant under transformations (represented by α) of the origins and units of the utility scales and also invariant under transformations (represented by β) of the commodity units.

Axiom 2 (invariance)

1 Let $\alpha : \mathbb{R}^2 \to \mathbb{R}^2$ be a strictly increasing affine transformation. Then

$$F(\alpha \circ \phi | \mathscr{E}) = F(\phi | \mathscr{E}).$$

2 Let $\beta : \mathbb{R}^2 \to \mathbb{R}^2$ be a strictly increasing linear transformation. Then

$$F(\phi | \beta \mathscr{E}) = \beta \circ F(\phi \circ \beta | \mathscr{E}).$$

Axiom 3 (efficiency).

For any trade $\mathbf{t}$

$$\phi(\mathbf{t}) > \phi(\mathbf{s}) \Rightarrow \mathbf{t} \notin \mathscr{E}.$$

Axiom 3 needs no comment. The next axiom, however, is of considerable significance since it is the analogue of the vital 'independence of irrelevant alternatives'. We require to begin with that the deletion of 'irrelevant' trades should leave the solution unchanged. Secondly, we require that, if both players' utilities for all feasible trades except the solution outcome and the no-trade point decrease, then the solution remains unchanged.

Axiom 4 (independence of irrelevant alternatives)

1 $F(\phi | \mathscr{E}) \in \mathscr{F} \subseteq \mathscr{E} \Rightarrow F(\phi | \mathscr{F}) = F(\phi | \mathscr{E})$.
2 If $\psi(\mathbf{0}) = \phi(\mathbf{0})$, $\psi(\mathbf{s}) = \phi(\mathbf{s})$ and, for any $\mathbf{t} \in \mathscr{E}$, $\psi(\mathbf{t}) \leqslant \phi(\mathbf{t})$ then

$$F(\phi | \mathscr{E}) = F(\psi | \mathscr{E}).$$

Our final axiom requires that both players and commodities be treated symmetrically.

Axiom 5 (symmetry).

If $\tau : (x_1, x_2) \mapsto (x_2, x_1)$, then

$$F(\tau \circ \phi \circ \tau | \tau^{-1} \mathscr{E}) = \tau \circ F(\phi | \mathscr{E}).$$

Theorem 1. Let $F : \mathscr{P} \to \mathbb{R}^2$ satisfy axioms 1–5 inclusive. Then, with the assumptions of section 2,

$$F(\phi | \mathscr{E}) = \boldsymbol{\gamma}$$

where $\boldsymbol{\gamma}$ is the competitive equilibrium of G (assumed unique and feasible in section 2).

Proof. We begin by using axiom 2 to transform the preference structure $\phi | \mathscr{E}$ to the configuration indicated in figure 11.3.

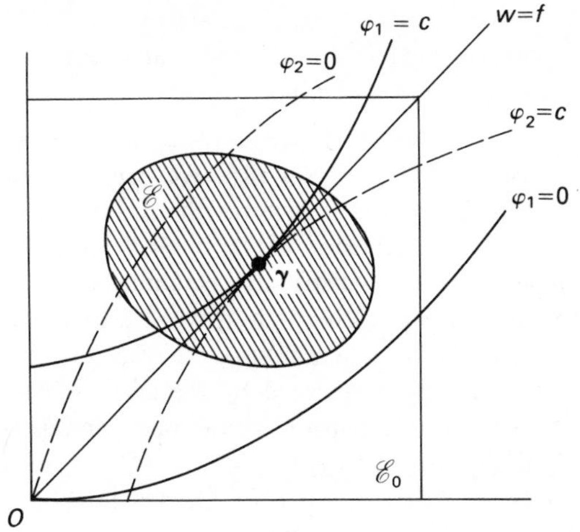

Figure 11.3

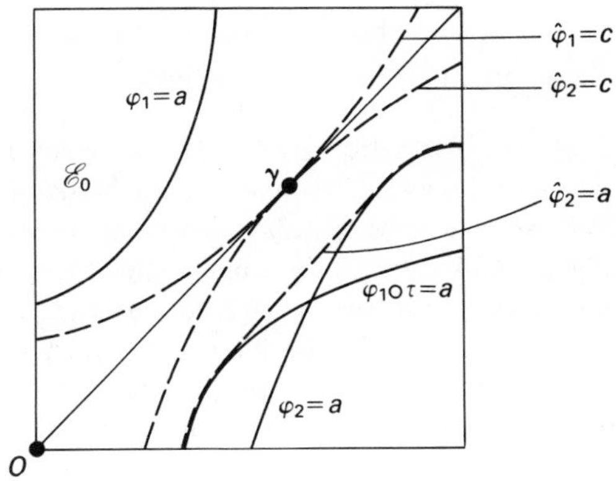

Figure 11.4

The square $\mathscr{E}_0$ is chosen so as to include the set $\mathscr{E}$. We next symmetrize by defining $\hat{\phi}_1$ to be the smallest quasi-concave function which dominates both ϕ_1 and $\phi_2 \circ \tau$. Similarly $\hat{\phi}_2$ is the smallest quasi-concave function which dominates both ϕ_2 and $\phi_1 \circ \tau$. Obviously $\hat{\phi}_2 = \hat{\phi}_1 \circ \tau$ (see figure 11.4).

Observe that $\boldsymbol{\gamma}$ remains a Pareto-efficient point for the symmetric game $G(\hat{\phi} | \mathscr{E}_0)$. It follows from axiom 5 that $F(\hat{\phi} | \mathscr{E}_0) = \boldsymbol{\gamma}$. Appealing successively to axiom 4(1) and (2) we obtain that

$$F(\hat{\phi} | \mathscr{E}_0) = F(\hat{\phi} | \mathscr{E}) = F(\phi | \mathscr{E})$$

as required.

Note 1. We discuss the problems which arise when several competitive equilibria exist in section 5. The proof given above would, of course, locate $F(\phi|\mathscr{E})$ at each competitive equilibrium inside $\mathscr{E}$ and hence lead to an inconsistency in the case of multiple equilibria.

Note 2. Note that the proof works equally well if ϕ_1 and ϕ_2 are required to be continuous or differentiable on $\mathbb{R}^2_{\circ}$. We cannot however insist in general that ϕ_1 and ϕ_2 be concave. To force the proof through in this case one needs to use axiom 2 to arrange matters so that not only is it true that $\phi_1(\gamma) = \phi_2(\gamma)$ but also that the normals to the endographs of ϕ_1 and $\phi_2 \circ \tau$ point in the same direction at γ. The point γ will then remain a competitive equilibrium when ϕ_1 is replaced by the smallest concave function $\hat{\phi}_1$ dominating ϕ_1 and $\phi_2 \circ \tau$ and ϕ_2 is replaced by the smallest concave function $\hat{\phi}_2$ dominating ϕ_2 and $\phi_1 \circ \tau$. Unfortunately one cannot simultaneously guarantee that $\hat{\phi}_1(\mathbf{0}) = \phi_1(\mathbf{0})$ and $\hat{\phi}_2(\mathbf{0}) = \phi_2(\mathbf{0})$. Of course, one can drop the provision in axiom 4(2) that $\psi(\mathbf{0}) = \phi(\mathbf{0})$ if this is thought acceptable.

Note 3. In the traditional context, Kalai and Smorodinsky (1975) replace axiom 4 by a 'monotonicity axiom'. A corresponding replacement here still leads to the competitive equilibrium and ϕ_1 and ϕ_2 may be concave.

4 MISREPRESENTATION CONTEST

In this section we analyse a specific bargaining game which is offered as a typical example of those to which the axiomatic analysis of the previous section applies. This game will be called a misrepresentation contest. We begin by describing a simpler bargaining contest which has essentially the same strategic structure. In both cases we restrict our attention to the case when $\mathscr{E}$ is an Edgeworth box - i.e. $\mathscr{E} = [0, W] \times [0, F]$ and $\boldsymbol{\gamma} \in \mathscr{E}$.

Suppose that, in the bartering problem described in section 2, the negotiation procedure consists simply of both players simultaneously announcing a commitment. These commitments are assumed to consist of two non-negative real numbers: a price p_i and a quantity q_i. The announcement of the price p_i commits the player in question to agree only to trades which yield him at least p_i units of the opponent's commodity for each unit of his own commodity surrendered at that trade. The announcement of q_i commits the player in question to agree only to trades in which he surrenders at most q_i units of his own commodity. We suppose that the trade which results from these commitments is that trade $\mathbf{t} \in \mathscr{E}$ which simultaneously maximizes the amount of wheat and the amount of fish exchanged subject to the constraints represented by the players' commitments. If $\mathbf{t} = \mathbf{0}$ we regard the commitments as incompatible.

Figure 11.5 illustrates the situation. Each negotiation strategy may be identified with a region like those shaded. The players' commitments exclude an agreement on trades in these shaded regions.

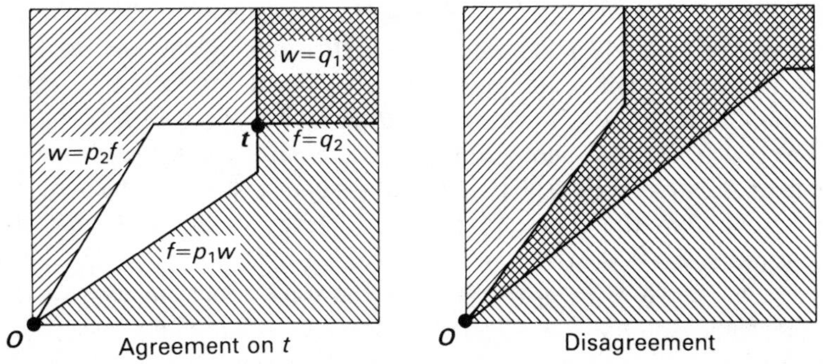

Figure 11.5

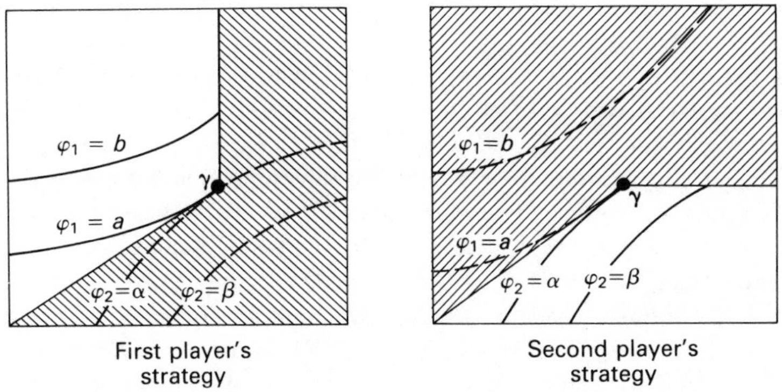

Figure 11.6

It is fairly evident that a Pareto-efficient, Nash equilibrium is obtained with $\mathbf{t} = \boldsymbol{\gamma}$ provided that the negotiation strategies are chosen in the manner indicated in figure 11.6.

It is also fairly evident that no trade other than $\boldsymbol{\gamma}$ can result from Pareto-efficient, Nash equilibrium strategies (provided the contours of ϕ_1 and ϕ_2 are smooth as in the figures).

We have therefore constructed a simple bargaining contest whose solution outcome is the competitive equilibrium $\boldsymbol{\gamma}$ as in the axiomatic analysis. This bargaining contest is rather special and so we continue by describing another bargaining contest for which the same result is true but which has an ostensibly more general structure.

The negotiation procedure begins with each player independently announcing a 'phony' utility function Φ_1 and Φ_2. The announcement of such a phony utility function is to be regarded as committing a player to act during the remainder of the negotiations as though his phony utility function were his true utility function. Once Φ_1 and Φ_2 have been announced, we assume that the players then select negotiation strategies for a secondary 'phony' bargaining contest $B_0 = B_0(\Phi|\mathscr{E})$ for which the status quo remains the no-trade point $\mathbf{0}$. In view of their commitments, the players are forced to choose the negotiation strategies for B_0 which are optimal for players whose utility functions actually are the phony utility functions announced. If the solution outcome of $B_0(\Phi|\mathscr{E})$ is $F_0(\Phi|\mathscr{E})$, then this outcome is therefore the outcome which results from the choice of the phony utility functions Φ_1 and Φ_2. One possible realization of this situation would be to take B_0 to be a Nash demand game and to take $F_0(\Phi|\mathscr{E})$ to be the trade which implements the 'Nash bargaining solution'. We do not however need to be very specific on the nature of B_0. In fact we shall only need to assume that the outcome $F_0(\Phi|\mathscr{E})$ lies in the core of the game B_0 (i.e. the players' phony preferences assign at least as much utility to $F_0(\Phi|\mathscr{E})$ as to the no-trade point $\mathbf{0}$ and there is no other trade to which both players assign a higher phony preference than $F_0(\Phi|\mathscr{E})$).

As a result of this discussion, we are left with a game H in which the players' strategies consist of phony utility functions. The choice of phony utility functions Φ_1 and Φ_2 results in the outcome $F_0(\Phi|\mathscr{E})$ which the players evaluate, of course, using their *true* utility functions ϕ_1 and ϕ_2. We call H a misrepresentation contest. The word contest is used to indicate that H is to be analysed as a non-cooperative game. This analysis requires certain technical assumptions which are described below.

We shall continue to assume that all the conditions of section 2 are satisfied. In addition, we shall require that ϕ_1 and ϕ_2 are strictly concave and differentiable. The former assumption ensures that we need not be concerned with the possibility that the use of mixed strategies might yield a Pareto-efficient outcome. The differentiability however is essential to the argument (see note 7). The analysis is not sensitive to the shape of $\mathscr{E}$ but we shall make a concrete choice and take $\mathscr{E} = [0, W] \times [0, F]$ (i.e. $\mathscr{E}$ is the Edgeworth box). As in section 2 we require that the competitive equilibrium $\boldsymbol{\gamma} \in \mathscr{E}$.

Next we need to state our assumptions on the admissible phony utility functions Φ_1 and Φ_2. It would be natural to make Φ_1 and Φ_2 subject to the same requirements as ϕ_1 and ϕ_2 but a technical difficulty intervenes. The game H is an infinite game and hence the theorem of Nash on the existence of equilibria does not apply. To guarantee the existence of equilibria in an infinite game one normally needs to embed the given strategy sets in larger compact sets. In our specific case, if we were to insist that Φ_1 and Φ_2 satisfy the same properties as ϕ_1 and ϕ_2, we could only deduce the existence of 'approximate' Nash equilibria. We get round this difficulty by widening the class of available Φ_1 and Φ_2. A player without access to this wider class would seek to approximate to functions in this class but would never be satisfied with the closeness of his approximation.

We shall continue to insist that the phony preferences represented by Φ_i be convex and that more of a commodity is always preferred to less. However, we shall allow Φ_i to take the value $-\infty$ and require that $\Phi_i : \mathbb{R}^2_\circ \to \mathbb{R}$ have the property that, for some closed, convex set D_i containing the no-trade point $\mathbf{0}$,

$$\Phi_i(x) = \begin{cases} -\infty & (x \notin D_i) \\ \psi_i(x) & (x \in D_i) \end{cases}$$

where ψ_i is admissible as a 'true' utility function. It may be helpful to think of $\Phi_i(\mathbf{x}) = -\infty$ as meaning that the ith player is committed to act as though $\mathbf{x}$ is not feasible.

It is fairly obvious that the competitive equilibrium $\boldsymbol{\gamma}$ is attainable as a Pareto-efficient, Nash equilibrium outcome of the game H. Figure 11.6 illustrates this remark. Appropriate Nash equilibrium strategies Φ_1 and Φ_2 may be chosen to take the value $-\infty$ in the shaded regions of figure 11.6 and to be equal to the true utility functions in the unshaded regions. Observe that the core outcome of $B_0(\Phi_1, \Phi_2)$ is then the single point $\boldsymbol{\gamma}$ and hence $F_0(\Phi_1, \Phi_2) = \boldsymbol{\gamma}$. No contour $\phi_2(\mathbf{x}) = c$ with $c > b$ has a point of intersection with $\{\mathbf{x} : \Phi_1(\mathbf{x}) > -\infty\}$ and hence has no point of intersection with $\{\mathbf{x} : \Phi_1(\mathbf{x}) \geqslant \Phi_1(\mathbf{0})\}$. Thus the second player cannot improve his payoff by deviating from Φ_2. Similarly for the first player.

It is not so obvious that the converse result is also true and so we quote this as a theorem.

Theorem 2. Any trade $\mathbf{t}$ resulting from a Pareto-efficient, Nash equilibrium of the game H is necessarily the competitive equilibrium $\boldsymbol{\gamma}$ of the game G.

Proof. Suppose that a (genuinely) Pareto-efficient trade $\mathbf{t}$ results from the play of a Nash equilibrium pair Φ_1 and Φ_2 of strategies for H. We assume that $\boldsymbol{\gamma} \neq \mathbf{t}$ and seek a contradiction.

Let $S = \{\mathbf{x} : \Phi_1(\mathbf{x}) \geqslant \Phi_1(\mathbf{0})\}$ and $T = \{\mathbf{x} : \Phi_2(\mathbf{x}) \geqslant \Phi_2(\mathbf{0})\}$. Then $\mathbf{t}$ lies in the closed, convex set $S \cap T$. Let the endpoints of the largest line segment in $S \cap T$ which contains $\mathbf{0}$ and $\mathbf{t}$ be $\mathbf{0}$ and $\mathbf{e}$. Denote the line segment itself by $[\mathbf{0}, \mathbf{e}]$. We then define the points $\mathbf{a}$ and $\mathbf{b}$ by

$$\Phi_1(\mathbf{a}) = \max_{x \in [\mathbf{0}, \mathbf{e}]} \Phi_1(\mathbf{x})$$

$$\Phi_2(\mathbf{b}) = \max_{x \in [\mathbf{0}, \mathbf{e}]} \Phi_2(\mathbf{x})$$

Figure 11.7 illustrates two possible configurations.

Since Φ_1 is strictly concave on $[\mathbf{0}, \mathbf{e}]$ we have that Φ_1 is strictly increasing on $[\mathbf{0}, \mathbf{a}]$ and strictly decreasing on $[\mathbf{a}, \mathbf{e}]$. Similarly for Φ_2 and the line segments $[\mathbf{0}, \mathbf{b}]$ and $[\mathbf{b}, \mathbf{e}]$. Since $\mathbf{t}$ lies in the core of $B_0(\Phi_1, \Phi_2)$ it is a Pareto-efficient point of this game and it follows that either $\mathbf{a} = \mathbf{b} = \mathbf{t}$ or else $\mathbf{a} \neq \mathbf{b}$

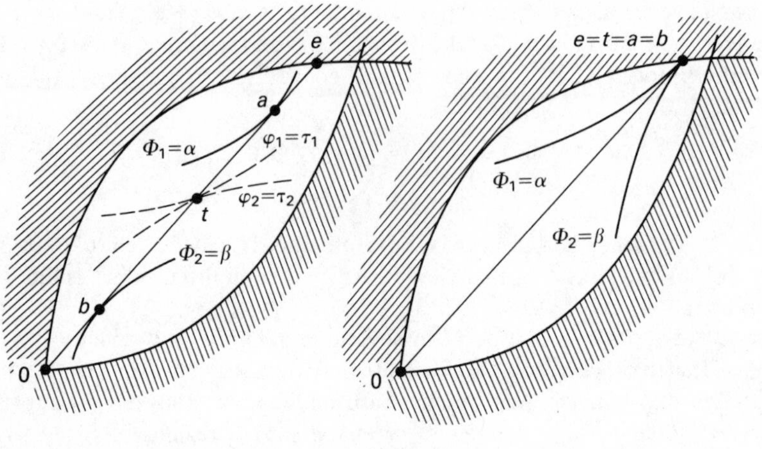

Figure 11.7

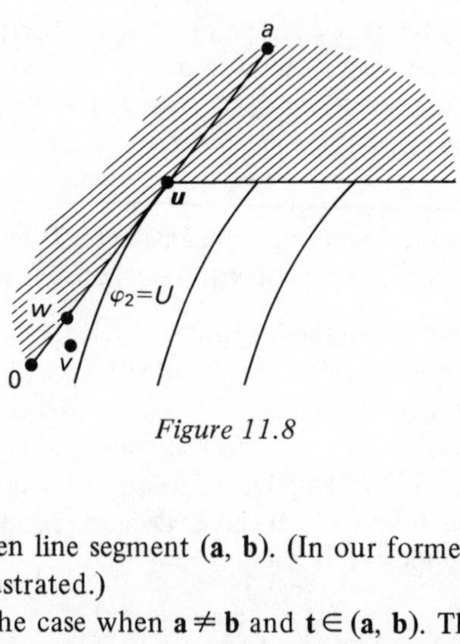

Figure 11.8

and **t** lies in the open line segment (**a**, **b**). (In our former case it need not be true that **t** = **e** as illustrated.)

We begin with the case when **a** ≠ **b** and **t** ∈ (**a**, **b**). Then, either **a** ∈ (**e**, **t**) or else **b** ∈ (**e**, **t**). We assume the former. Since **t** ≠ **γ**, the strictly concave function ϕ_2 does not achieve a maximum on (**0**, **a**) at the point **t**. Hence there exists a point **u** ∈ (**0**, **a**) for which $U = \phi_2(\mathbf{u}) > \phi_2(\mathbf{t})$. But the second player can then assure himself of a (genuine) payoff of at least U by changing his strategy choice from Φ_2 to the phony utility function Ψ_2 obtained from ϕ_2 by changing its value to $-\infty$ in the region shaded in figure 11.8.

Provided the first player does not alter his strategy choice, the result of such a strategy switch by the second player will be a point in the core of

$B_0(\Phi_1, \Psi_2)$. Consider a point $\mathbf{v}$ in the region where Ψ_2 is finite but $\phi_2 < U$. The second player prefers $\mathbf{u}$ to $\mathbf{v}$ according to the utility function Ψ_2. Also the first player prefers a point $\mathbf{w}$ on $[\mathbf{0}, \mathbf{u}]$ to $\mathbf{v}$ and hence prefers $\mathbf{u}$ to $\mathbf{v}$ according to Φ_1 (which is strictly increasing on $[\mathbf{0}, \mathbf{a}]$). Thus $\mathbf{v}$ is not Pareto-optimal according to the phony preferences and so the core of $B_0(\Phi_1, \Psi_2)$ lies inside the region where $\phi_2 \geqslant U$. The second player's switch is therefore profitable. But this is a contradiction since we assumed Φ_1 and Φ_2 to constitute a Nash equilibrium. It follows that $\mathbf{t} = \boldsymbol{\gamma}$.

The case when $\mathbf{a} \neq \mathbf{b}$ and $\mathbf{b} \in (\mathbf{e}, \mathbf{t})$ is dealt with similarly. There remains the case when $\mathbf{a} = \mathbf{b} = \mathbf{t}$. Here it is only necessary to observe that, because $\boldsymbol{\gamma} \neq \mathbf{t}$, there is a point $\mathbf{u} \in (\mathbf{0}, \mathbf{t})$ which is (genuinely) preferred *either* by the first player *or* by the second player. The preceding construction (for Ψ_1 or Ψ_2) can then be employed to the same effect. This completes the proof.

The interpretation of the misrepresentation contest H requires some thought. The naive view is that the players really do not know the true preferences of their opponents. Nash carefully distinguishes between this case (which he calls 'haggling') and the case in which the true preferences of the opponent are known (which he calls 'bargaining'). Only in the latter case of course are we dealing with a game in the strict sense of von Neumann and Morgenstern. It remains true, however, even in the former case that there exists a unique Pareto-efficient, Nash equilibrium outcome and this is the competitive equilibrium $\boldsymbol{\gamma}$. But, if the players are totally ignorant about the preferences of their opponents, they will have no way of determining what the optimal strategies are and so the analysis is not too helpful. It may be of course that they know the prices which sustain the competitive equilibrium $\boldsymbol{\gamma}$. This information is adequate for the computation of optimal strategies. Such an assumption about the information available to the players makes some sense especially in the n player context since one can tell a tâtonnement story of the type familiar to economists. As a matter of principle, however, it would seem more satisfactory to deal with the haggling case by regarding it as a 'game of incomplete information'. The game as described here would then be prefixed with a chance move which selects the players for the game H according to given probability distributions from populations with known characteristics. It seems unlikely that an analysis of such a game would yield as simple a result as theorem 2.

In the preceding paragraph we considered possible interpretations of a misrepresentation contest in contexts other than that of a straightforward two-person game. (Strictly speaking, the 'game of incomplete information' contemplated has as many players as there are members of the population.) But it remains possible to sustain interpretations even when both players are assumed to be familiar with the preferences of the other. The simplest such interpretation postulates an arbiter equipped with a predetermined arbitration scheme who is assumed, unlike the players themselves, to be unfamiliar with their preferences. He therefore takes as his input for the arbitration scheme the preferences which the players report to him.

A more sophisticated version of this notation replaces the arbiter by a lawyer. In this version one supposes that the choice of phony utility functions Φ_1 and Φ_2 by the players is an idealization of a more complicated bargaining procedure in which the opening moves consist of the players simultaneously exchanging messages in which they *commit* themselves to initial bargaining positions which then serve as a 'baseline' for the subsequent negotiations. Such an opening commitment will specify, for example, certain trades which the player designates as unacceptable under all circumstances. Other trades will be designated as unacceptable *unless* the opening commitment of the opponent satisfies certain properties. An outside observer (e.g. a lawyer or a judge considering possible damages for contract violation) may well regard such a battery of unilateral and conditional commitments as revealing 'preferences' of the player. Since, in our context, all communications take place directly in terms of quantities of commodities, the domains of these 'revealed preferences' will be the set of feasible trades. In the misrepresentation game H we shortcircuit this chain of ideas by supposing that the players summarize their opening bargaining positions with a direct announcement of a phony utility function defined on the set of trades.

With this interpretation, of course, neither player should be seen as attempting to deceive his opponent about his preferences. Both players will be fully aware that the other is playing a role but, since the players have *committed* themselves to their roles, this will make no operational difference to their subsequent play. Our contract lawyer on the other hand is not assumed to know the players' true utility functions (otherwise these could be incorporated in the contractual statements). In so far as he has to take the players' preferences into account in deciding whether or not the players have honoured their contract, he will have nothing to go on except the players' own statements during the negotiations. It is in this sense that the announcement of Φ_1 and Φ_2 can be said to 'misrepresent' the players' preferences.

In all the above discussion we have taken it for granted that one can draw a genuine distinction between 'true' and 'phony' preferences. One could however regard the discussion as negating this proposition by contradiction. If the players produce, consume and trade in commodities as part of an ongoing process, it would seem clear that certain preference structures would never be revealed by the players' actions because, if they held these preferences, it would be in their interests to act as though they held different preferences. But if a player systematically chooses to reveal a certain set of preferences, then, from an operational point of view these *are* his preferences. It is not at all evident, indeed, that a player himself will be able to distinguish between his 'true' preferences and those he has chosen to 'reveal' over a long period. It is a familiar fact that we tend to become what we pretend to be. The point of these remarks is that, in an economic context, it might be a helpful simplification to restrict attention to those preference structures which are stable, in the sense that, if held, they would not be misrepresented.

Note 4. What can be said of the competitive equilibrium $\boldsymbol{\gamma}$ in so far as the 'phony' game $G(\Phi|\mathscr{E})$ is concerned? It can be shown that either $\boldsymbol{\gamma}$ is a competitive equilibrium for $G(\Phi|\mathscr{E})$ or else at least one of the sets S or T of theorem 2 has the line segment $[\mathbf{0}, \boldsymbol{\gamma}]$ as part of its boundary. In the latter case it can happen (as in figure 11.9) that $G(\Phi|\mathscr{E})$ has no competitive equilibrium.

Note 5. Under what circumstances can the theorem be proved if we insist that Φ_1 and Φ_2 satisfy all the same requirements as ϕ_1 and ϕ_2? Under appropriate hypotheses, it can be shown that the theorem remains true provided that the solution outcome $F_0(\Phi|\mathscr{E})$ of the secondary bargaining game $B_0(\Phi|\mathscr{E})$ is the competitive equilibrium of this game. This is a satisfying result but is not, of course, very helpful if one's aim is, as in section 4, to seek to justify the axiomatic analysis of section 3.

Note 6. Questions of interchangeability and equivalence (see Luce and Raiffa 1957, p. 106) are relevant in the case of the game H since theorem 2 does not assert the uniqueness of the *strategies* which sustain $\boldsymbol{\gamma}$. Equivalence, however, is guaranteed. On the other hand, figure 11.9 shows that interchangeability may fail. The fact that interchangeability fails however is not significant in this context since it is clearly the right-hand case which should be regarded as typical of a 'solution strategy'.

Note 7. Figure 11.10 illustrates that the theorem may fail if the Nash equilibrium is not Pareto-efficient or if the functions ϕ_1 and ϕ_2 are not differentiable.

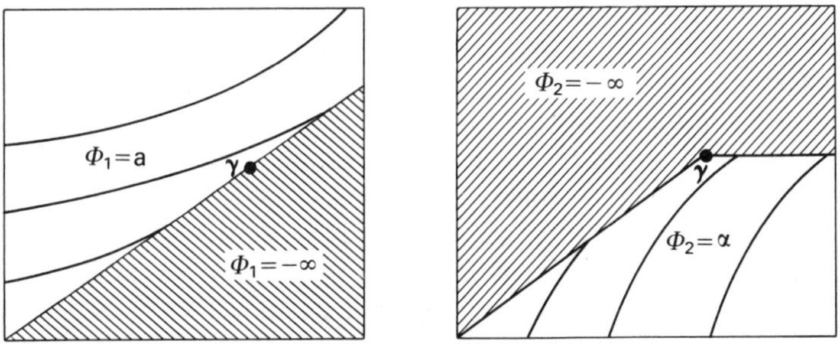

Figure 11.9

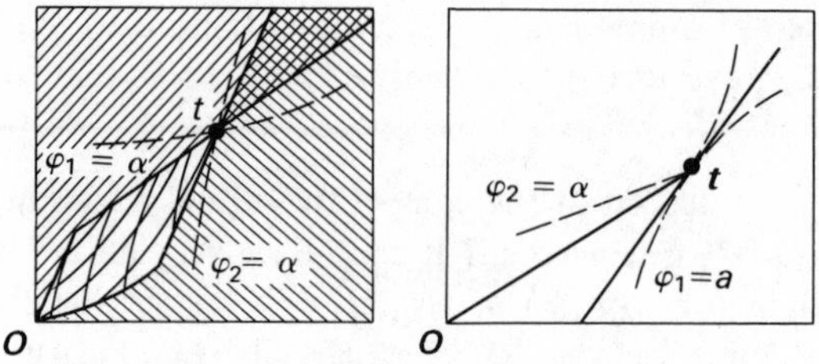

Figure 11.10

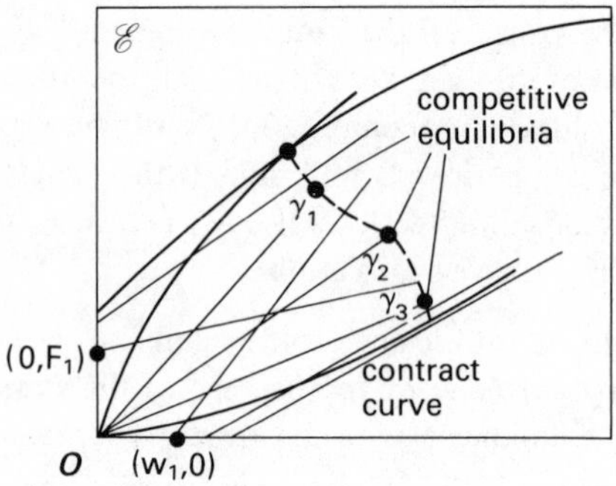

Figure 11.11

5 MULTIPLE COMPETITIVE EQUILIBRIA

There are various ways in which multiple competitive equilibria may occur. Figure 11.11 indicates one of these possibilities. In this figure, the straight lines emanating from the contract curve are the common tangents to the indifference curves at the points on the contract curve where these touch.

In the misrepresentation game based on this preference configuration each of the three competitive equilibria is attainable as a Pareto-efficient, Nash equilibrium outcome. A convention is therefore needed to distinguish among them and the convention which selects the middle one (i.e. $\boldsymbol{\gamma}_2$) seems natural. Certainly symmetry would otherwise have to be abandoned. But note that the first player would then be better off if he could induce the second player to accept a preliminary gift of a quantity W_1 of wheat. Similarly, the second player would be better off if he could induce the first player to accept a preliminary gift F_1 of fish. In so far as an analysis of this situation as a bargaining contest is concerned, it would therefore seem that this would be useful only in so far as it forms part of a more general

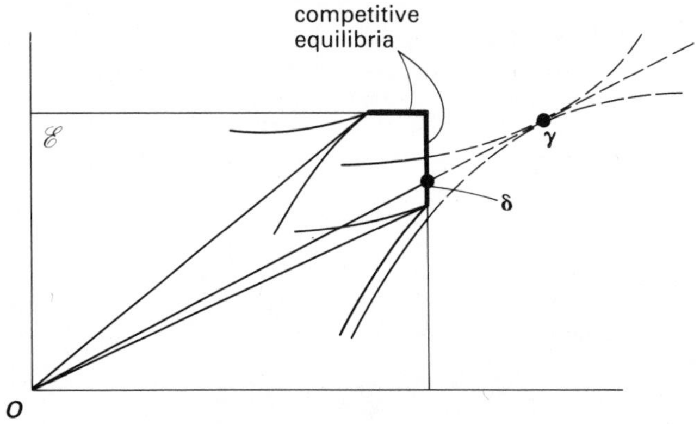

Figure 11.12

analysis in which a status quo is not regarded as given. Possibly the proper framework is that of a Nash threat game as described in Nash (1953).

In figure 11.12 a whole continuum of competitive equilibria lie on the boundary of the Edgeworth box. In the notation of section 2, this will occur when $\boldsymbol{\gamma} \notin \mathscr{E}$. Such a situation is not in the least pathological.

Each of these competitive equilibria is attainable as a Pareto-efficient Nash equilibrium outcome of an appropriate misrepresentation contest. Again we have need of a convention to distinguish among these but, in this case, none of the possible choices for a convention destabilize the bargaining set-up as in the preceding example. On the other hand, the number of possible conventions is embarrassingly large (as in the traditional Nash bargaining problem studied in chapter 2). Rather than embarking on a long discussion of the issue, we shall simply draw attention to the attractiveness of the outcome $\boldsymbol{\delta}$.

REFERENCES

Harsanyi, J. C. 1977: *Rational Behaviour and Bargaining Equilibria in Games and Social Situations.* CUP, Cambridge.

Hurwicz, L. 1979: On allocations attainable through Nash equilibria. *J. Econ. Th.*, **21**, 140–65.

Kalai, E. and Smorodinsky, M. 1975: Other solutions to Nash's bargaining problem. *Econometrica*, **43**, 513–18.

Luce, R. and Raiffa, H. 1957: *Games and Decisions.* Wiley, New York.

Nash, J. F. 1950: The bargaining problem. *Econometrica*, **18**. 155–62.

Nash, J. F. 1951: Non-cooperative games. *Annals of Mathematics*, **54**, 286–95.

Nash, J. F. 1953: Two-person cooperative games. *Econometrica*, **21**, 128–40.

Schmeidler, D. 1980: Walrasian analysis via strategic outcome functions. *Econometrica*, **48**, 1585–93.

Index